I0213273

RETURN TO LAUNCH

Return to Launch

Florida and America's Space Industry

Stephen C. Smith

UNIVERSITY OF FLORIDA PRESS

Gainesville

Cover: Photo by Julia Bergeron

Published in the United States of America

31 30 29 28 27 26 6 5 4 3 2 1

DOI: http: http://doi.org/10.5744/9781683406563

Library of Congress Cataloging-in-Publication Data
Names: Smith, Stephen C. (Space policy commentator) author
Title: Return to launch : Florida and America's space industry / Stephen C. Smith.
Description: Gainesville : University of Florida Press, 2026. | Includes index.
Identifiers: LCCN 2026006252 (print) | LCCN 2026006253 (ebook) | ISBN 9781683406563 hardback | ISBN 9781683405856 pdf | ISBN 9781683405986 ebook
Subjects: LCSH: John F. Kennedy Space Center—History | United States. National Aeronautics and Space Administration—History | Florida—History | Brevard County (Fla.)—History—20th century | BISAC: SCIENCE / Space Science / General | TECHNOLOGY & ENGINEERING / Aeronautics & Astronautics
Classification: LCC TL4027.F52 S65 2026 (print) | LCC TL4027.F52 (ebook)
LC record available at https://lccn.loc.gov/2026006252
LC ebook record available at https://lccn.loc.gov/2026006253

UF PRESS

University of Florida Press
PO Box 140239
Gainesville, FL 32614
floridapress.org

GPSR EU Authorized Representative: Mare Nostrum Group B.V., Doelen 72, 4831 GR Breda, The Netherlands, gpsr@mare-nostrum.co.uk

For Carol

CONTENTS

INSTITUTIONAL ABBREVIATIONS AND ACRONYMS

Bureaucracies, public and private, love abbreviations and acronyms. The aerospace industry, government and commercial, uses these as its own lexicon. This table lists the terms used in this book. You may wish to flag this page for reference.

See the "Abbreviations for Sources" for media abbreviations listed in the endnotes.

AADC	Alaska Aerospace Development Agency
AFL-CIO	American Federation of Labor and Congress of Industrial Organizations
AMF	Astronauts Memorial Foundation
ARM	Asteroid Retrieval Mission
ARPA	Advanced Research Projects Agency (later DARPA)
ASA	Aerospace States Association
AST	Office of Commercial Space Transportation
ATK	Alliant Techsystems
BEDC	Brevard Economic Development Council (later Corporation)
C3PF	Commercial Crew and Cargo Processing Facility
C3PO	Commercial Crew/Cargo Project Office
CAIB	Columbia Accident Investigation Board
CASIS	Center for Advancement of Science in Space
CCAFS	Cape Canaveral Air Force Station (until December 2020)
CCSFS	Cape Canaveral Space Force Station (starting in December 2020)
CEV	Crew Exploration Vehicle

CIT	Center for Innovative Technology
COMSAT	Commercial Satellite Corporation
COTS	Commercial Orbital Transportation Services
CSA	California Space Authority
DARPA	Defense Advanced Research Projects Agency (originally ARPA)
DOD	Department of Defense
EDC	Economic Development Commission of Florida's Space Coast
EELV	Evolved Expendable Launch Vehicle
EIS	Environmental Impact Statement
ELV	Expendable Launch Vehicle
ESA	European Space Agency
FAA	Federal Aviation Administration
FDOT	Florida Department of Transportation
FIT	Florida Institute of Technology (Florida Tech)
FSA	Florida Space Authority
FSI	Florida Space Institute
FSRI	Florida Space Research Institute
FWS	US Fish and Wildlife Service
GAO	Government Accountability Office
HALO	Habitation and Logistics Outpost
HIF	Horizontal Integration Facility
HLS	Human Landing System
IAM	International Association of Machinists
IAMAW	International Association of Machinists and Aerospace Workers
IBEW	International Brotherhood of Electrical Workers
ICBM	Intercontinental Ballistic Missile
IGY	International Geophysical Year
ISS	International Space Station
JPL	Jet Propulsion Laboratory

JSC	Johnson Space Center
KSC	Kennedy Space Center
KSCVC	Kennedy Space Center Visitor Complex
LC	Launch Complex (Sometimes SLC, or Space Launch Complex)
LEO	Low Earth Orbit
LLF	Launch and Landing Facility
Lockmart	Lockheed Martin
LUT	Launch Umbilical Tower
MARS	Mid-Atlantic Regional Spaceport
MINWR	Merritt Island National Wildlife Refuge
MSTO	Multiple Stage to Orbit
NACA	National Advisory Committee for Aeronautics
NAS	Naval Air Station
NASA	National Aeronautics and Space Administration
NextSTEP	Next Space Technologies for Exploration Partnerships
NOAA	National Oceanic and Atmospheric Administration
NOTU	Naval Ordnance Test Unit
NSDC	National Spaceport Development Corporation
O&C	Operations and Checkout Building
OIG	Office of the Inspector General
OMB	Office of Management and Budget
OPF	Orbiter Processing Facility
OPPAGA	Office of Program Policy Analysis and Governmental Accountability
OSP	Orbital Space Plane
OTTED	Office of Tourism, Trade and Economic Development
PPE	Power and Propulsion Element
REACH	Regional Economic Action Coalition
RLV	Reusable Launch Vehicle
RLVH	Reusable Launch Vehicle Hangar

RpK	Rocketplane Kistler
SAA	Space Act Agreement
SCA	McNamara-O'Hara Service Contract Act
SCDC	Space Coast Development Commission
SFA	Spaceport Florida Authority
SFF	Space Frontier Foundation
SFPP	Space Flight Participant Program
SIP	Spaceport Improvement Program
SLS	Space Launch System
SLSL	Space Life Sciences Lab
SNC	Sierra Nevada Corporation
SOB	Save Our Beaches
SRB	Solid Rocket Booster
SRF	Space Research Foundation
SRI	Space Research Institute
SSI	Space Studies Institute
SWS	Strategic Weapons System Ashore
TRDA	Technological Research and Development Authority
TWU	Transport Workers Union
UE&C	United Engineers and Constructors, Inc.
ULA	United Launch Alliance
ULV	Unmanned Launch Vehicle
US	United States
USA	United Space Alliance
USAF	United States Air Force
USBI	United Space Boosters, Inc.
USDOT	US Department of Transportation
USOS	United States Orbital Segment
USSR	Union of Soviet Socialist Republics (Soviet Union)
VSE	Vision for Space Exploration
WCSC	Western Commercial Space Center

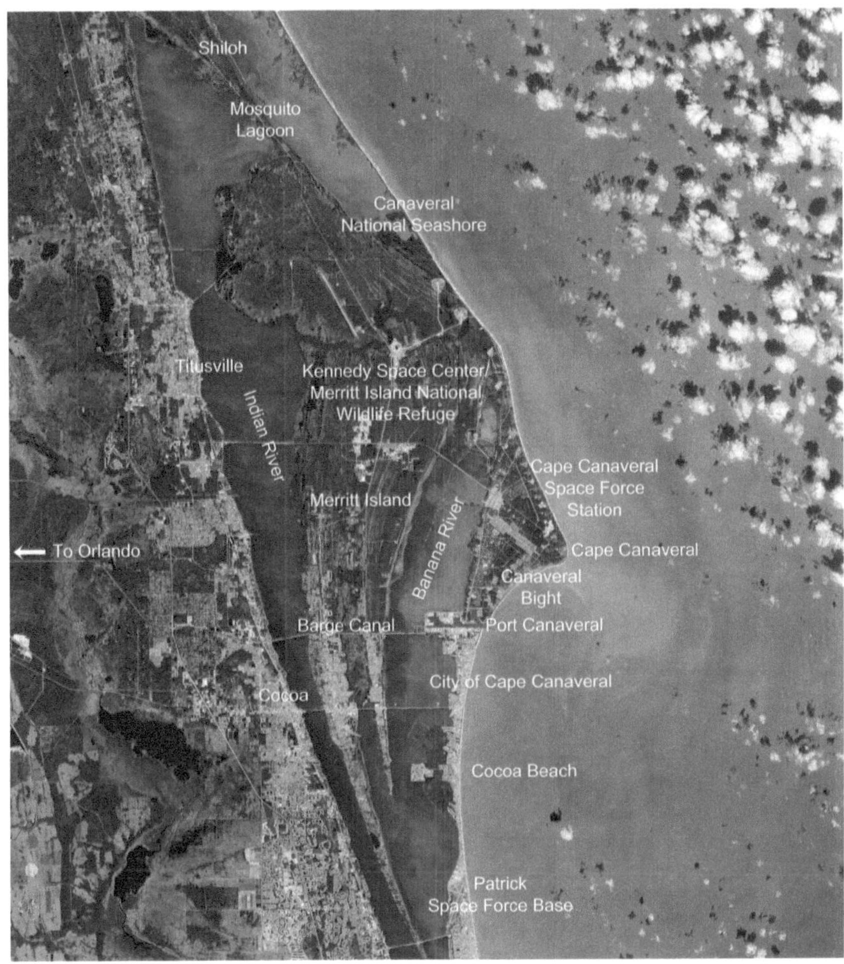

The Space Coast on April 16, 2021. Image source: European Space Agency Copernicus Sentinel-2 satellite. Labels added by the author.

1

Space Canaveral

The Nazi weapon hissed steam as it stood on a concrete slab carved into the Cape Canaveral swamp. Tending to the missile this day were Americans, not Germans, military and civilian. They were about to change history.

Less than ten years before, after the United States entered World War II, German U-boats hunted merchant ships in the waters off Canaveral. Sea battles could be witnessed from the beach. An enemy missile standing near where they once watched must have bemused the few remaining Cape residents.

Had World War II lasted a little longer, this missile might have been launched at London, Antwerp, or some other Allied target. But the war ended as history determined, so the Germans hid this weapon, along with about a hundred others, in an underground slave-labor factory near Nordhausen. The Nazi propaganda ministry called these missiles *Vergeltungswaffe Zwei*, "Vengeance Weapon 2," or V-2 for short.

At the end of the war, the US Army confiscated these V-2s and took them, along with rocketeer Wernher von Braun and his designers, to a test site near White Sands, New Mexico. The German engineers taught the Americans what they knew, but now it was time for the Americans to show they could launch missiles themselves.

On this day, July 24, 1950, the vengeance weapon was ready. Florida summers are notoriously hot and humid. Today was no exception. The launch was scheduled for 8:00 a.m., before the heat and humidity peaked. The Army evacuated the locals to safety in case the missile went off course, which sometimes happened at White Sands. The only people present were Army personnel and employees of General Electric, the company contracted to conduct the tests. Multitudes of mosquitoes pestered the poor pad workers.

This missile was part of a program called Project Bumper. It would be used as a booster for an upper stage called the WAC Corporal, a small missile "Without Attitude Control." The first six Project Bumper tests had been at White Sands. Two more were left; these were moved to Cape Canaveral so the V-2s could launch over the ocean. The New Mexico range was limited, so launches went more vertical than horizontal. At Cape Canaveral, the missiles could travel far down range over the ocean for hundreds of miles, much farther than White Sands. This particular missile test was dubbed Bumper 8.

After some delay, Bumper 8 launched at 9:28 a.m. The V-2 went somewhat off course, veering down range to a near-horizontal attitude. The WAC Corporal successfully separated from the V-2, but then broke up due to aerodynamic pressure. An Army colonel called the launch "a complete success in every way," which was charitable, but it showed what was possible.

Bumper 8 was the first launch from Cape Canaveral. Thousands more were to follow.[1]

* * * * *

For millennia, humans have lived on Cape Canaveral. Not all of them were welcome, as some sought to supplant and enslave those already living there. The Cape's strategic attributes often attracted military interests, welcome or not.

Cape Canaveral and its neighboring communities are part of Brevard County, in east-central Florida. By the late 1960s, these communities had been nicknamed the "Space Coast."[2]

The Space Coast topography is a mosaic of rivers, lagoons, lakes, swamps, sands, shoals, shores, channels, canals, ports, inlets, and outlets. Some were made by humans. Most were not.

Nature's ebbs and flows have carved, erased, and reshaped Cape Canaveral. Over tens of millions of years, as the polar ice cap grew and shrank, the ocean advanced and retreated.[3] Each cycle of advance and retreat left unique deposits defining that epoch; the Cape is believed to have eleven distinct ridges of limestone, clays, sands, and coquina rock.

The Canaveral Peninsula forms an outer beach barrier to the Atlantic Ocean, separated from Merritt Island by the Banana River, and Merritt Island is separated from mainland Florida by the Indian River. The current peninsula is believed to have begun forming about 7,000 years ago. Surface deposits today are largely sand and sandy coquina.[4] The peninsula formed a natural harbor to its south, called a bight. The deep water of the harbor created a relatively safe haven for ships at sea during a storm.[5]

The earliest artifacts suggest that Indigenous peoples inhabited the Cape about 5,000 years ago. They collected shellfish, fished in nearby waters, and hunted for game. The Banana River estuary to the west of the Cape, and its adjacent marshes, provided a rich wildlife habitat as a food source. A University of Central Florida student dig in 2022 found stone spear points estimated to be 2,000 to 6,000 years old.[6]

European discovery of the Cape is generally credited to Spanish explorer Juan Ponce de León, whose ship is believed to have anchored offshore in 1513.

Bumper 8 launches from Cape Canaveral on July 24, 1950. Image source: NASA/DVIDS.

The Spaniards gave Canaveral its name; *Canaveral* is believed to mean a place overgrown with canes or reeds; it's sometimes translated as "canebrake."[7] European occupation is considered largely responsible for the Indigenous diaspora of the eighteenth century.[8]

The name *Merritt Island* might have evolved out of a Spanish land grant that referred to the island by many names. One of those names was "Isla de Marratt," perhaps after Captain Pedro Marratt, the head surveyor of the Spanish governor in East Florida at the end of the eighteenth century. By the early nineteenth century, the term *Merritt's Island* was in use. The possessive was eventually dropped by the turn of the twentieth century.[9]

The fledgling United States in 1822 acquired Florida from Spain. Florida's territorial legislative council in 1824 included the future Space Coast in a newly formed county, aptly named Mosquito County. In 1854, the county was divided into smaller counties. Brevard County was named after Florida comptroller Theodore W. Brevard. By the 1850s, pioneers began to trickle into the area, although it remained sparsely settled until the twentieth century.[10] A lighthouse was established on the tip of Canaveral, around which a small settlement grew. The second lighthouse, assembled from cast-iron plates lined with brick, has been operational at its current location since 1894.[11]

As the twentieth century began, homesteaders started to occupy the Cape's interior. New dirt roads were plowed to connect the homesteads. A church and a store opened. Boat transportation was the primary means of reaching the peninsula from Merritt Island and mainland shoreline towns such as Cocoa and Titusville.[12]

In the 1920s, bridges replaced steamers as the primary access from the mainland, east across the Indian River to Merritt Island, Cape Canaveral, and the barrier islands. A toll bridge funded by a Titusville bond issue connected that city in 1921 to north Merritt Island.[13] Ten miles south of Cape Canaveral was a town called Oceanus, also known as Cocoa Beach. By 1922, Cocoa Beach was connected to the mainland by a bridge that began on the mainland in the City of Cocoa, connected to Merritt Island, and then continued to the township.[14]

A pier was built on the Atlantic side of the Cape; Pier Road still exists at today's Space Force station, although the pier itself is long gone.[15] Local businesspeople advocated upgrading the Canaveral Bight, the natural harbor south of the peninsula. Not only did they want the bay dredged for larger ships, but they also foresaw improved access from the mainland by road and rail, and possibly a cross-state canal.[16]

Land speculators formed companies to purchase—and sell at a profit—properties near Canaveral Bight. The Merritt Island Realty Company published a two-page advertisement in the October 14, 1925, *Miami Herald.* They declared the harbor to be "Another Miami in the Making!" The advertisers illustrated a proposed Orlando to Canaveral railroad line, and a proposed causeway the railroad would use to cross the rivers. The ad declared Canaveral Harbor to be "the deepest harbor between Jacksonville and Miami" and "the logical point for harbor development and the accommodation of the largest ships in ocean transportation."[17]

The local real estate boom went bust along with the rest of the nation in 1929 after the stock market collapsed and the Great Depression began. Financing dried up. The Florida state legislature in 1929 created a Canaveral Harbor District, but it went unfunded.

The Atlantic Peninsula Corporation, formed in Titusville, received a permit from the US War Department in 1933 to build a pier, breakwater, and seawall, and to dredge a channel out to deep water. The Corporation failed to raise funds for the project. An investor group called the Port Realty Company, later revealed to be a subsidiary of the Corporation, announced in February 1938 that it had bought 8,000 acres on the Cape and Merritt Island, hoping to "create a harbor at Canaveral for a direct trade and passenger route to Europe, and to develop all adjacent industries."[18] The group even submitted a proposal to the federal government to purchase the lighthouse reservation.[19] With war clouds looming in Europe, the corporation's founder argued that a seaport at Canaveral would relieve wartime congestion in American Atlantic ports and accommodate US naval forces.[20]

The State of Florida in 1939 created the Canaveral Port Authority to replace the earlier Canaveral Harbor District. The Corporation and the Authority partnered to request the US Army Corps of Engineers study how to build the port and dredge a channel through the Cape peninsula to the Banana River. This channel would have a turning basin with terminal facilities. The channel would then cut through Merritt Island to connect to the Indian River, part of the Intracoastal Waterway.[21] Placing the harbor on the Banana River side of Cape Canaveral, rather than in the Canaveral Bight, was considered safer for berthed ships than building jetties for a mile out to sea.[22]

In March 1941, the Army Corps of Engineers announced its approval for the project, estimating a cost of $1.6 million with half paid by federal funding, the rest by the locals. The division engineer's recommendations would become the footprint for today's Port Canaveral, servicing Space Coast government and commercial launch operations. The report included a twenty-seven-foot deep channel protected by stone and granite concrete jetties at its entrance, a turn basin, and a barge canal with a lock system that would connect the Cape, Merritt Island, and the Banana and Indian Rivers, which are part of the Intracoastal Waterway.[23]

Celebrations were premature. War was nigh. The federal government was disinclined to fund the project unless it was found necessary for the national defense.[24] The proposed fifty-fifty funding split continued through World War II, with the stipulation that no money would be appropriated until after the war unless the chief of Army engineers certified the project essential for national defense.[25] It wasn't until March 1945 that a rivers and harbors bill with funding for the Canaveral project managed to pass both houses of Congress and was signed by President Franklin Delano Roosevelt.[26]

Federal funding for Canaveral Harbor work was finally allocated by the Army engineers in July 1946.[27] Now it was time for the locals to do their part.

The Canaveral Port Authority began meetings that month to discuss how to raise the money.[28]

In October 1947, the Authority announced a plan to issue $1,365,000 in revenue bonds—$940,000 to match government funds, $425,000 for harbor facilities. The bonds would be repaid by revenue from long-term rentals of Authority land, and ad valorem taxes against taxable properties in the district.[29] The scheme was approved in a November 25, 1947, special election, with a 93 percent yes vote cast by Canaveral Harbor district property owners.[30]

The Atlantic Peninsula Corporation and the Port Realty Company investors had acquired their lands on Cape Canaveral and Merritt Island because they envisioned a privately built and integrated barge canal, port, and city that they would develop. It was only after English-backed funding was frozen in 1939 due to the war that the investors chose to pursue government financing of the harbor, hoping to reap a profit from the sale of the harbor land and the development of the remaining properties.[31]

What they hadn't anticipated was the Authority using the government power of eminent domain to condemn and seize their lands. A court ruled in April 1949 that the Authority could use condemnation to acquire 300 acres intended for port construction.[32] A jury in July 1949 awarded compensation largely in line with what the Authority had offered.[33]

* * * * *

American military presence on the Canaveral peninsula traces back to the late 1930s, when the Roosevelt administration prepared for a second world war.

Aggression by Nazi Germany in Europe, and imperial Japan in the Pacific, led to 1938 congressional legislation directing the secretary of the Navy to create a board to report on the need for "additional submarine, destroyer, mine and naval air bases on the coasts of the United States." Named the Hepburn Board after its chair, Admiral Arthur J. Hepburn, the committee traveled the nation finding facts and touring sites, including those in Florida. The report recommended Jacksonville, Florida, as the site for a new major Atlantic naval base but felt it wasn't optimal for seaplane patrols. The board identified "a suitable site . . . in the lower reaches of the Banana River near Cocoa Beach" as an adjunct to the Jacksonville base.[34]

Plans for Canaveral Harbor ran in parallel to, but were separate from, the Navy's plans for an air station at Cocoa Beach. The Atlantic Peninsula Corporation's 1939 request for the Army engineers to study the proposed seaport was considered independent of the naval station, although the assumption was that the Navy fleet would benefit from the harbor.[35]

By April 1939, Congressional legislation had authorized the selection of Jacksonville for the major base with an auxiliary seaplane station at Banana River, despite efforts by a Miami contingent to have Congress dictate their city for the base instead of Jacksonville.[36] In May, Jacksonville realtors descended on Brevard County to choose the final site for the naval station and appraise its value.[37] They were soon followed by a survey team. According to an *Orlando Morning Sentinel* report, "The northern boundary of the area to be surveyed lies about four miles south of Cocoa Beach on the Canaveral peninsula and the proposed site is between two and four miles in length, lying along the east shore of Banana River."[38]

As would happen a decade later with Port Canaveral, the federal government in October 1939 began condemnation proceedings to acquire 890 acres of land for the seaplane station. The site lacked a primary landowner, as did Canaveral with the Atlantic Peninsula Corporation.[39] A judge approved the condemnation by the end of the month.[40]

Naval Air Station (NAS) Banana River was commissioned on October 1, 1940.[41] Practice bombing ranges were set up on Cape Canaveral; an October 19, 1941, *Orlando Sunday Sentinel* article warned that a bombing target had been erected at "the head of the Banana River," northwest of the Canaveral lighthouse, and that practice would be held daily from 8:00 a.m. to 4:00 p.m.[42] Pilot trainees dropped inert practice bombs on their targets.[43] A separate tow target range for bombers out of Orlando Air Base was established east of Canaveral in the Atlantic Ocean.[44]

After the declaration of war against the Axis in December 1941, the practice bombs were replaced with real ones. Training flights were supplemented with active duty patrols. Coastal patrols were replaced with anti-submarine patrols, as German U-boats attacked commercial shipping lanes. An August 13, 1942, *Miami Daily News* article detailed convoy patrols from the Banana River naval station, with planes ranging from scouts to patrol bombers. The reporter wrote of witnessing the hulks of torpedoed ships sunk along the Florida coastline. After early losses due to "lone wolf merchantmen," the Navy herded them into convoys protected by surface and air craft, reducing losses.[45]

As the American military organized and strengthened its defenses, and with merchant shipping protected in convoys, U-boat attacks plummeted and fewer ships were lost. By one estimate, twenty-four ships were attacked off the Florida coast during World War II, but most of those were in 1942. Only four ships were torpedoed in 1943, and none from 1944 until the end of the war in 1945.[46]

* * * * *

World War II and the presence of NAS Banana River fundamentally changed the local economy of Brevard County. Before the war, the region was best known for Indian River citrus, commercial fishing, resort tourism, and an unspoiled coastline that speculators hoped to privately develop into a major harbor and tourist destination. The locals had relied on government to build and maintain infrastructure, but not as a significant source of employment.

In March 1945, near the war's end, 730 officers and 5,237 servicemen were stationed at Banana River.[47] The federal government had become a significant employer in the area; newspapers throughout Florida ran articles on behalf of the US Civil Service Commission seeking civilian workers, such as aircraft mechanics, electricians, metalsmiths, woodworkers, and machinists.[48] Brevard's local economy was no longer diversified, relying instead on a transient government program as its economic engine. This pattern continued into the twenty-first century, with American aerospace at its core.

After Japan surrendered on September 2, 1945, to end the war, the Navy began discharging officers and servicemen. In early 1946, the Navy announced that NAS Banana River would be one of five Florida installations to maintain full status, assuring each would have more than one-half the personnel and facilities required during wartime, but other air stations would be considered surplus.[49] By August, the total number of personnel at Banana River was down to 2,200.[50] Although the Navy continued to assure locals that Banana River was a "permanent" facility tasked with training aviators, an evolution in global military strategy was about to assure its closure.

Relations began to deteriorate between the United States and its wartime ally, the Soviet Union. America had the nuclear bomb; it was suspected that the Soviets might soon have one too. Military strategists believed that the next threat to the American mainland would not sail across the Atlantic Ocean, but fly "across the roof of the world," enemy bombers crossing the North Pole and the Arctic Circle. The "polar concept" would deter Soviet aggression by transferring air power to the north, specifically to Alaska. This change in thinking meant that naval air stations such as Banana River were no longer needed.[51]

At the same time, the Navy faced significant budget cuts. In late May 1947, US Senator Claude Pepper (D-FL) broke the bad news to his Florida constituency that it was likely NAS Banana River's training squadrons would be a casualty.[52] The Cocoa Chamber of Commerce and local businesspeople met to strategize their appeal to save the air station.[53] It made no difference; the secretary of the Navy notified Banana River's commander that the station would close on August 1.[54] The immediacy of the closure came as "a great shock" to "those whose incomes would be affected by the closing," according to a July 14, 1947, *Orlando Evening Star* article. But in an era of congressionally mandated spend-

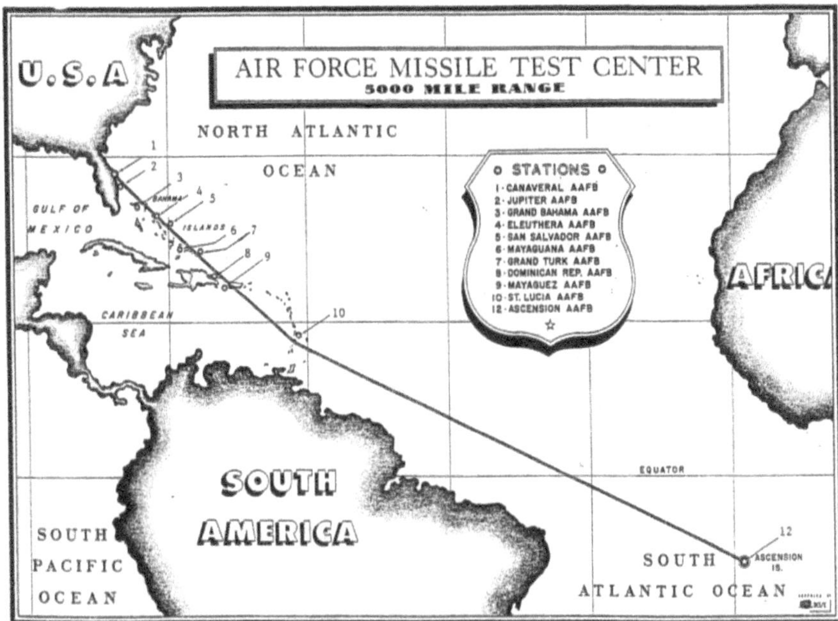

A map of the Atlantic missile range in the early 1950s. Cape Canaveral was selected as a test site so missiles could launch along an Atlantic island chain, sending data to tracking stations and ships at sea. Image source: NASA.

ing reductions, closing Banana River made sense to federal budget analysts. "The base is one of the most expensive to operate, per capita, of any base in the country," due to the unstable soil and the saltiness of the artesian well water.[55] A last-minute reprieve extended the closure date to October 1.[56]

It was around this time that Wernher von Braun and his German engineers were reaching the limits of their rocket range at the Army test site in White Sands, New Mexico. A Pentagon committee tasked with finding a new site selected El Centro, California, on the Mexican border. Rockets would launch southwest over Baja California, but Mexico President Miguel Alemán Valdés declined to grant the United States the right to fly over the Baja. The committee went with their second choice—Cape Canaveral.[57]

By spring 1948, rumors started to circulate in Florida newspapers that NAS Banana River might reopen. The tenant might once again be the Navy, or it might be the newly minted Air Force, or it might even be the Army. A Cocoa Chamber of Commerce secretary received a phone call on July 22 from an Army colonel asking "what the local reaction would be if a cleanup crew of Negroes were sent" to Banana River.[58]

The Air Force announced in early September 1948 that its branch would take over responsibility for NAS Banana River, although the station would remain inactive for now.[59] A Miami reporter in January 1949 figured out that the base would be a joint operation of the Air Force, Army, and Navy as "a test ground for many of the latest weapons, including large caliber guns."[60] The next day, an Orlando newspaper reported that 7,500 servicemen would be assigned to Banana River.[61] A congressional hearing on February 14 revealed that the military was looking for a broadened test range of at least 500 miles that might be extended one day to 5,000 miles.[62] In early March, Banana River's commanding officer confirmed that it was being prepared for guided missile experimentation, pending federal funding.[63]

The May 4, 1949, *Orlando Evening Star*'s front-page headline declared, "Cocoa AF Project Hailed." The news confirmed all the rumors of recent months. Brevard County Commissioner and Canaveral Port Authority Chairman A. Fortenberry told the paper, "It will mean a lot to Cocoa and Melbourne as well as this entire Central Florida." He predicted that "the missile base and harbor would be beneficial to each other."[64] The paper foresaw a "building boom" for the East Coast thanks to the new guided missile proving ground at Banana River.[65]

As had the Canaveral Port Authority, the Air Force used the government's power of eminent domain to condemn the land it wanted on Cape Canaveral. Condemnation proceedings began in April 1950.[66] As at Canaveral Harbor, the Port Realty Company, a subsidiary of the Atlantic Peninsula Corporation, was the primary landowner. The company had been working toward creating a Canaveral city, but its plans had been "wrecked by seizure of its land for the guided missile base," according to a September 1951 Orlando newspaper report.[67] Litigation continued for years; a final settlement was not reached until 1955.[68] In a 1962 retrospective it was estimated that, for every dollar the Corporation investors had put into their Canaveral project, they got back less than a quarter.[69]

* * * * *

Dependency on federal contracts for survival of the local economy was not foreseen when the Department of Defense (DOD) began launching rockets from Cape Canaveral in the early 1950s.[70]

Brevard County had a population of only 23,653 in 1950. Over the next decade, it grew to 111,435.[71] By 1970, the year after the first Project Apollo crewed lunar landing, the population was 230,006. The Cocoa Beach–Merritt Island division of the 1970 US Census grew from 19,320 in 1960 to 51,119 in 1970.[72]

In the 1950s, the first jobs were at what would later be called Cape Canaveral Air Force Station (CCAFS).[73] That's also where Project Apollo began under President Dwight Eisenhower in 1960. But as Apollo evolved under President John F. Kennedy into a crewed lunar program, the civilian National Aeronautics and Space Administration (NASA) expanded its operations to Kennedy Space Center (KSC) on north Merritt Island, on land adjacent to and generally northwest of the Cape. The local labor pool supported both the military operations on Cape Canaveral and the civilian operations on Merritt Island. Largely because of the NASA buildup at KSC, employment in Brevard County more than doubled, from 33,500 in 1960 to 68,400 in 1965. The county unemployment rate in 1965 was only 1.8 percent.[74]

By one account, the number of KSC jobs peaked at 26,000 during Apollo 7 operations in 1968. The US space program that year provided over 40 percent of Brevard County's employment.[75]

Government programs come and go, and with them jobs. Project Apollo was no exception.

Apollo's new goal, articulated by Kennedy on May 25, 1961, was to land a man on the moon by the end of the decade and return him safety to the Earth. Once that goal was in sight, KSC began to shed jobs. Between 1968 and 1970, over 10,000 jobs were lost. Economic growth reversed and plunged. Some businesses couldn't survive the downturn and closed. Many aerospace workers, unable to find other work, left the county.[76] Most of the lost jobs were contractor employees, which are by definition temporary, while federal government civil servants by law have certain protections.[77] By 1974, the number of KSC jobs was down to 9,246.[78] Some of those job losses were due to the end of the NASA Skylab space habitat project; the county unemployment rate rose from 7.0 percent in 1973 to 10.2 percent in 1974.[79]

Jobs also declined across the Banana River at CCAFS. According to a 1975 article in a local newspaper called *Today*, Cape jobs peaked at 27,447 in 1964, then began a slow decline. (*Today* was the original name for Gannett's current publication, *Florida Today*.) The number of Air Force station–related jobs in 1974 went down to 13,906. Brevard County at the time had the second highest unemployment rate in Florida, at 14.4 percent. The county's Division of Health and Social Services blamed the rate on "the county's heavy dependence on Air Force Eastern Test Range and Kennedy Space Center employment."[80]

The decline in military contractor employment coincided with a series of arms limitation agreements signed by the United States and the Soviet Union. The objective was to stabilize US-Soviet relations by limiting the development of strategic defense systems.[81] Development and testing of long-range ballistic launch systems at Cape Canaveral ended by 1970. Operational tests of intercontinental ballistic missiles (ICBMs) relocated to Vandenberg Air Force Base in California.[82]

The job losses at KSC and CCAFS depressed local real estate. In 1970, the Brevard County housing market was flooded with homes abandoned by their owners. A November 21, 1971, *Orlando Sentinel* article compared the Space Coast economy to a gold or oil boom going bust. The author wrote, "Brevard County's housing market has rocketed in 15 years from famine to feast and back again to famine." When the space boom went bust, "Foreclosures on FHA and VA loans became commonplace, and the huge subdivisions sprouted a forest of 'for sale' signs." But by the end of 1971, the housing market had begun to recover, as realty agencies placed advertisements in publications targeting military and federal retirees and veterans. One realtor estimated that 70 percent of the homes sold in the last year were to retirees.[83]

This boom-and-bust cycle would repeat itself time and again in future years. Another boom began as the Space Shuttle program came online in the late 1970s. The number of KSC jobs increased from 8,441 in 1976 to 16,067 in 1985.[84] An October 27, 1980, *Florida Today* article predicted that Brevard County would move into the new decade "riding the crest of an economic revival." The Space Shuttle program was seen as "stabilizing employment in the north and central areas of the County. Brevard County during the 1980–1985 period will enter the most active five years growth period experienced since the space race of the sixties."[85]

The loss of the orbiter *Challenger* started another "bust" in the cycle. *Challenger* was destroyed 73 seconds after launch on January 28, 1986. In the year that followed, more than 2,000 KSC jobs were cut. This time, though, the job losses were temporary, and by the time the Space Shuttle program resumed operations in September 1988 about 500 more people were employed at KSC than before the accident.[86]

After *Challenger,* the space shuttle was phased out as a launch vehicle for uncrewed NASA payloads such as orbital satellites and interplanetary probes. There was no reason to risk lives for payloads that didn't require people to deploy them. The uncrewed Soviet *Progress* cargo ship had been delivering payloads to Russian space stations since 1978.[87] The United States could have developed a robotic cargo ship, but chose instead to have people deliver payloads. That decision cost NASA seven lives with *Challenger,* and another seven with *Columbia* in 2003.

Commercial payloads such as communications satellites were also phased off the shuttle. They would now launch on expendable commercial rockets, meaning the rocket would be used only once. After the payload separated and headed into space, its booster fell back into the ocean. The booster was destroyed on impact, like an egg dropped onto concrete, with the remains sinking to the bottom of the ocean.

At the time, no one knew how to recover a rocket and use it again.[88]

With commercial customers now flying on expendable boosters, the Space Shuttle program no longer had that revenue stream to defer expenses. An independent review team issued a report in February 1995 that recommended "disengaging NASA from the daily operation of the space shuttle" by "consolidating operations under a single-business entity."[89]

By August 1995, Lockheed Martin and Rockwell International had joined forces to form United Space Alliance (USA), a company that would take over all Space Shuttle program operations from NASA. Rockwell COO Kent Black told a *Florida Today* reporter, "The only way to reduce costs is to reduce head count, and I expect that to happen across the board. We'll be scrutinizing every area."[90] Boeing acquired Rockwell in August 1996 and assumed the old company's fifty-fifty USA partnership with Lockheed Martin.[91] USA eventually consolidated thirty heritage contracts that supported the Space Shuttle program.[92]

The number of KSC Space Shuttle program jobs declined throughout the 1990s. The total was 19,088 in 1991. When USA took over the contract in 1996, the total was 16,208. In 2001, the first year of the twenty-first century, the total was down to 13,499.[93]

The orbiter *Columbia* was lost on February 1, 2003, breaking up over eastern Texas as it glided toward a KSC landing. "Economic Effect May Be 'Scary,'" read the headline of a *Florida Today* article the next day.[94] In the weeks and months that followed, local leaders expressed cautious optimism that the *Columbia* tragedy would not trigger the economic downturn that impacted the Space Coast after the end of Apollo and the loss of *Challenger.* As it turned out, the loss of *Columbia* didn't impact the job numbers at KSC, nor did it impact the Brevard County economy. In the years since *Challenger,* the Brevard County economy had diversified. In the mid-1980s, aerospace accounted for a fourth of Brevard's economy. By the time of *Columbia,* it was a tenth.[95]

The post-Apollo recession, nonetheless, was ingrained into the collective psyche of the KSC workforce, and the nearby communities where they lived. These fears persisted through the *Challenger* and *Columbia* disasters, and would emerge yet again as the Space Shuttle program came to an end in the early twenty-first century.

* * * * *

Cape Canaveral's first space age began that hot and humid summer day in July 1950, when the converted German V-2 rose above the Florida scrub palmetto to veer over the Atlantic shoreline and out to sea.

For the next six decades, Cape launches by and large were driven by one government program or another—first military, then military and civilian. Rocket technology was considered too magical and mystical for entrepreneurs to attempt, much less succeed. Brevard County's economy therefore became dependent on federal spending. Time and again, the boom went bust. When it did, businesses shut, jobs were lost, and homes were foreclosed.

Was there a way out of the boom-and-bust cycle? The answer is yes, and this is that story.

Cape Canaveral's second space age began with a group of local visionaries who convinced a politically unpopular Florida governor to "capture the future" by taking Canaveral's destiny into their own hands. Their vision was to free Florida of the boom-and-bust cycle by nurturing a commercial space industry.

This vision paralleled a few entrepreneurs, some of them billionaires, who were willing to risk their fortunes on founding their own rocket companies—and to compete with the government's subsidized contractors.

The visionaries and the entrepreneurs would intersect at Cape Canaveral. As the Space Shuttle program came to an end, they faced opposition from Florida politicians of all partisan stripes, representing constituencies who wanted to continue their government contracts in perpetuity. In a county traditionally conservative in its political beliefs, many nevertheless embraced and even demanded a continuance of the federal-funding model as a significant driver for their local economy, despite the many historical lessons that such a model was vulnerable.

For the first time since the 1960s, American ingenuity in space would capture the imagination of the world and reestablish the United States as the preeminent global space power. As this new space industry flourished, jobs returned, launches were frequent, and American astronauts once again launched from American soil. This comeback was symbolized by one company's reusable rocket booster that could launch, land, and launch again.

To tell the story of Cape Canaveral's second space age, we must return to its first, when a vigorous young president challenged his nation to go to the moon.

2

Something Old, Something New

> I believe that this nation should commit itself to achieving the goal, before this decade is out, of landing a man on the moon and returning him safely to the earth. No single space project in this period will be more impressive to mankind, or more important for the long-range exploration of space; and none will be so difficult or expensive to accomplish.
>
> —President John F. Kennedy, "On Urgent National Needs," May 25, 1961[1]

Popular mythology tells us that President Kennedy strode down the House of Representatives aisle one day, challenging Congress and America to race the Soviet Union to the moon, in a great quest to defeat communism. The nation united behind him and, after his assassination in Dallas on November 22, 1963, Project Apollo became a monument to his legacy.

That's not quite what happened.

Kennedy had been inaugurated president four months earlier. Space was not a priority for him until the Soviets launched the first human into space, Yuri Gagarin, on April 12, 1961.[2]

Five days later, Cuban exiles trained by the Central Intelligence Agency failed in their attempt to invade Cuba and overthrow Fidel Castro. Kennedy's closest advisers recalled that the president on April 19 was "anguished and fatigued," and "in the most emotional, self-critical state." His brother Robert reportedly told the president's top appointees, "You got the president into this. We've got to do something to show the Russians we are not paper tigers."[3]

The next day, the president sent a memo to Vice President Lyndon Johnson, whom Kennedy had appointed to chair the National Aeronautics and Space Council.[4] He charged Johnson with "making an overall survey of where we stand in space." He asked, "Do we have a chance of beating the Soviets by putting a laboratory in space, or by a trip around the moon, or by a rocket to go to the moon and back with a man. Is there any other space program which promises dramatic results in which we could win?"[5]

Johnson responded on April 28 with an eight-page memorandum.[6] It said that "the Soviets are ahead of the United States in world prestige attained through impressive technological accomplishments in space." Johnson wrote, "Dramatic accomplishments in space are being increasingly identified as a major indicator of world leadership."

> Manned exploration of the moon, for example, is not only an achievement with great propaganda value, but it is essential as an objective whether or not we are first in its accomplishment—and we may be able to be first.

The goals of achieving "prestige" and "world leadership" became a recurring justification for the moon program during the Kennedy administration. NASA Associate Administrator Robert Seamans, for example, used prestige as a justification in speeches and other public commentary. In a December 1961 speech, Seamans said that "the prestige of a nation" depended on "mastery and control of man's physical environment."[7] A United Press International article around the same time quoted Seamans as saying that prestige "is one of the most important elements of international relations." He said that people tend "to equate space and the future."[8]

Kennedy's May 25, 1961, speech was not just about the space program. In fact, the moon part was near the end of a forty-seven-minute delivery that was largely a shopping list of new programs Kennedy wanted Congress to fund to show strength in the face of Soviet achievements.

The *New York Times* report the next day noted that while members of Congress generally liked what Kennedy had proposed, they were skeptical about the cost. Support for the moon initiative was tepid even within his own party; some liberals feared the moon program might divert funding from programs such as aid to the aged.[9]

A Gallup poll conducted around the time of the speech showed the American public was also reluctant. When asked if the United States should spend $40 billion—about $225 per person—to send a man to the moon, 58 percent said no. Only 33 percent said yes. When asked why they opposed it, many of the respondents replied it wasn't necessary or it wasn't worth the cost.[10]

What was the actual cost? At the time, no one really knew. Estimates were anywhere from $20 billion to $30 billion to $40 billion or more. A 1990 NASA estimate put it at about $25.4 billion ($95.0 billion in 1990 dollars), or about $236.0 billion in 2025 dollars.[11]

Space historian Roger D. Launius in 2003 published a monograph that looked at 1960s Apollo polls. He concluded that the only time a majority of Americans thought Apollo was worth the cost was in July 1969, at the time of the Apollo 11 crewed lunar landing. Consistently throughout the 1960s, about 45 to 60 percent of Americans thought their government was spending too much on space.[12]

Americans weren't opposed to space exploration, per se. They mostly questioned the return value from the enormous expense. Launius found that polls over the decades have shown NASA is one of the most respected government agencies in the United States, and Americans believe that work is important, even if they're not sure what that work is. He wrote, "The American public is notorious for its willingness to support programs in principle but to oppose their funding at levels appropriate to sustain them."[13]

Americans may have been reluctant about paying for it, but in retrospect they cite the Apollo moon program as one of the nation's great achievements. A Pew Research Center nationwide survey conducted at the end of the twentieth century found respondents cited "the space program . . . as the country's single greatest achievement of the century." Almost one in five cited the exploration of space as America's greatest feat.[14]

At the end of the space shuttle era in July 2011, CNN in partnership with ORC International conducted a telephone poll of about 1,000 adult Americans. Although half thought the end of the Space Shuttle program was bad for the United States, when asked, "Do you think the US should rely more on the government or more on private companies to run the country's manned space missions in the future," by 54 percent to 38 percent the respondents preferred the private sector.[15]

Until the end of the Space Shuttle program, the government controlled both the military and civilian space programs. The private sector built the launch vehicles and the payloads, but only after receiving a government contract to design, build, launch, and operate them to the government's specifications.

How did the American space industry arrive at such a dichotomy? Why did government bureaucrats and politicians guard the gates to space, even though the American public wanted otherwise?

The answers for the military and civilian programs differ, although they run in parallel and often overlap.

* * * * *

Federal government agencies have several options for purchasing launch services, falling into two general categories: cost-plus and fixed price.

With a cost-plus contract, the contractor is compensated by the government for all allowable incurred costs. The contractor may not exceed the contract's total amount, except at its own risk, without approval of the contracting officer. Cost-plus contracts are typically used when labor and material costs are highly uncertain to perform the contract. Contractors might not bid on high-risk programs if they're not guaranteed to make a profit, such as the research and development projects common among the military and NASA. The government therefore assumes the risk that the contract might not be completed on time or for the expected cost. The "plus" refers to an additional payment to assure the contractor makes a profit.[16]

With a fixed-price contract, the government acquires commercial products and commercial services from a contractor, much like we might purchase a commercial airline ticket to fly from one place to another. The price can only be adjusted by the terms of the contract. Unless specified, the contractor assumes all the risk. A fixed-price contract is used when the requirement is well-defined and financial risks are insignificant. The pre-negotiated fee paid to the contractor typically provides no incentive for performance or cost savings.[17]

The US military began after World War II to research and design long-range rockets and missiles. Because of their experimental nature, cost-plus contracts were the typical means of selecting a contractor. The need for concurrent research, development, and production of new aerospace technologies exploded at the beginning of the Cold War. It became impossible to introduce competition into the process. The military branches turned to sole-source acquisition through cost-plus contracts that covered the contractor's costs and allowed an additional margin for profit. Thomas McNaugher, who at the time was a research associate at the Brookings Institution, estimated in 1987 that "cost growth during the Cold War era was 200 to 300 percent on average" due to uncertainty during high-risk development. McNaugher wrote that the weapons acquisition process in the United States "injects powerful, technically divisive political incentives into the development of weapons systems." Politicians seek to ensure that their districts and states get "their fair share, or more" of contracts.[18]

The history of NASA's predecessor followed a different path. In the early twentieth century, airplane development in the United States was largely dependent on amateur inventors, such as Samuel Langley, Glenn Curtiss, and the Wright Brothers. In Europe, where war would soon begin, governments recognized the potential of the "aeroplane" as a new weapon. European governments encouraged scientists, engineers, and industrialists to dedicate their skills and resources to developing this new technology. The United States made no similar effort. When World War I began in 1914, it was estimated that France had 1,400 airplanes, Germany 1,000, Russia 800, Great Britain 400, and the United States only 23.[19]

To finally compete with Europe's aviation industry, Congress approved the creation of the National Advisory Committee for Aeronautics (pronounced "N-A-C-A") in March 1915. Advances in aeronautics were viewed as helping to advance the science in other fields.[20]

The War Department in October 1916 purchased land near Hampton, Virginia, for joint use as an aircraft proving ground by the Army and Navy. It was named Langley Field, after Samuel Langley. The NACA chose Langley as its first research center. Lacking its own facilities, NACA contracted with others to perform its research. After the war ended in 1919, the Army and Navy continued to maintain their own separate aeronautical research sites elsewhere in the country, while the NACA remained with its own wind tunnel at Langley.[21]

The NACA built a competent engineering staff at Langley, and earned a reputation for independence, but would also contract with other facilities elsewhere, typically at universities. The agency's capabilities were demonstrated by the development of unique wind tunnels where aeronautic hypotheses could be tested. One significant advance was the construction of a tunnel where full-scale airplane propellers could be tested under conditions of flight. NACA's research into engine nacelle design increased airplane speeds.[22]

NACA contributed to the advent of airmail service in the United States, from which evolved commercial aviation. The agency first "examined the problem of the carriage of mail by air" in 1916. It recommended to Congress that experimental airmail routes be created as technology demonstrators; Congress appropriated $100,000 in 1918 to establish the first delivery routes between Washington, Philadelphia, and New York using military airplanes flown by Army pilots. The Air Mail Service became the pathfinder for commercial aviation. NACA recommended to President Calvin Coolidge that only the US Post Office should operate airmail service until commercial aviation had matured to assume the responsibility.[23]

By the mid-1930s, NACA's work was internationally known and respected. Except for findings applicable to military security, all NACA research was freely available to the public and to the world. NACA won the United States international "prestige" and "leadership" long before the Cold War. Additional research facilities opened at Moffett Field, California, in 1939 and Cleveland, Ohio, in 1940.[24]

As the military began to explore the sciences that would support rocketry, NACA participated as well. The agency researched various exotic fuels that not only could be used by aircraft, but also by rockets and missiles. NACA determined that a blunt-nose shape would help create a very high-pressure drag on reentry that would dissipate heat in the region of the shock wave. The initial use for this knowledge was to redesign existing ballistic missile warheads, but NACA's "blunt body" research also led to the capsule-shaped spacecraft design still used today, and winged vehicles such as the space shuttle orbiter and the Sierra Space Dream Chaser.[25]

At its end in 1958, most of NACA's research was conducted in-house. The agency's final budget for fiscal year (FY) 1958 was $76 million, but of that only $500,000 was for research contracts with educational institutions or government agencies.[26] When it ceased to exist on October 1, 1958, NACA was unencumbered by bloated high-risk development contracts, and the attending political interests that afflicted their siblings in the military.

* * * * *

The tale of how NASA came to be has often been told in print and in film, in fact and in fiction. Our interest is in how an academic culture largely free of political influence came to be an instrument of economic policy and an exercise of soft power in foreign relations.

Just as European superiority in aeronautics during the first world war prodded Congress into action, so did Sputnik and subsequent events in 1957 force Congress and the White House into a more assertive aerospace policy. The launch of Sputnik 1 by the Soviet Union on October 4, 1957, and Sputnik 2 a month later with a dog named Laika, panicked not only the American public but also the politicians who represented their interests.

Sputnik (Russian for "satellite") wasn't a secret. It was the Soviet contribution to the International Geophysical Year, a global event planned since 1952 pooling the resources of scientists around the world to study the planet. On July 28, 1955, the Eisenhower administration announced American participation in the IGY by launching "small earth-circling satellites . . . This program will for the first time in history enable scientists throughout the world to make sustained observations in the regions beyond the earth's atmosphere." Four days later, the Soviet press announced that the USSR would also launch a satellite for IGY.[27]

On September 30, 1957, at an IGY conference held at the National Academy of Sciences in Washington, DC, a Soviet delegate delivered a presentation on Sputnik. Although he was circumspect about when it might launch, he provided the radio frequencies Sputnik would use to broadcast. The presentation was reported by major newspapers and wire services.[28]

Sputnik 1 transmitted a "beep beep" heard around the world. Eisenhower's press secretary James Hagerty told the media that the launch "did not come as any surprise; we have never thought of our program as in a race with the Soviets."[29] But to the average American, it was concerning, if not downright frightening.[30] Members of Congress called for investigations; Senator Henry Jackson (D-WA) declared that Sputnik was "a devastating blow to the prestige of the United States as the leader in the scientific and technical world."[31]

The American contribution, called Project Vanguard, was a nominally civilian program, not a military priority. Vanguard was designed and operated by the Naval Research Laboratory. Their proposal was chosen over an Army Ballistic Missile Agency (ABMA) design using a modified Redstone booster proposed by Wernher von Braun.[32]

Vanguard TV-3 attempted to launch on December 6, 1957, from Pad 18A at Cape Canaveral. Two seconds into the launch, four feet off the pad, TV-3 fell back to the ground and exploded live on American television. The problem was traced to a loose connection in a fuel line above the engine.[33]

While Project Vanguard struggled with technical problems, ABMA revived its plans to launch a satellite using a modified Redstone booster called a Jupiter-C. Formal approval was granted on November 8, 1957, to proceed with planning for a launch. The mission, called Explorer 1, launched into orbit on January 31, 1958. Its payload was a satellite designed by the Jet Propulsion Laboratory (JPL) in Pasadena, California.[34]

American media reports the next day declared the partial restoration of American prestige. Associated Press reporter John M. Hightower, the 1952 Pulitzer Prize winner for international reporting, commented that the United States had "regained . . . some of the prestige it lost to Russia last October." Hightower speculated that "the launching should strengthen this country's hand in negotiating with Russia for a summit conference."[35] An Orlando newspaper quoted one Brevard County resident as saying, "It'll help our prestige a lot."[36]

The Vanguard TV-3 failure and Explorer 1 launch preparations occurred while emergency hearings were being held by the US Senate to investigate the US supposed inferiority in rocketry. That wasn't the truth, as Republican President Eisenhower knew, but he couldn't say so publicly because he wanted to preserve the secrecy of certain missile and spy satellite projects, a vulnerability the Democratic majorities in both houses exploited. Senate Majority Leader Lyndon Johnson chaired a special "preparedness subcommittee" of the Senate Armed Services Committee to investigate US missile programs. Both houses of Congress created special committees in early 1958 to create a new American space policy.[37]

Senator Clinton Anderson (D-NM) suggested that if immediate propaganda results were the top priority, Congress should "turn von Braun loose" developing his "Super-Jupiter" booster (which eventually became the Saturn program). There was also some debate about whether any new agency should be military or civilian. In late November 1958, Eisenhower created the Advanced Research Projects Agency (ARPA) within the Pentagon to coordinate and direct all government space projects.[38] (Today, it's DARPA, the *D* is for *Defense*.) But after advice from Vice President Richard Nixon, scientists, and other advisers, Eisenhower decided it made sense to separate military and civilian space efforts.[39]

By spring 1958, legislation was working its way through Congress to create the agency that would become NASA. Although the new agency was to be nominally civilian, it was acknowledged that initially it would have to rely on military hardware and expertise. In an era when partisan rivalries were friendlier than today, bipartisan resolution of legislative differences was customary. On July 7, 1958, Johnson went to the White House to meet with Eisenhower and resolve any differences regarding the legislation.[40] The final version was called the National Aeronautics and Space Act.[41] (It's commonly known as the Space Act.)

NASA began on October 1, 1958. The old NACA disappeared into history; its assets and personnel transferred to NASA.[42] Among the programs reassigned to NASA were Project Vanguard,[43] the X-15,[44] an ARPA cislunar probe program called Pioneer,[45] the Army's contract with JPL,[46] and a military program for crewed spaceflight that would soon be called Project Mercury.[47] NASA was to be responsible for the nation's civilian space activities; "development of weapons systems, military operations, or the defense of the United States" remained the purview of the DOD. Where responsibility was uncertain, the president made the final decision.[48]

The Space Act did not require NASA to engage in any specific activities, only to "contribute materially to one or more" of a list of objectives.[49] Nowhere in the Space Act does it require NASA to explore other worlds, to own its rockets, or to fly people into space, although one objective allows NASA to "contribute materially" to the "development and operation of vehicles capable of carrying instruments, equipment, supplies and living organisms through space." The Space Act also authorized NASA to enter into contracts to obtain various services, such as "aeronautical and space vehicles." The contracting authority was rather broad, allowing NASA to enter into contracts "or other transactions as may be necessary in the conduct of its work and on such terms as it may deem appropriate." This language would become known as a *Space Act Agreement* (SAA) and would become politically controversial in future decades, when NASA administrations entered into SAAs to stimulate private sector development of space services.[50]

NASA was intended, as the NACA before it, to be a technology incubator, not an economic stimulus. All that changed with Project Apollo.

* * * * *

John M. Logsdon, founder of the Space Policy Institute at George Washington University, wrote in his 2010 book *John F. Kennedy and the Race to the Moon* that "Kennedy was not at all a visionary in the sense of having a belief in the value of future space exploration; rather, his vision was that space capability would be an essential element of future national power."[51]

Kennedy's NASA administrator, James Webb, had different plans for Apollo.

Webb's college degree was in education. After serving his country in the 1930s as a pilot for the Marine Corps, he went on to earn a law degree. He worked in both the public and private sectors, including a stint with President Harry S. Truman as director of the Bureau of the Budget. (Today, it's called the Office of Management and Budget, or OMB.) He'd also been a deputy governor with the World Bank and the International Monetary Fund, as well as a director for an oil company, and for McDonnell Aircraft.[52]

Walter A. McDougall, a historian and international relations professor, wrote in 1985 that Webb's career experiences had taught him "the virtues of thinking big. By 1960, he had also learned the potential of progressive big government from his service . . . Webb was not a scientist, but a lawyer and manager who administered the power inherent in applied science."[53]

By May 1961, around the time that Kennedy was preparing to deliver his "On Urgent National Needs" speech, Webb was already articulating his agenda for what his biographer, W. Henry Lambright, called in 1995 a "grand mix of noble vision and pork-barrel politics." He envisioned using universities and research institutions across the nation, but particularly the Southwest, as NASA centers or partners to stimulate science and technology industries.[54]

For all his "noble vision," Webb also acknowledged the reality of "pork-barrel politics." In a May 23, 1961, memorandum to Vice President Lyndon Johnson, Webb wrote, "Considerable interest has been expressed in this program by members of the Congress" and proceeded to detail how to curry the favor of certain representatives and senators by investing research funds in their districts and states. "I am convinced, and believe you should consider very carefully, that will attract the kind of strong support that will permit the President and you to move the program on through the Congress with minimum political in-fighting."[55]

McDougall wrote that Webb, Johnson, and others "viewed the political distributivism not as palm-greasing but as the economic and intellectual component of the Second Reconstruction . . . Technological infusion was to call to life a New South, and the space program thus addressed several large items on the national agenda all at once."[56]

Journalist Piers Bizony, another Webb biographer, described "the strange relationship that existed between government and private industry in the 1950s and 1960s." The relationship was somewhere between capitalism and socialism—private companies relied on government contracts for survival, yet these companies remained privately owned and were guaranteed profits. "The thousands of employees . . . were, by any sensible definition, workers for the State," Bizony wrote.[57]

In his December 1961 speech, NASA Associate Administrator Robert Seamans reflected Webb's thinking. Seamans said, "The primary significance of the national space program is that it is a powerful instrument of governmental policy, such that the social and economic impact of the new technology can be channeled to desirable ends."[58]

While Webb thought big, Kennedy constantly worried about the cost of the spending spree he'd unleashed on the American economy. In his "On Urgent National Needs" speech, Kennedy warned Congress that this "new course of action" would "last for many years and carry very heavy costs . . . If we are to go only half way, or reduce our sights in the face of difficulty, in my judgment it would be better not to go at all." During the thousand days of his administration before his death on November 22, 1963, Kennedy ordered several studies questioning the wisdom of the enterprise.[59] Throughout his term, Kennedy sought to convince the Soviets to unite their human spaceflight program with the Americans, partners in a joint mission to the moon. One prominent example is Kennedy's speech to the United Nations on September 20, 1963:[60]

> Why, therefore, should man's first flight to the moon be a matter of national competition? Why should the United States and the Soviet Union, in preparing for such expeditions, become involved in immense duplications of research, construction, and expenditure? Surely we should explore whether the scientists and astronauts of our two countries—indeed of all the world—cannot work together in the conquest of space, sending some day in this decade to the moon not the representatives of a single nation, but the representatives of all of our countries.

The contrast in priorities between Kennedy and his NASA administrator is illustrated by a meeting between the two in the White House Cabinet Room on November 21, 1962. Kennedy had the meeting recorded. Kennedy and Webb, among others, discussed requesting from Congress a $400 million supplemental appropriation for NASA.[61]

The meeting is famous for a lengthy, almost heated, dispute between Kennedy and Webb about what was Apollo's top priority. Webb argued that the science collected in preparing for the first crewed landing was the top priority, because it helped the agency understand the environment in which the astronauts would operate. This would also be consistent with Webb's grand vision for a national technological stimulus.

Kennedy, however, made it clear that he was not interested in "spending six or seven billion dollars to find out about space . . . we shouldn't be spending this kind of money because I'm not that interested in space."

> I think it's good; I think we ought to know about it; we're ready to spend reasonable amounts of money. But we're talking about these fantastic expenditures which wreck our budget and all these other domestic programs and the only justification for it, in my opinion, to do it in this time or fashion, is because we hope to beat [the Soviets] and demonstrate that starting behind, as we did by a couple years, by God, we passed them.

Kennedy recorded another meeting with Webb on September 18, 1963. Webb spoke of "a basic need to use technology for total national power" as a benefit from Project Apollo.[62]

By one NASA account, Project Apollo at its peak employed 400,000 Americans across the United States, and partnered with 20,000 industrial firms and universities.[63] As mentioned earlier in this chapter, it cost about $25 billion in 1960s dollars, or about $236 billion in 2025 dollars. Most of those jobs were with private sector contractors, which by definition meant that the jobs were temporary.

In those early days of Apollo, no one seems to have considered the long-term consequence of so many jobs depending on a federal program with a predefined deadline "before this decade is out." Apollo had a specific goal—a man on the moon by the end of the decade. Apollo had an end date. It was never intended to keep going, to build lunar bases and stations, to send humans elsewhere in the solar system. It was to demonstrate that the United States was technologically superior to the Soviet Union.

When the inevitable layoffs came at Cape Canaveral and elsewhere, those responsible for this shortsighted vision were no longer in office. John F. Kennedy lay in Arlington Cemetery. Lyndon Johnson was in political retirement at his Texas ranch. James Webb left NASA in October 1968, just before Richard Nixon was elected president. Only Nixon, Kennedy's political rival, was in office to reap the rewards of "prestige."

* * * * *

Ralph Cordiner, a far-right conservative and the chief executive officer at General Electric, was invited by the University of California Los Angeles in 1960 to participate in a lecture series under the general theme *Peacetime Uses of Space.* The title of his May 4 lecture was, "Space and Competitive Private Enterprise."[64]

During his time, Cordiner's remarks had no meaningful impact on American national space policy. This author could locate no evidence that his opinions influenced either of the presidential candidates, Kennedy or Nixon.[65]

But history has been kind to Cordiner. He correctly predicted what Webb would attempt with Apollo and its consequences. He also correctly predicted what a commercially oriented space policy could do for American prestige and leadership. His predictions are significant because, in the early twenty-first century, Cordiner's views would inspire a progressive NASA deputy administrator to attempt dismantling the model Webb had created a half-century before. Hers would be a historic political battle for the future of NASA, unseen since the agency's origins.

In 1961, McGraw-Hill published a collection of essays by the UCLA *Peacetime Uses of Space* lecturers. Cordiner's twenty-eight-page chapter, titled "Competitive Private Enterprise in Space," expanded on the ideas he'd introduced the year before.[66]

> As we step up our activities on the space frontier, many companies, universities, and individual citizens will become increasingly dependent on the political whims and necessities of the Federal government. And if that drift continues without check, the United States may find itself becoming the very kind of society that it is struggling against—a regimented society whose people and institutions are dominated by a central government.

Cordiner's warning was evocative of President Eisenhower's January 17, 1961, farewell speech. Eisenhower warned about a "military-industrial complex" that had the "potential for the disastrous rise of misplaced power." He also warned about the increasing influence of federal funding in misdirecting priorities:[67]

> The prospect of domination of the nation's scholars by Federal employment, project allocations, and the power of money is ever present and is gravely to be regarded.

Cordiner's essay just as easily could be viewed as warning us about a "space-industrial complex" comprising NASA, its cost-plus contractors, and the members of Congress with a vested interest.

His personal views notwithstanding, Cordiner and General Electric were major benefactors from Project Apollo, and without competition. NASA awarded GE a contract in February 1962 to play "a major supporting role," responsible for assuring overall reliability of the space vehicle, and for developing and operating a checkout system.[68] In May, The *New York Times* reported that the contract was a no-bid contract, and had created a significant conflict of interest. GE was acting in an advisory and management role, but could also bid later on contracts to produce hardware. The *New York Times* reported that, after GE failed to win the Apollo capsule contract, Cordiner called Webb to say his company "had invested considerable time and money in Apollo design studies," and expressed his desire for GE to obtain "some additional work." The *New York Times* confirmed that GE had received a no-bid "sole source procurement" contract, which NASA had not admitted when the contract was awarded in February.[69] In the end, Cordiner had appealed to Webb's "grand vision" to win an Apollo contract, after competition had failed.

By the time the McGraw-Hill book was published in 1961, Cordiner was facing price-fixing allegations leveled at GE and its senior management. GE executives, and those at other electrical equipment companies, were fined, and some were sent to prison. Cordiner was not charged, but, by the end of 1963, he had retired, and GE had agreed to pay $75 million in settlement of various lawsuits brought by public and private electric utilities alleging overcharges on heavy electrical equipment.[70]

It appears that Cordiner never spoke out again publicly about the consequences of a government-controlled space economy. He retired to a cattle ranch near Clearwater, Florida.[71]

* * * * *

On February 13, 1969, a little more than three weeks after being sworn into office, Nixon issued a memorandum that created a Space Task Group to recommend "on the direction which the U.S. space program should take in the post-Apollo period." Nixon appointed Vice President Spiro Agnew, by law the chair of the National Aeronautics and Space Council, to lead the review.[72]

The group's report, issued in September 1969, recommended that "the United States adopt as a continuing goal the exploration of the solar system, with men and machines." It presented a grand vision of a crewed space station in "near-earth orbit," a "new and truly low-cost space transportation system" to service the station, crewed exploration of the moon, and crewed missions to Mars as early as 1981. Robotic explorations would be sent to the other planets in the solar system; a "key scientific objective of this exploration will be the search for extraterrestrial life."[73]

The report didn't say how to pay for all this, nor did it discuss reform of NASA's procurement system, nor did it consider a transition to a more competitive free enterprise model. It invoked the same arguments made nearly a decade before, in the early days of the Kennedy administration. "Will America continue to lead, or will some other nation build on what we started?"[74]

The report instead concluded with a chapter titled, "Values of the Space Program," listing "tangible benefits" in science and applications. The intangibility of "prestige" was invoked. "The existence of this space capability contributes to our national security and maintains our prestige and standing in the world."[75]

The aerospace industry's addiction to government funding was viewed by the group just as Webb had envisioned—an ongoing stimulus to the economy. The report noted that total employment by NASA programs had peaked mid-decade at about 420,000 people, and would be down to about 190,000 by the end of FY 1970.[76] This "highly labor-intensive" dependency was championed as "the most efficient flow of economic benefits," if perpetuated.[77]

> Over ninety percent of space funds has been spent in the private sector, impacting over 20,000 firms in the first levels of prime and subcontracts. These contracts have gone to all states in the nation. Twenty states have received over $100 million each in prime contracts alone. The twenty-five largest contractors actually perform their contracts in 30 states and 177 cities.

Nixon received the report but did not act on it. His first NASA budget proposal reduced the agency's prior year budget by 15 percent. In his 2015 book, *After Apollo? Richard Nixon and the American Space Program,* Logsdon wrote that Nixon concluded "an ambitious space program was not something that would gain support." Nixon's space policy statement issued on March 7, 1970, ended NASA's singular exceptionalism within the government bureaucracy. The crewed lunar landing had been achieved. Its prestige had been reaped. Time to move on to other, less costly priorities.[78]

The only element of the Space Task Group report to survive was what eventually became known as the space shuttle, "for the routine transfer of men and materiel to and from Earth orbit . . . the earth-to-orbit shuttle would be brought into operational use at the same time as the space station, in order to provide economic logistics support."[79]

But Nixon didn't approve funding for a space station, only the shuttle. Employment in California was one of the president's motivations. Logsdon wrote of a November 24, 1971, taped conversation between Nixon and his chief domestic adviser, John Ehrlichman, "Jobs—right, John? Do it in terms of jobs. It ought to be in California." Nixon told his political adviser Charles Colson, "In Florida and California this is a big deal. It will save the aerospace industry." He later told a group lobbying for Grumman, "This is jobs. I mean that is really what is at stake here, jobs."[80] Nixon directed NASA to proceed with the Space Shuttle program on January 5, 1972, the beginning of his presidential reelection year.[81]

* * * * *

Project Apollo was the "prestige" centerpiece for three presidential administrations, from Kennedy through Johnson to Nixon. All three administrations created an inadvertent "highly labor-intensive" dependence on NASA by the aerospace economy. No thought was given to how aerospace corporations might one day be weaned off government funding, to stand and compete on their own. But there was one effort in the early 1960s to midwife a modest commercial industry.

In October 1958, Bell Telephone Laboratories engineers John Robinson Pierce and Rudolf Kompfner proposed "to interest some agency" in launching a passive communications satellite test. In January 1959, Bell, NASA, and JPL negotiated an agreement for a test communications relay satellite called Project Echo. The satellite was a balloon that would passively relay a signal bounced off the object from one Earth station to another.[82]

The first commercially functional communications satellite was *Telstar 1,* developed by Bell under a contract for AT&T, which had signed an agreement with NASA. Launched on July 10, 1962, from Cape Canaveral, Telstar was the first privately funded satellite. The satellite relayed live television signals between Europe and North America.[83]

Congress passed the Communications Satellite Act of 1962 "to provide for the widest possible participation by private enterprise" in establishing "a commercial communications system, as part of an improved global communications network."[84] The act created a Commercial Satellite Corporation (COMSAT), owned fifty-fifty by the government and private companies. COMSAT is one of the earliest examples of a new way for the US government to stimulate the aerospace private sector. The government assumed much of the early risk so that private enterprise could invest in developing technologies that they could market to the rest of the world.[85] In today's space industry, this would be called a public-private partnership.

Commercial satellite payloads had one significant vulnerability—the companies relied on a government launch vehicle to reach orbit.

The typical boosters were intermediate or intercontinental ballistic missiles converted from military use—the Thor-Delta (which became known as just *Delta*) and the Atlas-Centaur. Each had an upper stage designed to boost the payload into orbit, instead of plunging it back into the atmosphere with a nuclear warhead.[86]

Allied nations, lacking their own launch vehicles, had to rely on the United States to place their communication satellites in orbit. A September 17, 1965, memorandum approved by President Johnson directed, "It is the policy of the United States to support the development of a *single global communications satellite system* to provide common carrier and public service communications . . . Assistance to any of these foreign governments in the development of communications satellite systems can potentially develop competitors seeking to divert traffic from the single global system being developed by the international consortium."[87] (*emphasis in the original*).

This policy had the unintended consequence of encouraging European nations to develop their own launch vehicle, the Ariane. The European Space Agency in 2009 recalled, "When ESA came into being in 1975, one of its first objectives was to build a European launcher. The reason was simple: no launcher—no independent access to space—no space programme."[88] Ariane quickly emerged as an affordable and reliable competitor to the space shuttle.

Early budget estimates showed that the space shuttle could not be cost-effective unless it flew a routine schedule, so NASA concluded that it would have to be used for all US payloads, both civilian and military. One DOD/NASA report projected the space shuttle would fly 30 to 70 flights a year, maybe up to 140 missions a year. This certainly strained credulity, but it was the only way on paper to show a projected low cost for space shuttle operations.[89]

President Jimmy Carter in May 1978 issued a directive that the space shuttle was available for "all authorized space users—domestic and foreign, commercial and governmental—and will provide launch priority and necessary security to military and intelligence missions while recognizing the essentially open character of the civil space program."[90] This meant that military missions would be encouraged to transition from proven multiple uncrewed expendable launch vehicle (ELV) options to a single experimental system that required people to operate it.

The corporations building boosters such as Atlas (General Dynamics), Delta (McDonnell Douglas), and Titan (Martin Marietta) faced an uncertain future, as they were now in competition with a government delivery service. NASA derived revenue from using the space shuttle to launch commercial payloads, reducing the cost of flying a mission. Ariane also competed for contracts.[91]

President Ronald Reagan had a more favorable view of commercialization. His administration in May 1983 issued a directive that "fully endorses and will facilitate the commercialization" of US ELVs. The government would "encourage free market competition" but would not subsidize the industry.[92] Reagan signed in July 1984 an amendment to the 1958 Space Act, requiring NASA to "seek and encourage, to the maximum extent possible, the fullest commercial use of space."[93] Congress declared in the act its acknowledgment that "the private sector will continue to evolve as a major participant in the utilization of the space environment."[94]

The course of commercial space policy during the Reagan administration is its own long and complex topic, documented and explored by Logsdon's 2019 book, *Ronald Reagan and the Space Frontier.* Logsdon concluded:[95]

> The emergence of a dynamic private space sector in recent years can trace its heritage to the Reagan administration encouraging the first-generation space entrepreneurs of the 1980s. The development of a commercial space sector can appropriately be considered a Reagan space policy.

NASA saw commercialization as a threat to the space shuttle's payload revenue stream. Those running the Departments of Defense and State worried about the consequences of exposing American technology to foreign customers. DOD leadership, dissatisfied with the space shuttle's low launch rate and worried about exclusive reliance on one launch vehicle, tried to purchase more ELVs but encountered fierce political headwinds from NASA's congressional supporters.[96] A compromise reached in February 1985 allowed the DOD to purchase two ELVs per year, while committing to flying on one-third of upcoming space shuttle flights over the next ten years.[97]

The loss of *Challenger* and its crew on January 28, 1986, proved the folly of relying on one launch system. Congress increased the DOD's budget to purchase more ELVs for national security payloads. *Challenger* finally provided DOD leadership with the justification to phase out the space shuttle for military missions.[98] A second space shuttle launch site, built at Vandenberg Air Force Base in California, had been intended for polar orbit missions. In July 1986, USAF Secretary E. C. "Pete" Aldridge directed that the space shuttle's launch pad at Vandenberg Air Force Base be placed in "operational caretaker" status. Vandenberg was never used for a space shuttle launch.[99]

* * * * *

In the months after the *Challenger* accident, the Reagan administration decided that the private sector would take responsibility for launching commercial satellites from American soil. NASA would concentrate on payloads important to defense, foreign policy and science. Launch companies could now compete for customers who wanted to send nonhuman payloads into space.[100]

To ensure that these companies would no longer face competition from the government's space shuttle, Reagan in December 1986 issued a directive. "NASA shall no longer provide launch services for commercial and foreign payloads unless those spacecraft have unique, specific reasons to be launched aboard the Shuttle."[101]

Most of these commercial launches would be from Cape Canaveral. Mothballed launch sites were reopened and updated for Atlas, Delta, and Titan.

While the commercial launch industry struggled to mature, the United States and the USSR began to sign various peace agreements that limited the number of nuclear weapons. US defense spending began to decline.[102] To remain competitive and solvent, aerospace companies began to merge. After the fall of the Soviet Union in 1991 and the end of the Cold War, the DOD in the early 1990s decided to subsidize the industry's consolidation so that some launch vendors would survive. In 1992, the DOD had over sixteen prime contractors capable of producing major combatant weapons systems. By 1998, only three remained.[103]

The DOD sought to create a new generation of boosters, called evolved expendable launch vehicles (EELVs), to ensure affordable access to space for government payloads. At first, the DOD wanted to select only one contractor, but then it decided to select two to maintain competition. It was believed that a commercial launch market would emerge to support those two providers; in 1998 the Pentagon selected Lockheed Martin (often called "Lockmart") to build the Atlas V to succeed the Titan IV, while Boeing would build the Delta IV to complement the existing Delta II.[104]

Economic forecasts in the 1990s generally suggested that a robust commercial satellite industry was about to emerge. A May 1995 analysis by the Office of Commercial Space Transportation (AST) produced an encouraging forecast for the number of future launches to deliver commercial satellites to low Earth orbit (LEO), but acknowledged the "dynamic and somewhat uncertain nature of this market segment."[105]

Reality didn't comply. The number of American commercial launches for the period peaked at twenty-two in 1998, and then declined to just three by 2001; 2002 and 2003 had just five each, six in 2004, and just one in 2005.[106] A 2017 review of these past forecasts observed they were "inherently over-optimistic in their prediction of future commercial launches." The authors blamed "the combination of self-reporting bias from industry as well as the overarching goals" of AST.[107]

With so few launch opportunities, Boeing and Lockmart fought to survive. In June 2003, Lockmart sued Boeing, claiming that Boeing employees had illegally obtained Lockmart proprietary information, including "extremely sensitive and detailed cost and technical data regarding Lockmart's EELV proposal."[108] The Department of Justice filed a criminal complaint against two former Boeing company managers, charging them with "conspiring to steal Lockheed Martin trade secrets concerning a multi-billion rocket program for the United States Air Force."[109]

The USAF in July 2003 punished Boeing by reallocating planned launches to Lockmart, an estimated $1 billion revenue loss to Boeing. Under the original allocation, Boeing received nineteen launches to Lockmart's seven. The new allocation was Lockmart fourteen and Boeing twelve. The Air Force also suspended three Boeing space subsidiaries from winning new government contracts for an unspecified period.[110]

The suspension didn't last long. The Air Force reinstated Boeing on March 4, 2005, declaring the company "a full partner on our national security space team."[111]

Two months later, on May 2, 2005, the two rivals buried their legal hatchets to form a joint partnership, called United Launch Alliance (ULA).[112] The press release estimated that the combination would save the government $100–$150 million a year.[113]

With the collapse of the commercial launch market, DOD's interest was "assured access to space." Its new EELV acquisition strategy, according to a later review by the Government Accountability Office (GAO), was to acknowledge "the government's role as the primary EELV customer, and the need to maintain assured access to space by funding two launch vehicle families." When the Federal Trade Commission initially opposed the ULA merger, the DOD argued that having two launch vehicle families "presented unique national security benefits that outweighed the loss of competition," according to the GAO report. The EELV program was focused on mission success—not cost control—but by 2010 the DOD realized the strategy was not sustainable.[114]

Because ULA enjoyed a government monopoly, commercial payload customers went overseas. A 2018 calculation by the US Department of Transportation (USDOT) showed that, during the ten-year period from 2006–2015, 224 commercial space payloads were launched worldwide. Russia was the primary launch provider (93, or 41.5 percent), followed by Europe (53, 23.7 percent) and then the United States (46, 20.5 percent). Of the 46 US launches, only 17 were by ULA, and none of those were in 2011–2013.[115]

The promise and potential of real competition envisioned by Ralph Cordiner a half-century before seemed dead, a victim of the "central government" institutions he had warned about.

* * * * *

Inspired by NASA's exploits, classic science fiction literature, and television shows such as the 1950s Disney *Tomorrowland* episodes and the 1960s *Star Trek* series, several citizen space advocacy groups sprung up in the 1970s and

1980s. Wernher von Braun helped form the National Space Institute in 1974, an educational nonprofit organization.[116] Princeton University physics professor Gerald K. O'Neill published an article in the September 1974 *Physics Today* suggesting a rotating cylindrical habitat could be constructed in space to create an off-world human colony.[117] O'Neill's vision inspired the creation of a grassroots group in 1975 called the L5 Society.[118] The organizations eventually merged into the National Space Society in 1987.[119] O'Neill in 1977 founded a space research and education nonprofit called the Space Studies Institute (SSI).[120]

The Public Broadcasting System science series *Cosmos,* hosted by Cornell University professor Carl Sagan, aired in fall 1980. *Cosmos* became the highest-rated public television series of its time.[121] Sagan, JPL director Bruce Murray, and JPL spacecraft engineer Louis Friedman in 1979 created The Planetary Society to demonstrate public support for robotic planetary exploration. Their first issue of *The Planetary Report* was published in December 1980, as *Cosmos* reached its finale.[122]

By 1988, another advocacy group had emerged, the Space Frontier Foundation (SFF). Unlike the National Space Society and The Planetary Society, the SFF saw the existing government space bureaucracy as a hindrance. According to its website, "At the time of its founding, the Space Frontier Foundation stood apart from other public space groups who were merely promoting the current paradigm in space policy." Those groups "were trying to promote the current space program," while the SFF was "dedicated to opening the space frontier to human settlement as rapidly as possible." They intended to "wage a war of ideas in the public sphere for a wholly new and ecumenical space agenda." The group was granted the right to market O'Neill's vision of massive industrialization and settlement of the inner solar system.[123]

As the organization grew, the SFF leadership concluded they needed a brand that described its philosophy and direction. What to call their movement? An early term was *alt.space*, but that seemed clunky and inadequate. The phrase *new space* was sometimes used as a shorthand for describing the emerging generation of commercial space development. In July 2004, SFF leadership settled on the term *NewSpace*. Six months later, their board of directors decided to build around the phrase *Advancing NewSpace*, and decided to change the name of the organization's annual meeting to the NewSpace Conference.[124] The SFF defined NewSpace as, "People, businesses and organizations working to open the space frontier to human settlement through economic development."[125]

The logical antithesis to NewSpace? OldSpace.[126]

For NewSpace to work, it would require entrepreneurs to have deep pockets filled with investment capital. It was easier to attract investors if federal funding was committed to the project, but that also meant the project was vulnerable to the whims of politics and bureaucracy.

The NewSpace community believed that reusability was the key to lowering costs. Reusability meant that the launch vehicle wouldn't be expended, but it would be recovered and flown again, which in theory should reduce cost.

A somewhat commercial attempt at reusability was the McDonnell Douglas DC-X, a proposed "Single Stage to Orbit" vehicle that could launch, land, and launch again. The program originated with the Strategic Defense Initiative Organization, a DOD agency created to develop advanced space weapon technologies. The contract was awarded to McDonnell Douglas in 1989. The company nicknamed the vehicle *Delta Clipper-Experimental,* evoking the memory of nineteenth-century "clipper" merchant sailing ships while attaching the company's Delta rocket moniker as an adjective. Test flights were at the White Sands Missile Range in New Mexico. After the end of the Cold War, the Pentagon eventually lost interest in funding the DC-X; the program found funding with NASA as a reusable launcher technology demonstrator called DC-XA. On July 31, 1996, the Delta Clipper completed its demonstration flight, but a landing strut failed, causing the vehicle to topple over and explode.[127]

NASA passed on the DC-XA in favor of a Lockmart proposal for a project NASA designated X-33. The project was to demonstrate a suborbital reusable space plane that would lower the risk of building and operating a full-scale reusable vehicle fleet. The X-33 had a delta-shaped lifting body design, similar to the space shuttle orbiter, but it relied on a cutting-edge aerospike engine fueled by liquid oxygen and liquid hydrogen.[128] After five years of development and unresolved technical challenges, NASA canceled the X-33 program in March 2001. NASA had spent $912 million on the project, while Lockmart had spent $356 million.[129]

Lockmart invested in X-33 because it had plans for a fully reusable orbital vehicle that it would market to launch commercial customers. Named VentureStar, the company viewed the spacecraft as a one-stop commercial vehicle that would launch vertically like a rocket, take payloads to the International Space Station (ISS) and other orbital destinations, and then land like an airplane. Jack Gordon, president of Lockmart's Skunk Works in Palmdale, told the *Los Angeles Times,* "This is high risk, and it'll be high payoff, not only for Lockheed Martin but also for the nation." VentureStar would be 100 percent privately funded, with the company projecting an investment of up to $5 billion for two ships and private spaceports.[130]

According to a 2011 retrospective by NASA engineer Gary Letchworth, "In hindsight, the program was simply trying to accomplish too much with inadequate cost and schedule margin." He noted the many shifting stakeholders within the program, both within the government and the private sector. NASA became more risk adverse during the program's lifetime, and the projected market for launching commercial satellites didn't materialize.[131]

Because NASA contributed about three-fourths of the project spending, the agency controlled the priorities. NASA was less interested in VentureStar than the evolution of new technologies such as the aerospike engine, a reusable cryogenic tank, and new composite materials.[132] NASA's birthright was to research and develop new aerospace technologies, just as its NACA predecessor had pushed aviation advances. But with a tight budget, NASA finally gave up and walked away, and the commercial satellite launch market wasn't there to make the business case for VentureStar to continue. After the loss of government funding, Lockmart simply wasn't willing to shift to a more nimble entrepreneurial culture, especially given the absence of customer demand, public or private. Unlike NASA, which wanted to invent new aerospace technologies, a strictly commercial R&D cycle in competition with other companies typically would have looked for the simplest, cheapest, fastest way to produce a solution for potential customers.

The company that finally figured this out was a California-based startup called Space Exploration Technologies Corporation—"SpaceX" for short. The SpaceX success story is well-known and oft told. One of the earliest profiles was *Elon Musk*, a biography about the company's founder, published in 2015 by author Ashlee Vance.[133]

Vance wrote that SpaceX wanted to prove what the NewSpace community had always contended: If costs could come down and launches were routine, the commercial payload market would emerge.[134] The company didn't have its hand out to NASA, the DOD, or any other government entity asking for a subsidy. It was funded by the money Musk made with Silicon Valley ventures, such as the one that evolved into the online payment company PayPal.[135] Musk wanted SpaceX to run "lean and fast and capitalize on the huge advances in computing power and materials that had taken place over the past couple of decades," Vance wrote. "As a private company, SpaceX would also avoid the waste and cost overruns associated with government contractors."[136] In an October 2012 interview with *Wired* magazine, Musk said that a rocket should cost about 2 percent of the typical price. Why was it more expensive? He cited "the incredible aversion to risk within aerospace firms. Even if better technology is available, they're still using legacy components, often ones that were developed in the 1960s."[137]

SpaceX accepts failure. COO Gwynne Shotwell told Makers.com in 2018, "You don't learn anything from success, but you learn a lot from your failures."[138] SpaceX has had some spectacular failures over the years, but those failures have led to innovations that have reduced costs while increasing reliability. When there was a failure, SpaceX didn't bill the government to solve the problem. After the first SpaceX Falcon 9 booster launched in December 2010, Shotwell, the company's president, told ABC News, "If we overrun this program, we have to come up with the money through investment to cover the cost, which is dramatically different from contracts where if the contractor overruns, taxpayers have to pay the overruns."[139]

SpaceX began with Musk investing $100 million of his own money.[140] The company began development with a small rocket called Falcon 1 (for one engine), launching from Kwajalein Atoll in the Pacific Ocean.[141] Despite three launch failures, in August 2008 SpaceX received a $20 million investment from Founders Fund, a technology venture capital firm. In a press release, a Founders Fund managing partner said, "Having reviewed their technology, outstanding engineering & business talent and the infrastructure they have built, we are highly confident in the future of the company."[142]

SpaceX wasn't above soliciting government money, but the money had to be earned.

In November 2005, NASA opened the Commercial Crew/Cargo Project Office to "manage orbital transportation capability demonstration projects that may lead to the procurement of commercial cargo and crew transportation services to resupply the space station."[143]

One of the office's first projects was called Commercial Orbital Transportation Services (COTS). Private industry would coordinate with NASA to develop and demonstrate transportation capabilities to low Earth orbit. At the end of that phase, NASA had the option to competitively procure transportation services to resupply the ISS with cargo and, one day, perhaps, crew. Participants were to successfully demonstrate a series of "milestones." A participant did not receive a milestone payment until the task was successfully demonstrated. Participants had to invest their own money up front and throughout the demonstration process; in exchange, NASA made no claim to their intellectual property rights. If a participant failed to demonstrate a milestone, NASA could terminate their COTS agreement.[144] Then–NASA Administrator Michael Griffin justified COTS by citing the US Post Office Department airmail precedent of a century before.[145]

In August 2006, NASA announced that two companies had been selected to participate in the COTS project. One was SpaceX, which was eligible for $278 million, to help develop a cargo ship named Dragon. The other was Rocketplane Kistler (RpK), which could receive $207 million to help complete a reusable rocket called K-1 that would launch a cargo module.[146] RpK had chosen a launch site in Australia, and was planning a second spaceport at the Nevada Test Site.[147]

RpK managed to complete only three of its milestones before NASA ended their agreement. The company was unable to demonstrate adequate funding to continue, their fourth milestone requirement.[148] NASA had paid RpK $32.1 million in milestone payments to that point. The remaining funds, about $170 million, were awarded to Orbital Sciences.[149]

One could argue that the taxpayer was out $32 million, but the idea behind the program was for the government to assume risk up front so that private entrepreneurs would invest their own capital to grow a vibrant commercial launch industry. By that standard, COTS was an unqualified success.

By the early 2020s, the SpaceX Falcon 1 had evolved into the Falcon 9, then a Falcon 9 that could land, and then a Falcon 9 that could land and launch again and again. The Falcon 9 evolved into the Falcon Heavy, essentially three Falcon 9 boosters connected side-by-side, offering the most thrust of any launch vehicle since the Saturn V. Musk and his SpaceX shareholders risked their investments because NASA gave the company a contract to develop the Dragon spacecraft for delivering cargo, and one day people, to low Earth orbit.

In September 2021, SpaceX launched four civilians in a crew Dragon on the Inspiration4 mission, a three-day orbital flight that achieved the highest altitude by any human since the Apollo era. A private booster launched a private spacecraft with four private passengers.[150] The Falcon 9 had flown twice before.[151] The Dragon capsule was the same spacecraft used for the Crew-1 flight that had taken four astronauts to the ISS in September 2020.[152] A private crew in a reusable private spacecraft had been launched by a reusable private rocket.

NewSpace had arrived.

* * * * *

The Project Apollo moon landings are often cited as one of the great technical achievements of the twentieth century. But President John F. Kennedy's challenge to land a man on the moon by the end of the 1960s had the inadvertent consequence of turning NASA into a propaganda organ. No thought was given to turning it back. The objective was no longer science. It was prestige. When it was over, hundreds of thousands of people were dependent on the government for their jobs, and NASA was left with empty facilities across the nation, including two launch pads and a cavernous assembly building at Kennedy Space Center.

For the next forty years, NASA found itself "in the steel trap of regimentation" of a nationalized space industry, as Ralph Cordiner had once warned. Sporadic attempts to escape the trap were thwarted by the space-industrial complex, including some career civil servants within NASA itself. Politicians representing NASA centers and their legacy contractors fought in the halls of Congress to protect jobs in their districts and states.

The loss of *Columbia* and its crew in February 2003 set in motion an unforeseen and unanticipated sequence of events. President George W. Bush convened an investigation board that declared the space shuttle "a complex and risky system."[153] The board expressed its "conviction that operation of the Space Shuttle, and all human spaceflight, is a developmental activity with high inherent risks."[154] The report also blamed a "lack of an agreed national vision for human space flight."[155]

On January 14, 2004, Bush delivered a speech at NASA Headquarters in Washington, DC, to unveil his Vision for Space Exploration (VSE).[156] The space shuttle would be retired once the space station's assembly was finished around 2010. Once ISS was complete, the Space Shuttle program would end, the orbiters would be retired, and the contractor employees would be laid off. The countdown had started.

During Bush's second term, his administration evolved a new NASA human spaceflight program that came to be called Project Constellation. Although no one promised that Constellation would save the contractor jobs, many assumed so. Politicians—federal, state, and local—acknowledged the Space Shuttle program's denouement but did little to prepare the Space Coast community for its arrival. And when it did, it coincided not only with the election of the first African American president in US history, but also the housing and banking crises that led to the Great Recession, the worst downturn for the American economy since the Great Depression of the 1930s.

The VSE focused primarily on transition of NASA crew missions in the next decade from low Earth orbit to deep space. But if one dug into the details, the first seeds were being planted for a revolution in the way the United States acquired its space technologies.

A Bush-appointed presidential commission issued a report in June 2004 that included recommendations for creating a "robust space industry" that would "become a national treasure."[157] NASA would no longer own its LEO transportation systems, but "acquire" them as one might purchase a package delivery or an airplane ticket.[158] Other than perhaps a few commercial space evangelists, at the time no one seemed to realize the implications, or the consequences.

The COTS program may have seemed innocuous, no threat to the space-industrial complex and the next big "prestige" project. Constellation was the big dollar, big job program. COTS was one small component, robotic deliveries in the 2010s to the ISS while NASA transitioned to beyond Earth orbit.

But within a decade, the way the government did business in space would change. The "robust space industry" vision rippled first through NASA and then the Pentagon, each with its own traditions and bureaucracies, and with opponents entrenched in the status quo. Many politicians attacked this NewSpace vision; they perceived it as not only threatening the jobs of local voters, but also the generous contracts of the companies that employed them. Those companies spent hundreds of thousands of dollars each election cycle in campaign contributions, and millions each year lobbying members of Congress to preserve OldSpace.

The Space Coast would be at the eye of this political hurricane. And a Florida state agency, with roots in an effort twenty years before to open a commercial spaceport at the Cape, would ride the storm.

3

Spaceport Florida

Stephen Morgan described himself as a "space nut."[1]

After a four-year stint in the US Air Force, Morgan returned home in 1980 to Brevard County, where he found work in the aerospace industry as a software engineer with the Harris Corporation and, briefly, McDonnell Douglas. Within three years, he'd been promoted by Harris to senior engineer.[2]

But space evangelism was his destiny.

Morgan joined the Space Studies Institute, Gerald K. O'Neill's space research and education nonprofit. Unlike other organizations such as the National Space Society, The Planetary Society, and the Space Frontier Foundation, SSI was not politically active; they only sought to advance the state-of-the-art.[3] In 1984, the SSI chose Morgan to lead their international expansion—from Palm Bay, in south Brevard County. He already headed SSI's Florida Support Team. He was profiled by *Florida Today* in a July 4, 1984, article, in which he described SSI's research into developing a mass driver foreseen by O'Neill, "kind of an electromagnetically operated catapult," and solar-powered satellites.[4] With perhaps a little audacity, Morgan offered a complete set of *Omni* magazines to the first person who donated $1,000 to his chapter.[5]

By the summer of 1987, Morgan was program manager for a project to create a "space institute" connected to the Florida Institute of Technology and other east central Florida universities. He told the *Orlando Sentinel,* "In a nutshell, we're trying to establish an independent research institute loosely affiliated with FIT and the other schools to do research, especially in space management and commerce, such as commercial launch and payload processing."[6]

Morgan worked on another idea, a space development "chamber of commerce." He was the "acting executive director" of the East Central Florida Space Business Roundtable, described by the *Orlando Sentinel* as "a group of aerospace and general business executives he hopes to bring together to promote the commercialization of space in Central Florida."[7]

The Roundtable was affiliated with yet another space advocacy group called the Space Foundation, founded in 1983 in Colorado, "to establish an organization that could, in a non-partisan, objective and fair manner, bring together the various sectors of America's developing space community and serve as a credible source of information for a broad audience—from space professionals to the general public," according to their website.[8]

Much of the Space Foundation's early activities targeted the Houston, Texas, space community. The Foundation's leadership saw the need to reach into other space communities. The "roundtable" was a means to build support for commercial space. Morgan wrote about the need to obtain strong support from the business and financial communities for what he termed *space enterprise,* defined as "space-based or space-oriented commercial activity."[9]

By the end of 1986, the Florida Roundtable had a fourteen-member board of trustees, with Morgan now its permanent executive director. Morgan's photo appeared on the front page of the December 19, 1986, *Florida Today* business section, along with an article detailing the Roundtable's plans for a direct mail recruitment campaign. He told the newspaper, "In this area there is a surprising lack of awareness on the part of business as to what is going on in space and what it means to them as far as opportunities."[10]

Frustration with local ignorance, if not apathy, was an ongoing problem for Morgan and the Roundtable founders. He told *Florida Today* about a business friend who was "laughed out of the offices of several Brevard County investment banking firms when he approached them for a loan" to pursue commercial space ventures. When the friend went to Texas, the Houston Economic Development Foundation offered to assist him—if he brought his company to Houston. Morgan noted that six other space roundtables had been established across the nation, including Houston, and each was promoting itself as a space business center. The Houston group had nearly 400 members, while the Florida roundtable had only 89.[11]

In September 1986, the Roundtable's Board of Trustees directed Morgan to approach government leaders in Florida's state capital, Tallahassee, about creating a state-sponsored "commission on space." The idea was inspired by President Reagan's National Commission on Space, which had issued a report earlier that year titled *Pioneering the Space Frontier.*[12] Perhaps with more audacity, Morgan sent a letter directly to Republican Florida governor Bob Martinez outlining their proposal for a similar state-sponsored commission. Morgan emphasized that other states such as Texas, Virginia, Hawaii, and California had their own commercial space initiatives. If Florida did not act, the state could be left behind in this nascent national competition. A series of letters and telephone calls led to Morgan and Roundtable member James Vevera meeting in April 1987 in Tallahassee with Florida Commerce Secretary Jeb Bush, a son of US Vice President George Bush.[13]

Based on the Roundtable's recommendations, Governor Martinez signed an executive order creating the Florida Governor's Commission on Space, on May 28, 1987, at Kennedy Space Center's Spaceport USA. (In 1997, the attraction was renamed the Kennedy Space Center Visitor Complex.) "For almost 40 years, Florida has been America's gateway to the stars," Martinez said. "The space industry has become a critical component of our economy. The commission will help us understand how to maintain a competitive edge." The commission's purpose was to complete a comprehensive report on the Florida aerospace industry and recommend how the state could attract more space companies and investment.[14]

Three months later, Governor Martinez announced the twenty-five Space Commission members. Martin Marietta Electronics and Missiles Group President and CEO Thomas Young was named commission chair. Florida Commerce Secretary Jeb Bush was named vice chair. Several Brevard County businesspeople were named to the panel, including Morgan, as well as State Representative "Bud" Gardner (D-Titusville) and State Senator John Vogt (D-Cape Canaveral).[15]

While the Space Commission members deliberated, the Reagan administration on February 11, 1988, released a summary of the President's Directive on National Space Policy. Although the presidential directive was classified, an unclassified summary was released. One of its principles stated, "The United States shall encourage and not preclude the commercial use and exploitation of space technologies and systems for national economic benefit without direct Federal subsidy. These commercial activities must be consistent with national security interests, and international and domestic legal obligations."[16]

The policy stated that American space transportation capabilities would rely on both the shuttle and "unmanned launch vehicles." It directed that government agencies were to facilitate these commercial launch vehicles, and to encourage "the use of its launch and launch-related facilities for United States commercial launch operations." The government would not subsidize the commercialization of these uncrewed vehicles, but "will price the use of its facilities, equipment, and services with the goal of encouraging viable commercial ULV activities."

The policy perked the collective ears of Florida's commercial space advocacy community. Morgan told *Florida Today*, "Anyone reading that will notice that there are a lot of opportunities and new avenues through which commercial enterprise in space can be pursued."[17]

On February 9, 1988, two days before the release of the National Space Policy, Florida's Space Commission voted to recommend a six-month, $500,000 study to consider the creation of a state space development authority. The authority would "address the development of an industrial base in proximity to the nation's primary launch complex, and . . . provide a mechanism for future economic development," Morgan later wrote.[18]

Governor Martinez endorsed the proposal. At Cape Canaveral's Launch Complex 26, the site of the first US satellite launch, Martinez on March 25, 1988, announced he would ask the state legislature for $500,000 to study the creation of the nation's first commercial spaceport, a "Spaceport Florida."[19]

Rep. Bill Nelson (D-Melbourne), a Brevard County native whose district included the Cape, told *Florida Today* that he supported the Republican governor's proposal. "The study of Florida's commercial space potential will help the state get ready to capture much of the commercial space business in coming years."[20] Nelson had already acknowledged his interest in running for governor in 1990, positioning him as a potential rival and impediment to Martinez's reelection plans.[21]

Some Democrats opposed the proposal. State Rep. Michael Friedman (D-Miami Beach) said the money was better spent on children's programs, and suggested that the private sector should "put a package together."[22] Rep. Bud Gardner of Titusville, a Space Commission member, said, "I think it is hard to justify spending $500,000 on something like trying to build another space center in Brevard County in addition to the one we already have," and was confident that Space Coast firms would pay for the research if interested.[23] But after a meeting with Space Coast Development Commission[24] Executive Director Bob Allen and others, Gardner modified his position to ask that a qualified advisory board oversee the process.[25]

The Space Commission's final report, titled *Steps to the Stars,* was delivered to Governor Martinez on July 7, 1988, with a cover letter signed by Jeb Bush.[26] The forty-eight-page report cited a 1988 study, which found that, statewide, "space-related activities have not been a significant part of the state's overall economy. However, in Brevard County, approximately 38 percent of the jobs stem from Kennedy Space Center operations."[27]

The commission surveyed nineteen Florida space company senior executives to identify key factors that enhance the success and growth of aerospace industries in Florida. The respondents ranked, on a 1–10 scale, both the importance and Florida's rating for each factor. The results (in the following table) exposed several perceived inadequacies.[28]

The report offered a series of recommendations to address these deficiencies.

The commission saw in recent developments new "opportunities for Florida to build a permanent space-related industrial complex."[29] The report cited the federal policy decision to support a mixed fleet of space vehicles; the release of

Factor	Importance Rating	Florida's Rating
Work Ethic of Population	8.4	7.3
Labor Skills and Availability	8.4	6.2
K–12 Education	7.9	5.8
Proximity to a University	7.9	6.6
Highway Facilities	7.2	4.5
Availability of Good Local Vendors	6.8	5.5

the National Space Policy; and the administration's intent to support commercial space activities. The commission supported the governor's $500,000 proposal to conduct the study for a commercial spaceport somewhere in Florida. It also recommended "a permanent office for space programs" within the state's Department of Commerce. The department should "reorganize their economic development programs to include the development and location of space enterprises within the State," and should become more active in "promoting Florida as a world center for space commerce."[30]

This one sentence perhaps best reflects the authors' message to the people of Florida:[31]

> The goal of Florida's space industry initiative must be a long-term commitment to making Florida the center of space commerce in the United States.

* * * * *

Upon delivery of the final report, the Space Commission disbanded, its purpose completed. A new Office of Space Programs in the state's Department of Commerce was created, led by Chris Shove, a Commerce economic analyst who had a doctorate in urban and regional planning. He had been assigned to the Space Commission, so he was already familiar with their proposals. His job was to define how to implement the Space Commission's recommendations.

In a 2002 essay for the *Journal of the American Planning Association*, Shove wrote, "Initially, there were two basic organizational models for the commercial spaceport concept: state government operated and private company operated." Governmental agencies, while well-intended, "generally produce unintended outcomes due to political influences, aversion to risk, and lack of clear market signals." Shove believed that the significant challenge for the emerging US commercial space industry was "not technical but revolved around government policy."[32]

Government had to be involved as an oversight to any private operations, which might not view safety as a top priority. A government agency would be subject to more public scrutiny due to public records laws. A government operation also assured that one company could not establish a monopoly over commercial launch sites.[33]

A new model was required.

In the late 1980s, US launch vehicles still failed from time to time. Insurance was all but unavailable for commercial spaceport operations. In his 2002 essay, Shove wrote, "Sovereign immunity was one clear method of potentially insuring space flight activity, in that unlimited liability could be capped if the state would extend its sovereign immunity to the spaceport." A state-operated spaceport would also have nonprofit tax status, eligibility for federal grants, and could issue tax-exempt bonds. "A review of government organization types led to the decision to create a single-purpose, quasi-independent entity of the state."[34]

In a 2022 email to this author, Shove wrote, "I proposed establishing a state-enabled authority to implement reuse of launch pads based on the NY Port Authority that developed airports, seaports etc. The background thinking was that we did not want another government agency but we wanted sovereign immunity in case of an accident."[35]

Shove's first task as the office's director was to begin the next phase—hiring a company to plan the proposed Spaceport Florida.[36] He told *Florida Today*, "I feel like the guy who was the beginner of the Federal Aviation Administration, when air transportation reached a point of vitality that it had to be dealt with."[37]

Eight teams submitted proposals—four from Brevard County, four from outside the area.[38] On June 20, 1988, the office announced that United Engineers and Constructors, Inc. of Denver, Colorado, had been selected to conduct the study. The UE&C report was targeted for delivery in February 1989.[39]

Space Coast advocates assumed their county was the obvious choice. Much of the infrastructure was already in place, including the Eastern Range, but these facilities were on government land. The National Space Policy directed federal agencies to cooperate with commercial space enterprises, but all that still needed to be negotiated. It might be easier to go elsewhere, where NASA or the Air Force wouldn't be the landlord.

Another concern was the local transportation infrastructure. Aerospace industrial parks were being planned in the Titusville area, but the roads were considered congested and "unsightly." State Roads SR-3 and SR-405 in particular were cited as requiring attention. "If something is not done about (SR-3) it will stop this development," Shove told the Titusville City Council. Bob Allen of the Space Coast Development Commission (SCDC) added, "The United

States' global competition is not guaranteed and it is not guaranteed for Florida because the competition is out there."[40]

Shove expressed concern about the lack of overt political support for Spaceport Florida. "The governor has not received one letter of support for Spaceport Florida," he said. "The Legislature has not received one letter of support. I don't know how long a politician can run without support. The governor is running on intuition on this and vision, that's all." The Titusville City Council agreed to prepare a resolution supporting the project, and encouraging support from other Brevard leaders.[41]

The hesitancy by some elected officials to openly support the project may have been its close public identification with the governor. Bob Martinez was politically unpopular; after his mishandling of a proposed state services tax, registered voters in a September 1987 Florida Opinion Poll disapproved of Martinez by 12 percent to 53 percent.[42] His numbers had improved by August 1988; the Florida Opinion Poll found that, by 34 percent to 44 percent, registered voters disapproved of how Martinez was handling his job. Less than one-fourth said they would vote for Martinez again.[43] A poll taken in late March 1989 by Mason Dixon Opinion Research showed that 46 percent rated his performance "excellent" or "pretty good," while 51 percent rated him "fair" or "poor." In a hypothetical matchup against Rep. Bill Nelson, Martinez would win 43 percent to 36 percent, but if retired US Senator Lawton Chiles ran against Martinez, Chiles would win 47 percent to 39 percent. Chiles said he was not a candidate—but that would change.[44]

The study was due for release by Governor Martinez on February 13, 1989, but somehow a draft found its way to a member of the press, forcing the Department of Commerce to release it to the media on February 9. This irritated Bob Allen, who was planning a press conference for Titusville.[45] It was the first hint that scheming and infighting among Space Coast stakeholders might impede the project.

The complete UE&C report was about 5,000 pages.[46] A shorter version was released to the public titled *Spaceport Florida Feasibility Study: Executive Summary*.[47]

The study identified three potential launch sites—two orbital and one suborbital.

The two orbital sites were in Brevard County, one being the obvious Cape Canaveral Air Force Station, while the other was an abandoned farm town called Shiloh, north of Kennedy Space Center, that crossed into Volusia County. The CCAFS site had the advantage of converting an existing facility that already had support infrastructure and utilities. The study recommended Launch Complex 15 for the commercial spaceport, with neighboring pads at 14 and 16 available for future expansion. The main obstacle would be the DOD as a landlord. Shiloh was part of the Merritt Island National Wildlife Refuge (MINWR), which shared borders with KSC. It would operate separate from CCAFS and KSC, but infrastructure and utilities would have to be built, as well as improving local transportation routes.

Shiloh was acquired by the federal government in the early 1960s. Acting on NASA's behalf, the US Army Corps of Engineers in late 1961 began using eminent domain to acquire private land on north Merritt Island, up through Shiloh into Volusia County.[48] The far north part of the property would not be needed immediately; NASA foresaw using it for Project Nova, a planned successor to the Saturn V.[49] Property owners near Shiloh were allowed to remain temporarily, but eventually they would have to relocate. In November 1963, Shiloh had six homes, a post office, a service station, and a few other buildings. Once NASA decided that the Saturn V would be the launch vehicle for Project Apollo, the plans for the next-generation Nova were shelved, but NASA still owned the abandoned land.[50]

On August 28, 1963, NASA and the US Fish and Wildlife Service entered into an agreement for the FWS to manage all lands within Kennedy Space Center not being used by NASA operations. These lands eventually became known as the Merritt Island National Wildlife Refuge. NASA retained the right to use any location within MINWR for a future space program activity, at the sole discretion of KSC.[51]

Shiloh was within KSC boundaries when MINWR was established in 1963, and became part of the MINWR in 1972. During the nineteenth century, the area was converted by landowners into some of the largest citrus groves in the state. The hydrology of the lands was altered by the excavation of canals to service the citrus industry. The ecosystem was impacted by non-native plants and invasive non-native species such as feral hogs. Starting in the early 1980s, the FWS began removing the citrus groves, replacing them with various species of oaks and pines, with the hope of one day restoring native scrub vegetation. The FWS managed about 1,500 acres within the Shiloh area for waterfowl, shorebirds, and wading birds.[52]

A suborbital site could be at Shiloh, at Cape Canaveral, or in the Florida Panhandle at Cape San Blas, originally used by Eglin Air Force Base to launch sounding rockets.[53] The Eglin Gulf Test Range Site D-3 on Cape San Blas had been operational since 1958.[54] The study authors believed that the site would require "minimal new permanent facilities" as small rockets required only a flat concrete slab and not an exhaust duct as do larger boosters.[55]

The report identified "complementary opportunities" such as tourism, a "space business incubator," and educational opportunities. The authors proposed a "Space Experience Attraction" that would include an "Analog Moon Base," an "Information Retrieval Center," and a "Commercial Applications Pavilion" exhibiting the benefits derived from space activities.[56]

A "building block" approach should be taken to development of the launch sites, depending on market potential. The authors recommended starting with a small pad before building a site for launching larger rockets a few years later.[57]

Up to this point, there was little public discussion of the cost for building and operating a commercial spaceport. The Space Commission's *Steps to the Stars* provided the vision, but not the cost. UE&C's task was to present options and costs. Depending on the scenario, the authors estimated that constructing small and medium pads at CCAFS would cost about $42.6 million, or $57.9 million at Shiloh. Annual operating costs were estimated to be $8.1 million for CCAFS, and the same for Shiloh. The costs for Cape San Blas were insignificant.[58]

The authors wrote, "The state's ultimate concern is whether the benefits of the Spaceport will be greater than the costs."[59]

> Using conservative launch rates and fees, the study's financial analyses indicate that the direct cash benefits from launch operations are not sufficient to recover the capital and operating costs completely. With a less conservative launch rate or fee assumption, the medium-vehicle pad should recover annual operating costs, but probably will not recover capital costs. The small-vehicle pad option would require a very optimistic launch rate to recover all operating costs.
>
> The State of Florida, or some other source of grant funds, will probably have to provide a "capital contribution" to the Spaceport's operations. This contribution will be returned only in the form of the indirect benefits.

The report defined "indirect benefits" as "higher employment, an increased tax base, economic growth, etc."[60]

Speaking at Spaceport USA on February 13, Governor Martinez said he would include $10 million in his proposed state budget to start Spaceport Florida, begin construction of the launch pads, and fund its governing body, the Spaceport Florida Authority. He directed the Department of Commerce to begin work on the Cape San Blas site, which he said would cost about $200,000 and should be ready in about nine months.[61]

The Martinez proposal was endorsed by US Secretary of Transportation Sam Skinner, just appointed by newly inaugurated President George H.W. Bush. Skinner told the *Orlando Sentinel,* "I am totally committed as secretary of transportation to do everything in Washington that I can do to help the efforts of Gov. Martinez and everyone in the state of Florida come to fruition.

We intend to make Florida the leader of leaders in the development of space commercialization, because it is an effort that will clearly pay dividends into the 21st Century."[62]

Some Florida state politicians not representing the Space Coast, Democrats in particular, expressed skepticism. State Senator Gwen Margolis (D-North Miami), who chaired the Senate Appropriations Committee, told *Florida Today,* "It seems like a PR thing. You will have to spend more than $10 million. I haven't heard what it will cost the state of Florida ultimately and that's what I want to know." State Rep. Mike Friedman (D-Miami Beach), who chaired the House's education committee, told the *Orlando Sentinel,* "I think you will hear others such as myself argue that we can't afford $10 million for this program when we don't even have the money for other things like the needs of our children."[63]

Local differences quickly emerged. North Brevard County Commissioner Truman Scarborough opposed the Shiloh site not only because it lacked utilities infrastructure, but also because he anticipated that Volusia County commissioners were gearing up to lobby for Shiloh. Volusia Council chair Clay Henderson told *Florida Today,* "I don't think we would see any economic benefits from the Cape Canaveral site."[64] Bob Allen's SCDC clashed with the county's Brevard Economic Development Corporation; Allen told *Florida Today* that his lobbying trips to Tallahassee were frustrated by county representatives telling state agencies to deal with them and not the Titusville group.[65]

Environmental groups began to organize protests against the Shiloh site. Residents in Brevard and Volusia Counties met with the Florida Department of Commerce; so did environmental advocates such as the Sierra Club, Audubon Society, Save the Manatees, Florida Wildlife Federation, and Florida Defenders of the Environment.[66] Environmentalists claimed that an errant rocket could contaminate the fragile Mosquito Lagoon, the project would threaten endangered species, and it would set a dangerous precedent of turning over national park land for commercial exploitation.[67] Shove wrote in 2002 that the environmentalists' main concern was that an exemption for MINWR would create a precedent allowing commercial oil drilling in the Arctic National Wildlife Refuge.[68]

Shove warned that the commercial space industry might not want to launch from Cape Canaveral, fearful of military bureaucracy, and therefore would prefer Shiloh. He cited an example of representatives from India having to wait forty-five days for the USAF to grant them permission to access their payload awaiting launch from the base. "The launch companies could leave Florida and the new firms coming on line refuse to operate at Cape Canaveral Air Force Station because of the military requirements," he told Brevard County commissioners.[69]

A week later, the commissioners voted to support the Cape Canaveral site. Stephen Morgan, still representing the Space Business Roundtable, told *Florida Today,* "The resolution is counterproductive because it pits Florida community against Florida community rather than serving to unite Florida against the true competition," other states and counties. Kay Jackson, executive director of the Brevard Economic Development Corporation, favored the Cape site because the utility infrastructure was already in place. Commission chair Roger Dobson said he favored the Cape site because most of Shiloh would be in Volusia County. Bob Allen of the SCDC feared the resolution jeopardized Brevard's leadership role in the project, claiming, "It's going to muddle up the legislation" in Tallahassee to fund the project.[70]

On March 14, at a state House Science and Technology Subcommittee hearing, representatives of the aerospace community expressed a preference for Shiloh. Business officials asked for an all-purpose launch facility, tax breaks, and legislation reducing their insurance burden. Environmentalists once again spoke out against the Shiloh option.[71]

Partisan politics also crept into the spaceport project. On April 3, 1989, *Florida Today* ran an article quoting the Democratic chair of the Florida House Budget Committee, T. K. Wetherell (D-Daytona Beach), as claiming that the proposal was a Martinez scheme to preempt Bill Nelson's space industry expertise. "You've got the governor's proposal for a spaceport and a congressman running for governor who's been an astronaut. I guess the governor is trying to be an astronaut, too."[72] The accusation was baseless; Nelson had flown on the shuttle in January 1986 as a congressional observer, at the invitation of the NASA administrator. Nelson's "astronaut" honorific was a technicality, as he was not a career astronaut. We'll discuss his flight in Chapter 4.

Nelson endeavored to stay above the partisan politics creeping into the spaceport project. He refrained from criticizing the spaceport proposal, but neither did he endorse it, saying he was not familiar enough with the project and that criticism would appear self-serving.[73]

State Democrats also accused the Martinez administration of mismanaging the proposal. Wetherell and Margolis complained they had not been briefed about the legislation even though their committees would have to approve the $10 million request. A Martinez representative assured that they would be briefed soon.[74] Wetherell told the *Orlando Sentinel,* "If they can show how it can be economically beneficial, then we'll take a look at it. But it's going to be a tough sell."[75]

After the Florida Audubon Society threatened to organize a nationwide campaign against Spaceport Florida, Martinez decided to drop Shiloh from the project. "While Spaceport Florida remains high on my list of priorities, I refuse to allow it to proceed in a location where the precious natural resources of our state are threatened," he said. In return, the Florida Audubon Society pledged to support "whatever legislation is needed to get through the bureaucratic roadblocks."[76]

Dropping Shiloh exposed Martinez to more criticism from state Democrats. House Speaker Tom Gustafson (D-Fort Lauderdale) claimed the UE&C report was "sloppy work" for including Shiloh. "I'm just now wondering whether the whole report has just the same kind of inconsistencies or failures of information," he said at a planning meeting. But Rep. Dixie Sansom (R-Satellite Beach) pointed out that the Shiloh land had been acquired by NASA back in the early 1960s "for the express purpose of keeping it available for future space needs."[77]

This was not the last time NASA or the spaceport authority would propose the use of Shiloh or MINWR for commercial use, nor would it be the last time that environmentalists would oppose them. In any case, the UE&C summary barely mentioned the site's sensitivities, only that the state should "begin the process to obtain use of Shiloh from NASA" and to begin "the required environmental studies" for both Shiloh and CCAFS.[78] The study was not a political document, but its failure to anticipate the firestorm kindled by proposing a launch site in a wildlife refuge—even if it were no longer in its natural state—created a political vulnerability quickly exploited by opponents. It would happen again in future decades, a topic for discussion in Chapter 10.

With Shiloh off the table, legislative deliberation turned to the composition of the authority's governing board. Martinez had proposed a seven-member panel, appointed by the governor, and approved by the Florida Senate. House Democrats objected to that, too, preferring a more inclusive board. A House subcommittee proposed a nine-member board, with three appointed by the governor, three by the Speaker of the House, and three by the Senate president. The board's composition would include three members from the aerospace or banking industries, three from academia, one from labor, and two members of the public not employed by aerospace firms.[79]

The proposed $10 million was another legislative target. By the end of April 1989, the proposed amount was reduced to $2.6 million, because it was acknowledged that not all the money would be needed for the upcoming FY 1989–1990.[80] Even that wasn't included in the original House Appropriations bill, which zeroed out the project.[81] The Senate version proposed only $1 million.[82]

On May 7, 1989, the *Orlando Sentinel* reported allegations that state Democrats were trying to sabotage the spaceport proposal to help Bill Nelson's campaign for governor. Rep. T. K. Wetherell, described by the *Sentinel* as "a staunch supporter" of Nelson, told the paper, "We just can't get a handle on this thing. They floated this deal out that sounded so wonderful and everybody just got fired up about it. But when you start looking at it, it doesn't look so good." The paper cited "several sources who asked not to be identified" as claiming Nelson was urging space contractors to oppose the governor's proposal.[83] In a December 2023 interview, Nelson confirmed his April 1989 statement that he had been unfamiliar with the Martinez proposal, therefore he had not urged anyone to oppose it.[84]

House Democrats also added pro-labor amendments that were unacceptable to the governor. The amendments would have required spaceport construction contractors to pay union scale, forced the authority to engage in collective bargaining with employees, and placed a labor representative on the governing board.[85]

Nineteenth-century poet John Godfrey Saxe famously said, "Laws, like sausages, cease to inspire respect in proportion as we know how they are made."[86] The adage is certainly true of the Spaceport Florida legislation. Bills evolved in parallel in both the state House and Senate. At one point, a version of the House bill deleted the Cape Canaveral site, leaving that for the future, while funding only the Cape San Blas site; House leaders were skeptical that a market existed for state-launched orbital payloads, as existing launch companies already had operational pads on the Cape. Environmentalists suspected this was a maneuver to restore the Shiloh option.[87] A Senate version adopted Martinez's board membership proposal and dropped the House version's requirements for union wages and board membership.[88]

The final version of the bill passed the state legislature on June 3, 1989. The bill officially created the Spaceport Florida Authority, governed by a board with seven members appointed by the governor, but approved by the legislature. The authority's offices would be located near Kennedy Space Center. The bill provided $1.9 million to start commercial pads at Cape Canaveral and Cape San Blas. Although it was far less than the originally requested $10 million, it was enough to fund a single commercial midsized launch pad to act as proof of concept.[89] The bill also targeted education, creating a council of state educators to establish at least one residential math and science high school for honor students. It provided $650,000 for a network of high-technology business startup centers at community colleges. Although the authority didn't receive the full funding Martinez requested, the bill allowed the agency to issue up to $210 million in bonds its first year.[90]

At Spaceport USA's rocket garden on July 5, 1989, Governor Martinez signed the bill into law. "Why let it escape to another state?" Martinez said in his remarks. "If we don't move fast to capture the future, someone else will."[91]

* * * * *

For many of its more prominent supporters, Spaceport Florida did not benefit their careers.

Governor Bob Martinez was defeated for reelection. As the polls had predicted, he was beaten by former US Senator Lawton Chiles, who came out of retirement after all. In the September 4, 1990, Democratic primary election, Chiles handily defeated Bill Nelson, 69.5 percent to 30.5 percent. Nelson carried only two Florida counties, one of which was Brevard.[92] In the November 6, 1990, general election, Chiles defeated Martinez, 56.5 percent to 43.5 percent.[93] Nelson's star would rise again; in 2000, he would be elected one of Florida's two senators in Congress, and in 2021 he would be appointed NASA administrator by President Joe Biden.

Once the Spaceport Florida bill passed the legislature in June 1989, Chris Shove left the state Department of Commerce to accept a vice president position with the Brevard-based Space Research Foundation.[94] Shove departed the SRF after only six months. The reason why is a lengthy tale of Florida politics and legislative malpractice, which we will detail in Chapter 5.

Shove now describes that departure as "a blessing." He was hired as a professor of city planning and economic development at the University of Oklahoma. Shove now runs a business focused on promoting youth science and engineering education, hoping to create for them a career path into NASA's Project Artemis crewed lunar landing missions later in the 2020s.[95]

The Commonwealth of Virginia courted Steve Morgan for a new commercial space research program. The Center for Innovative Technology was a rough analog to Florida's Space Research Foundation. In October 1988, the CIT made $500,000 available for commercial space research and development in joint university-industry projects.[96]

After fulfilling his US Navy Reserve commitments, Morgan joined the CIT in 1989 as their director of space industry development. As he had in Florida, Morgan created a Virginia Space Business Roundtable, organized through the CIT. He also started the Virginia Space Business Incubator to attract companies to the Commonwealth, and managed over twenty joint state and industry co-funded projects at Virginia universities.[97]

* * * * *

Governor Martinez, Chris Shove, Steve Morgan, and others were right to worry that rival states and locales would try to lure an emerging commercial space industry from Florida. Morgan's departure for Virginia showed that the Commonwealth was serious, and so were other states. Each entity had its own unique story. Each story taught its own lessons.

Governor Bob Martinez chairs the first meeting of the Spaceport Florida Authority on September 11, 1989. Image source: Mark T. Foley Collection, State Archives of Florida, Florida Memory.

While Florida had to create its agencies and foundations through a treacherous political process, Virginia used the existing Center for Innovative Technology as its springboard.

The CIT was originally envisioned in 1983 as a cooperative effort by the state's three major research universities. Virginia Governor Charles Robb appointed the commission that recommended creating the institute, backed their proposal, and warned that narrow parochial interests could sabotage the project.[98] After the political sausage was ground, the legislation was passed in March 1984, creating a public authority and nonstock corporation located in northern Virginia. Critics complained that the agency was exempt from the state's Freedom of Information Act laws, but the Robb administration promised to seek amending legislation to require some disclosure.[99]

By late 1989, the CIT was beginning to invest in aerospace research. The institute was one of several entities comprising the Virginia Space Grant Consortium. NASA awarded grants to two consortium members, Hampton University and the College of William and Mary, to attract students to space disciplines.[100]

Dennis Barnes, vice president for government relations for the University of Virginia (another consortium member), also acted as Governor Gerald Baliles's space business advocate. Barnes recruited Steve Morgan to join the CIT. In March 1990, Barnes became president of the Virginia Space Business Roundtable. Organized through the CIT, the Roundtable—like its Florida counterpart—was a forum for Virginia's space community to create new space enterprise opportunities.[101]

As did Florida, Virginia sought to create its own commercial launch pads. NASA's predecessor, the National Advisory Committee for Aeronautics, in 1949 purchased a barrier island called Wallops on Virginia's eastern shore as a test range for guided missiles. When the NACA became NASA in 1958, Wallops became an independent center. In 1981, the test site fell under the supervision of NASA's Goddard Space Flight Center in Maryland, and was renamed the Wallops Flight Facility.[102] A consortium of private businesses from Virginia, Maryland, Florida, and California in 1994 failed to win USAF funding for a commercial complex at Wallops to launch small to medium payloads.[103] The Commonwealth persisted and in 1995 passed legislation to create the Virginia Commercial Space Flight Authority.[104] The first orbital commercial launch from Wallops was on October 23, 1995; the Conestoga rocket was owned by EER Systems, but two-thirds funded by NASA. Conestoga exploded about forty-five seconds after launch.[105]

By April 1996, Spaceport Virginia had won its first contract, to help launch Air Force rockets from Wallops. The USAF planned to convert old Minuteman ICBMs to carry scientific instruments into space; NASA contributed the launch site and its expertise.[106] The authority negotiated an agreement to manage NASA's Wallops suborbital launch operations.[107] In March 1997, NASA agreed to allow the authority to create a Virginia Space Flight Center at Wallops that would launch commercial satellites.[108] By the end of 1997, Spaceport Virginia had received its federal license to operate as an independent commercial spaceport.[109]

By the early 2000s, the center had yet to launch a commercial payload. The authority competed with other commercial spaceports to host Lockheed Martin's VentureStar, but the company canceled the spaceplane in 2001 after NASA withdrew from the project. Operations still depended largely on NASA and defense contracts. Several satellite companies had gone out of business after the emergence of fiber-optic cable as a viable ground-based communications option. A proposal emerged to replace the existing bureaucracy with a Mid-Atlantic Regional Spaceport, governed by both Virginia and Maryland, believing the joint power of two states would enhance the launch site's attractiveness.[110] The two states reached agreement in December 2003, and MARS replaced the Virginia space center.[111]

In November 2004, MARS won a $49 million contract to launch satellites and spacecraft for the USAF.[112] The first launch was December 16, 2006, an Orbital Sciences Minotaur I rocket carrying a USAF TacSat-2 satellite and a NASA experiment. The sixty-nine-foot Minotaur was diminutive compared to the larger rockets launched at the time from Cape Canaveral and Vandenberg, but it was a start.[113] Orbital Sciences announced in June 2008 that MARS would be the launch site for its Cygnus robotic commercial cargo craft, launching atop its Taurus II rocket, later renamed Antares.[114] Orbital is now part of Northrop Grumman, which in March 2022 won another Commercial Resupply Services contract from NASA. Cygnus and the SpaceX cargo Dragon are projected to provide redundant robotic cargo delivery to the ISS through at least 2026—Dragon from the Space Coast of Florida, Cygnus from both the Cape and from the eastern shore of Virginia.[115]

Although MARS has yet to launch a payload to the Red Planet, Wallops in 2013 launched its first probe beyond Earth orbit. NASA's LADEE mission, launched atop an Orbital Sciences Minotaur V, was the first lunar mission from Wallops.[116]

Florida and Virginia arguably are the best success stories for state commercial space authorities. They had existing federal government launch sites. California has Vandenberg Space Force Base, but so far the Golden State has struggled to achieve a sustainable state or regional authority. They've been at it longer than any other state.

Back in the 1970s, *Chicago Daily News* syndicated columnist Mike Royko was fond of using the word *moonbeam* to describe those with unconventional progressive political philosophies. In 1976, Royko wrote that California Governor Jerry Brown was pursuing "the moonbeam vote," which led to the sobriquet "Governor Moonbeam" that haunted Brown the rest of his political career.[117]

Brown declared in 1977, "I'm really into space now." He lobbied Capitol Hill that year to restore $20 million in funding for a Jupiter probe managed by JPL in Pasadena. He also fought to protect $56 million in funding for the space shuttle, which would be assembled in Palmdale. "Small is beautiful on Earth," he said, "but in space big is better."[118] Brown saw space exploration as a solution to earthly problems such as overcrowding, energy depletion, and environmental damage.[119]

Brown hired astronaut Rusty Schweickart as his chief science adviser, to help prod the federal government into moving faster on contracts affecting the California space industry. Schweickart helped Brown organize a "Space

Day" in Los Angeles to propose that the State of California enter into a joint venture with NASA to build and launch a California satellite.[120] The event, held on August 11, 1977, at the California Museum of Science and Industry, was attended by Schweickart, astronomer Carl Sagan, underwater explorer Jacques-Yves Cousteau, and NASA Administrator Robert Frosch.[121] Brown said the satellite could monitor the oceans, reservoirs, crops, and forests, and would create jobs for California aerospace workers. Sagan remarked, "I am absolutely lost in admiration of Governor Brown's talk."[122] Brown also attended the first glider test flight of the space shuttle prototype orbiter *Enterprise*. The governor described the flight as "a really unusual mixture of poetry, profit, technology, and ecology."[123]

Brown doesn't seem to have proposed any legislation that would have led to a state space authority. The satellite became his focus, both to collect environmental data and to create California jobs.

Hughes Aircraft Company in the 1960s built for NASA a series of communications satellites called Syncom, short for synchronous communications satellite. The Syncom satellites were built at a Hughes factory in El Segundo, California.[124] In December 1977, Schweickart briefed California state legislators on Brown's plan to include in his next proposed budget a three-year California Satellite Project. It would use at least one of the communications links on a proposed Syncom 4 that would launch on a future space shuttle mission. The plan also included ground-based transceivers, and recruitment and training for the personnel to operate the ground stations. Brown's staff believed that California's share of the costs would be only $3 million per year for the ground stations, with NASA and Hughes picking up the rest of the costs.[125]

NASA didn't see it that way. In April 1978, NASA sent a letter to California members of Congress advising that the agency hadn't made a final decision to fly Syncom 4. The letter expressed concern that giving the satellite a free ride might be construed as "undue commercial or competitive advantages to Hughes."[126] The satellite manufacturer was also reluctant; Hughes was unwilling to commit its $14 million investment without a firm commitment from the federal government.[127]

After California voters passed the property tax limitation measure Proposition 13 in June 1978, Brown and the legislature were forced to cut the state budget. The state space satellite was one of the victims.[128] Syncom 4 was deployed by the crew of the STS-51I mission aboard the orbiter *Discovery* in August 1985; it failed to work and Hughes wrote it off as a total loss.[129]

A later effort to support commercial space began in the early 1990s, as the Cold War ended and the nation reaped the "peace dividend." Facing a force reduction of 25 to 50 percent at Vandenberg Air Force Base north of Lompoc, local officials in May 1992 created a nonprofit called the Western Commercial Space Center (WCSC). The organization sought funding to research potential commercial uses for Vandenberg.[130] In 1993, the WCSC borrowed Florida's idea to create its own state space agency. Called the California Space Authority (CSA), the agency became eligible for federal grants to help promote the industry and encourage aerospace to remain in California.[131] The authority planned to build a commercial launch pad at Vandenberg. To raise matching funds for that project, a separate for-profit corporation was created, called the California Commercial Spaceport, Inc.[132] The CSA dissolved itself in June 2011, after $5 million in expected federal funding failed to materialize.[133]

Local officials tried again, and appear to have succeeded. In 2018, civic and business leaders formed the Hourglass Project, to grow and diversify the local economy. Vandenberg partnered with the group to attract more commercial enterprises to the base. By 2020, the organization had renamed itself the Regional Economic Action Coalition, calling themselves REACH.[134] In June 2021, REACH and its partners released a Commercial Space Master Plan produced by Deloitte.[135] Small-rocket NewSpace startups Firefly Aerospace and Relativity Space have established launch presences at the base.[136] In a May 2024 release, REACH said that a recent California Space Innovation breakfast in Sacramento had as one of its topics, "Cementing a statewide space entity á la Space Florida."[137]

In April 2023, SpaceX announced plans to take over Space Launch Complex 6, formerly used by ULA, as a second Vandenberg launch site.[138] SpaceX has encountered resistance from local environmentalists, complaining that their increased launch pace would threaten wildlife, create noise, and leave marine debris.[139] In October 2024, the California Coastal Commission rejected a Space Force request to increase the number of launches at Vandenberg; some commissioners cited Elon Musk's political activities. Elon Musk and SpaceX sued the commission, alleging he was being punished for his political views.[140]

California's Mojave Air and Space Port is about 135 miles inland east of Vandenberg, near Edwards Air Force Base. The spaceport has evolved over the decades from a desert general aviation airstrip into an aerospace research center. In June 2004, Mojave was licensed by the FAA as a commercial launch site. It's been used for several NewSpace "firsts" such as the Scaled Composites suborbital SpaceShipOne flights in 2004, the Stratolaunch dual-fuselage Roc carrier aircraft test flights in the early 2020s, and Virgin Galactic's suborbital spaceplane test flights before the company relocated to Spaceport America in New Mexico.[141]

Spaceport America is its own cautionary tale.

New Mexico's pursuit of an operational commercial spaceport traces back to the early 1990s. McDonnell Douglas DC-X test flights at White Sands inspired visions of a space industrial and research complex.[142] A group of individuals and businesses in the Las Cruces area formed the Southwest Regional Space Task Force, hoping to create a commercial spaceport. Apathy among local business and community leaders was an early concern; a June 1992 community luncheon was poorly attended, with only five of twenty invited local leaders showing up.[143] State elected officials were more receptive. In August 1993, the State of New Mexico's Economic Development Department announced plans to open an Office of Space Commercialization in Las Cruces, along with other offices in Albuquerque and Santa Fe. NASA awarded $950,000 to New Mexico State University for spaceport studies. The USAF awarded the state a $1.1 million grant to study a spaceport at White Sands. The state legislature chipped in $250,000 to run the three offices tasked with attracting space enterprises to New Mexico.[144]

As happened with other states, New Mexico indulged in overly optimistic projections of the future. George Harris Jr., the original director of the Office of Space Commercialization, said in August 1994 that he had already signed customers and chosen a location for the spaceport. Harris estimated that the spaceport could be operational as soon as 1996; he believed that 60 percent of the $200 million construction cost would be covered by the private sector. The state would acquire the land and build the utility infrastructure.[145] By the summer of 1995, McDonnell Douglas was looking at New Mexico for test flights of their version of the X-33, a planned successor to the DC-X. A representative of Lockheed Martin, which won the X-33 contract in 1996, confidently expressed his opinion that commercial spaceflight would be common by 2005. "The (US) government is hoping to be out of it by then. Commercial spaceport will be the way space travel will be done by then."[146]

But a June 1995 study presented to Republican Governor Gary Johnson by his Technical Excellence Committee found that the spaceport would not be viable unless the aerospace industry developed a reusable commercial rocket. It would not be suitable for certain existing spacecraft, while others would not be economical. There was also the risk of expendable rocket stages falling on populated areas. Governor Johnson concluded that the state should proceed cautiously before heavily investing in the spaceport.[147]

NASA provided little encouragement. Administrator Dan Goldin addressed the New Mexico Spaceport Summit in August 1997 and told the 200 attendees that existing spaceports had an advantage. "The (federal) deficit is not going to get cut by having ten different spaceports in America, all getting subsidized by the government. And this group will have to work harder than a lot of the established launch facilities." Governor Johnson estimated that the state would have to invest about $65 million in the spaceport over thirty years.[148]

The Office of Space Commercialization courted Lockheed Martin to bring VentureStar to south New Mexico. Being a single-stage-to-orbit vehicle, it would not discard expendable stages on the populace, and it would be a reusable commercial rocket.[149] But when Lockmart canceled VentureStar in 2001, New Mexico lost its best hope for a commercial spaceport anchor tenant.[150]

Democratic Governor Bill Richardson took office on January 1, 2003, and revived the state's spaceport dreams. He dispatched Peter Mitchell, the executive director of the Office of Space Commercialization, to an FAA conference in Washington, DC, with a letter from him urging the FAA to continue support of inland spaceport initiatives.[151] The state sought an FAA license for a commercial spaceport at Upham, about 50 miles northwest of Las Cruces; Mitchell touted the site's 4,200-foot altitude as an advantage for launch companies to save fuel.[152] The office also pursued hosting the XPRIZE Cup, a $10 million award offered by multimillionaire Peter Diamandis. The prize would go to the first private sector team that launched three humans above the Kármán line, named after Caltech physicist Theodore von Kármán, who had suggested in the 1950s that "space" begins at 100 kilometers (62 miles).[153] New Mexico won the rights to the XPRIZE Cup in May 2004.[154]

In September 2004, billionaire Sir Richard Branson announced in London, England, the birth of Virgin Galactic, a suborbital tourism venture inspired by SpaceShipOne crossing the Kármán line to win the XPRIZE Cup. Branson hired Scaled Composites to build SpaceShipTwo, a more advanced version that would launch initially from Mojave. With the braggadocio of a showman, he claimed the first passenger flights would be within eighteen months. Drinks would be served on board.[155]

After a year of rumors, Governor Richardson and Sir Richard announced on December 14, 2005, at a Santa Fe, New Mexico, hotel press conference that the state and Virgin Galactic had reached a twenty-year agreement to build a twenty-seven-square-mile spaceport near Upham. A study by New Mexico State University projected that, by its fifth year of operation, the spaceport would generate $1 billion of local spending, a payroll of $300 million, and employ 2,300 people.[156] At the Farnborough International Airshow in Hampshire, England, in July 2006, state officials announced to the world that the site would be named Spaceport America.[157]

The estimated cost for the spaceport was $225 million. How to pay for it? Richardson's budget proposed:[158]

- $100 million from state capital outlay funds over three years;
- $25 million from a state road improvement project fund called GRIP 2 (Governor Richardson's Investment Partnership);
- $10 million from encumbered existing state funds; and
- $90 million over three years from federal funding and local taxes.

Richardson's proposal relied on a local-option gross receipts tax, which required a majority-vote approval from local electorates in Doña Ana, Otero, and Sierra counties.[159] State legislation quickly passed, authorizing local governments to impose the tax with voter approval, and shifting spaceport activities from the Office of Space Commercialization to a new state Spaceport Authority.[160] Doña Ana County commissioners scheduled a special election for April 3, 2007, asking voters to approve a 0.25 percent tax increase to pay for twenty-year bonds.[161]

Commissioners from Otero and Sierra counties awaited the Doña Ana results before scheduling their own elections. The referendum narrowly passed by 270 votes, 50.8 percent to 49.2 percent.[162] About two-thirds of Sierra County voters approved the tax in April 2008.[163] Otero County voters rejected the tax in November 2008, 52.3 percent to 47.7 percent, but the Otero tax was expected to generate only about 3 percent of the spaceport's cost.[164]

On the last day of 2008, Virgin Galactic and Governor Richardson signed a twenty-year Spaceport America lease agreement.[165] The building was dedicated on October 17, 2011, with Branson rappelling down the side of its glass windows holding a bottle of champagne. The mother ship VMS *Eve,* carrying the VSS *Enterprise* spaceplane, flew an exhibition flight around the spaceport.[166]

Those operational flights Branson once claimed would begin in 2006? Just another desert mirage.

Human error destroyed VSS *Enterprise* on October 31, 2014, during a powered test flight over the Mojave. The National Transportation Safety Board accident report concluded that the copilot prematurely unlocked the plane's feather system, causing *Enterprise* to suffer a structural failure. The copilot was killed, but the pilot survived with serious injuries.[167] The accident set back Virgin Galactic several years, until a new ship called VSS *Unity* could be built and complete its test flights.

VSS *Unity* and VMS *Eve* finally arrived at the spaceport in February 2020.[168] With six people on board, including Branson, *Unity* completed Virgin Galactic's first commercial mission on July 11, 2021—more than fifteen years after Branson and Governor Richardson announced their agreement.[169] On November 8, 2023, Virgin Galactic announced the company would lay off 185 employees in anticipation of pausing flights in mid-2024 until a new Delta Class spaceplane comes on line sometime later in the decade.[170] The company reported a net loss of $502 million in 2023, and $384 million in 2024.[171] The company nevertheless remained committed to Spaceport America, promising that the Delta Class ships would carry more passengers and fly more often, generating more revenue for the spaceport. Virgin Galactic signed a letter of intent to build a larger hangar that can hold multiple Delta Class ships.[172]

Other states and locales in recent decades have sought commercial spaceport authorities and licenses. Florida, Virginia, California, and New Mexico are the most prominent, and each offers us lessons that are applicable not only to startup spaceports but any entity interested in establishing a public-private partnership.

The primary lesson is to minimize the risk of political interference. Opposition may be partisan, or for other reasons. In Florida, it was a risk to the environment, in parallel to petty parochial interests. In New Mexico, some legislators called for the money to be spent instead on the poor. In California, Jerry Brown's state satellite lacked widespread political support, and therefore lacked funding.

Virginia found the path of least resistance, evolving out of an existing agency, the CIT. When it appeared in the mid-1990s that the commercial space boom was going bust, the Commonwealth partnered with the State of Maryland to create a regional partnership. Seeking protective partnerships, Virginia and New Mexico in 2000 joined seven other states to form a joint counterweight to the perceived political clout of Florida and California in federal space launch policy.[173]

Gubernatorial leadership often helped shepherd legislation through their state's sausage grinder, and where such leadership was absent, local efforts struggled to succeed. Executive political leadership spanning administrations is needed to assure funding for agency continuity, and support should cross partisan lines.

Where all states failed was their wide-eyed belief that the commercial space revolution was just around the corner. The George H.W. Bush administration did what it could to encourage that revolution, but it failed to find the formula to sustain the robust commercial satellite industry beyond the 1990s. After the end of the Cold War, the commercial launch industry contracted, and might have collapsed if not for the protectionism of the DOD. Few were willing to risk their investments without government subsidy.

State leaders nonetheless believed they were in a new space race, as if staking their claim in a modern-era gold rush before other states did, but often their pans yielded no more than a fool's gold. In the most egregious example,

New Mexico leaders agreed to build Spaceport America and its supporting utility infrastructure based on Sir Richard Branson's claims that he'd be ready to fly within just a few years. In June 2022, the former mayor of Roswell, New Mexico, published an opinion column in the *Albuquerque Journal* calling Spaceport America a "concept/scam" that is now "the world's most expensive roller coaster."[174]

When a government official is courted by a private entity or entrepreneur predicting fantastic returns in exchange for a government subsidy or tax break, caveat emptor is the best advice. Put another away, recall the lesson learned in the Popeye cartoons by victims of Wimpy's empty promise:

"I'll gladly pay you Tuesday for a hamburger today."

4

Scion of the Space Coast

Clarence William Nelson II, who everyone called "Bill" or "Billy," was born in Miami on September 29, 1942. It was a time of war, as German U-boats sank American ships off the Florida coast, not far from his birthplace.

Throughout his political career, Bill described himself as a fifth-generation Floridian. His lineage traced back to his great-great-grandfather, who in 1829 landed at Port St. Joe in the Florida Panhandle, near where sounding rockets would launch one day from Cape San Blas.[1]

The third generation, Bill's paternal grandparents, used the Homestead Act in 1915 to acquire 160 acres of land in north Merritt Island near a spot called Wilson's Corner. That land was a few hundred yards from what later became the north end of the Shuttle Landing Facility runway.[2]

Bill's parents were the fourth generation, Clarence William Nelson and Nannie Merle Nelson. Bill's five generations of Florida lineage trace back through Nannie, whose family members for decades were born, raised, and died in a small Panhandle town called Chipley.[3] Clarence and Nannie both lost their prior spouses to tragedy. Nannie's husband drowned at age thirty-five in an April 1937 boating accident near Destin in the Panhandle.[4] Clarence's wife died at age thirty-four in Miami in November 1937 after a brief illness.[5] Their marriage in Gainesville, Florida, in September 1940 was a fresh start for both.

Clarence was Bill's exemplar. In 1929, Clarence was part of the first graduating class at the University of Miami School of Law.[6] He participated in debate and public speaking competitions.[7] After earning his law degree, he completed his bachelor of science in agriculture with the University of Florida in Gainesville.[8] In 1932, he ran for a Dade County commission seat, but lost.[9] In 1934, Clarence announced his candidacy for the Florida state House, but lost that race too.[10] In 1938, Clarence and a physician purchased a light plane so they could learn to fly.[11]

Clarence relocated the family to Malabar in south Brevard County in 1949.[12] A December 1950 classified ad in *The Miami Herald* lists him as selling property in the area, with an address in Melbourne.[13] By 1953, he was not only developing home sites on the Indian River near Malabar, but also a "high producing pasture."[14] By 1955, Clarence had named the pasture Rock Point Ranch, and was using it to raise cattle.[15]

In his adolescence, "Billy" followed in his father's footsteps. He joined the 4-H Club and began appearing in local newspapers as the winner of cattle breeding competitions.[16] His father saw only Bill's earliest achievements in life; Clarence passed away in August 1957 at age fifty-nine, leaving Billy and his mother Nan.[17] He was not yet fifteen.

Bill quickly stood out among his contemporaries. In May 1958, he won a 4-H speaking contest. The topic? "Promise in the Cattle Business."[18] He went on to win the state 4-H speaking championship and a $100 scholarship.[19] At age sixteen, he was elected president of Key Club International (a service organization with almost 50,000 high school members in the United States and Canada) at their global convention in Toronto.[20] The next year, he was Melbourne High School's student body president.[21] Bill often spoke to public groups about the need for teenagers to have adult role models; he noted that few teens were delinquents, but adult leadership was necessary to transform them into "creative" citizens.[22] Melbourne High's principal selected Bill to deliver the commencement address for his senior class.[23] The principal predicted that, one day, Bill might be Florida's governor.[24]

In October 1959, Bill departed on a ten-day trip as part of a Crusade for Freedom tour to the Iron Curtain, the only teenager in the group.[25] When he returned, he told a local Kiwanis Club gathering about his experiences. "Freedom is easily taken for granted by those born to it," he said. "Its true meaning is best understood by those persons who have lived under a dictatorship." He described the barbed wire he'd seen erected by Soviet troops to keep eastern Europeans behind the Iron Curtain.[26]

Bill was a rising star in the community. He received the key to the city of Melbourne at age seventeen for recognition of his public service. *The Miami Herald* noted that Nan was at his side, and described her as his "constant companion."[27]

To fund his college education, Bill sold off his ranch operations.[28] He intended to become a lawyer like his father, although he also considered becoming a minister. An *Orlando Sentinel* profile quoted his assistant principal's impression of his speaking skills. "He can take an audience and hold its attention throughout an entire speech. He can do that with 3,000 and he has done it right here with his fellow students . . . I would equal him with most any speaker I've heard and I've heard quite a few." Reporter Logan Owen observed, "Bill uses an approach which is analogous to the narrative hook used by writers. He establishes rapport with his hearers by referring directly to them and their interests. This accomplished, he carries his listeners right down to the last Amen."[29]

* * * * *

Bill began his college studies at the University of Florida in Gainesville. He was immediately elected president of his freshman class.[30] Six months later, in March 1961, he was elected to an honor court, "which upholds the honor system followed on campus."[31] At the end of that first year, he transferred to Yale University in Connecticut and largely disappeared from public view in Florida, save for the occasional social engagement and Christmas events at Rock Point Ranch.

Nelson's roommate at Yale was Bruce Smathers, the son of Senator George Smathers (D-FL) who was a friend of President John F. Kennedy. Nelson interned for two summers with Smathers's office in Washington, DC. His senior thesis was titled, *The Impact of Cape Kennedy on Brevard County Politics.* Nelson graduated from Yale in 1965 with a degree in political science. His speaking skills won him election as class orator.[32]

Bill went on to the University of Virginia Law School, where he earned his law degree. He then served as a captain in the US Army, stationed in Indianapolis and Miami. In 1969, while on leave, Bill and Bruce traveled together to Europe. While standing on a hill outside Bucharest, Hungary, they listened to the BBC shortwave radiocast of the Apollo 11 launch. Four days later, as he stopped in London on his way back to his duty station in the states, Nelson watched a grainy black-and-white TV image of Neil Armstrong and Buzz Aldrin walking on the moon.[33]

After he returned home to Rock Point, Nelson opened a law practice in Indian Harbour Beach, on a barrier island east of Melbourne.[34] The newspaper gossip columns began to report that Bill was seen about town dating one local woman or another. A *Today* staff writer commented, "Bill Nelson is considered by many to be one of Brevard's most eligible bachelors. There are some who call him the most eligible, with no competition."[35] A young woman from Jacksonville named Grace Calvert won his hand in marriage; they met while she was dating Bruce.[36] Bill and Grace wed in Jacksonville on February 19, 1972.[37]

With a new bride on his arm, it didn't take long for Nelson to launch his political career. On April 18, 1972, he announced his bid as a Democratic candidate for the Florida House of Representatives. He sought a seat vacated by a retiring Republican. "I am announcing my candidacy in the hope that I may begin a life of public service in which I can give of myself with whatever ability God gave me," Nelson told a crowd of 200 at a Melbourne press conference.[38] Grace told a newspaper society columnist, "I knew from the first time I met Bill that he was interested in politics. His friends all told me he'd be governor someday."[39]

The local newspaper *Today* endorsed Nelson over his Republican opponent. "We feel confident Bill Nelson will be an all-around quality legislator. We see him as a bright young man with a long and distinguished political future ahead."[40]

Most of the House District 47 voters agreed. Nelson won by 70 percent to 30 percent.[41] The next day, Bill and Grace stood roadside in the district holding "Thank you!" signs for passing motorists.[42]

Nelson established a reputation as an honest politician. He gave a speech in October 1973 to the Kiwanis Club of Cocoa on the topic, "Morality in Politics." "Compromise should not mean sacrificing your principles," he said. "It should mean finding an equitable solution to any problem which does not have a clear-cut answer."[43] His penchant for compromise would be a trait throughout his career—praised by some, lamented by others.

Although space activities at the Cape were within his district, Nelson's time in the State House largely dealt with issues that were far more parochial—expanding a bridge, capping an artesian well, education reform, restoring marshlands and watersheds, a Veterans Administration hospital.

By early 1974, the post-Apollo job losses were driving up unemployment in Brevard County. In late February, Nelson announced he was organizing a conference with the Florida Department of Commerce to be held in Cocoa Beach on March 6. "We're bringing the top people in the Department of Commerce to Brevard County to show them our need for replacing the space industry, as it's phased out, with other jobs and industry," he told local newspaper *Today*.[44] But it appears that no legislation came out of the event.

The Brevard County unemployment rate reached 12.7 percent in February 1975. Nelson called the report "devastating" and told *Today*, "It should cause us to redouble efforts to provide jobs." Nelson amended legislation by state senator William Gillespie (D-New Smyrna Beach) proposing a Florida solar energy center, directing that the operation be located at Cape Canaveral to generate jobs.[45] It was a rare example of Nelson, as a state legislator, acting on a space-related issue, although it was no more than an attempt to transition some jobs into an emerging high-tech industry.

Nelson was politically popular in his district. He was returned to office without opposition from either party in his 1974 and 1976 reelection bids. He was not as popular with his party's leadership in Tallahassee. Nelson had opposed House Speaker Donald Tucker (D-Tallahassee), who intended to ignore fifty years of precedent by succeeding himself in the post. To protect members from retaliation, Nelson suggested that the secret ballots not be signed, so Tucker would not know how members voted, but the motion was defeated. Nelson was chair of the House appropriations committee, but in May 1976, Tucker denied him what traditionally had been an automatic appointment for that chair to a House-Senate joint budget committee. Tucker refused to say why, but one anonymous veteran lawmaker said Nelson wasn't "in" with Tucker's leadership team. The lawmaker told a reporter, "He's Mr. Nice Guy. He's worried about his image. He tries to please everybody . . . and you can't do it. He's too compromising."[46]

Tucker punished Nelson again in November 1976, removing him from the House appropriations chair. Nelson told *Today*, "If you are not a yes-man, you have retribution taken against you." Tucker also slashed Republican staffing and denied the Brevard County Republican delegation any important assignments to committees. Nelson said it was "an attempt to silence dissent."[47]

In February 1977, Nelson announced that he would seek a US Congress House seat being vacated the next year by Rep. Louis Frey (R-Winter Park). The district was split between Brevard and Orange Counties. "The real critical decisions in this country will be made in Congress," he said at a press conference in his Melbourne law office.[48]

Nelson's congressional bid coincided with what seemed to be an innocuous announcement by the USAF that the Eastern Test Range (ETR) would be placed under the command of the Space and Missile Test Center at Vandenberg Air Force Base in California. The 6550th Air Base Group at Patrick Air Force Base would report to Air Force Systems Command at Andrews Air Force Base in Maryland. The reorganization was to "improve management efficiency and operational support" for both the eastern and western ranges. No personnel reductions were expected at either base.[49]

But that wasn't the way Bill Nelson viewed it. Just a few days after his declaration of candidacy, he sounded the alarm about what he believed was an "economically devastating" phase-down at Patrick. "Our country must unite immediately to prevent a move by the federal government which would be such an economic hardship to Brevard," he declared. Nelson called upon Governor Reubin Askew and the county's federal representatives, including Rep. Frey, to do everything possible to prevent the cutbacks. But Frey knew of no such cutbacks; he told *Today* that Air Force officials in Washington had assured him there were no plans for at least ten years to reduce personnel at Patrick. *Today* reported that the study was to determine if the ETR could be phased out in the mid-1980s once NASA's space shuttle fleet was fully operational, but the study was being conducted only at the request of the GAO.[50]

"Gov. Askew Enters Fight to Save ETR," blared a *Today* headline on May 4, 1977. Nelson was the source for the story, telling the paper that 4,000 jobs could be lost at the Range if Florida officials didn't act. "We are ready now, and unless we act now to reverse the Air Force momentum on its plans to phase out the Eastern Test Range by the mid-1980s, we will lose those jobs." Nelson said that the February reorganization of command was proof that the phaseout had already begun.[51]

Askew wrote a letter to President Jimmy Carter asking about the rumored Range consolidation. He received a reply from a Carter administration aide, who told him that any such study would look at the economic impact on the area.[52] Nelson, nonetheless, gave speeches to local groups about the supposed threat, quoting from the administration letter. The number of potential job losses had now grown from 4,000 to 5,000, and said he had discovered language in a DOD memo that was evidence of a phaseout. *Today* quoted the relevant passage: "To determine the technical and economic feasibility of transferring the Eastern Test Range missions to other agencies, facilities, and-or locations." "I suspect that in the innards of DOD there is this movement going on (towards closing the facility)."[53] Nelson told an *Orlando Sentinel* reporter that the memo was "shot through" with the assumption that the ETR was closing, but the Patrick public information officer said it was no more than a feasibility study. "I foresee an active future for the Range at least in the foreseeable future of 10 to 15 years," the officer said. "Launches are launches and tracking support remain no matter who manages the Range."[54]

Both the Air Force and Navy had operations at the Cape, so each military branch completed a study. Their reports were deemed classified, but summaries were provided to candidate Nelson, who gave them to *Today*. The Air Force report was a collection of "what-if" suppositions, attempting to project what the department might do with the Range if certain efficiencies were to occur in the next ten years, such as the NASA launch of satellites to replace the ETR's functions, and the Navy abandoning the Cape. But the Navy report made no such assumptions, and concluded that its department required ballistic missile testing ranges in both the Atlantic and Pacific.[55]

Nelson declared victory:[56]

> I think that the way that Brevard has responded by contacting their congressional delegation and the Department of Defense about the possible ETR closing has had a positive effect. That effect is that we do not have the imminent shutdown of the ETR.

The Patrick public information officer declined comment. "We can't comment until it's released. You all went ahead and leaked it, now you have to live with it."[57]

* * * * *

Nelson was elected by his district's voters to the House in November 1978, 86,486 (62 percent) to 53,219 (38 percent).[58] Space was not a significant issue in the campaign, as both he and his Republican opponent advocated for more federal spending on military and civilian space.[59] As he had after previous campaigns, Nelson stood on street corners thanking his constituents for their votes.[60]

The freshman legislator was appointed to two House committees. The Budget Committee oversees the general parameters of the annual federal budget. The plum for Nelson was a seat on the Science and Technology Committee, which has jurisdiction over the American space programs. Nelson told the *Orlando Sentinel Star,* "Both Cape Canaveral and the Kennedy Space Center are in my district. This is for the home folks. It is one of our major industries."[61]

One of Nelson's earliest stands on a NASA issue was to send a letter demanding an explanation for why the orbiter *Columbia* lost hundreds of heat shield tiles during a California test flight atop its Boeing 747 carrier aircraft. Rockwell employees at Kennedy Space Center told *Today* that tile workers about to be laid off at the Palmdale, California, factory had been lackadaisical in attaching the tiles. The tile problem delayed *Columbia*'s KSC arrival by about two weeks, and cost the taxpayers about $1 million. NASA sent Nelson a one-page report assuring him that the tile system was reliable, but he wasn't satisfied. "After I receive a personal briefing from NASA and they go into detail as to exactly what happened, then I might be satisfied," he told *Today.*[62] Nelson got his meeting a couple weeks later, and told *Today,* "NASA says they failed to anticipate a problem with the tiles and that was the major cause of the problem."[63]

As happens with most NASA projects, the launch date for the first shuttle mission receded farther and farther into the future. In January 1980, NASA officials said the first launch on paper was targeting June, but the fall was considered more realistic and a slip into early 1981 was possible.[64] NASA Administrator Robert Frosch appeared before Nelson's House Science and Technology Committee to discuss $300 million in supplemental funding he was about to request to keep the program on schedule. Although other committee members were skeptical, Nelson indicated he would support the request when it came before his Budget Committee.[65] NASA claimed that up to 25,000 jobs could be lost if the supplemental appropriation wasn't approved, but that didn't convince the Senate Budget Committee, which voted 7–5 to turn down the request.[66]

Entering his 1980 reelection year, Nelson sounded an alarm, just as he had during his first congressional campaign. In 1978, it was the supposed pending closure of Patrick Air Force Base. This time, he told a chamber of commerce group that NASA officials had informed him the agency intended to close the main access road to Playalinda Beach once the Space Shuttle program was operational, due to extra security for military payloads.[67]

Playalinda Beach, to this day, remains a popular tourist destination within the Canaveral National Seashore. Its main access is by State Road 402, the "beach road" that runs east from Titusville past the north end of the shuttle's runway to the Atlantic shoreline. The beach is about three miles from KSC's Pad 39A, and a little more than a mile from 39B. Pad 39A was the first Project Apollo pad to be refurbished for the shuttle, to be followed by 39B circa 1986. At that point, NASA intended to permanently close SR-402 within its borders so long as the Space Shuttle program was operational, according to Nelson.[68]

To resolve the problem, Nelson proposed that the Department of the Interior spend $3.5 million to build a new access road that would bridge the southern part of the Mosquito Lagoon to the north of SR-402. Park superintendent Donald Guiton said that NASA officials had told him the seashore would be closed about 35 percent of the year. He also predicted that conservationists would oppose any new road, which would require approval from multiple government agencies and possibly conflict with presidential orders protecting wetlands. Nelson believed that any environmental impact would be minimal, but local conservationists disagreed. The Florida Audubon Society objected to both NASA's projected closures and Nelson's alternate route, calling the closure "arbitrary and excessive" and the road proposal "premature."[69]

In June, NASA extended the security zone another half mile, essentially ending any possibility of Nelson's access road through the Mosquito Lagoon. Attempting to find a compromise, he and KSC Director Dick Smith released a proposal for government buses to shuttle beachgoers between Titusville and Playalinda. Nelson would ask Congress for $500,000 to lease buses from the General Service Administration, but no one had figured out how to accommodate surfboards or where the estimated 1,000 cars per day might park.[70]

By July, a citizens group had formed in Titusville, calling themselves Save Our Beaches (SOBs). Their goal was to eliminate NASA's proposed shuttle security perimeter and continue public access to Playalinda. The group's attorney dismissed Nelson's bus proposal, calling it a nuisance and a waste of taxpayer money.[71] Nelson arranged for the SOBs to meet with KSC Director Smith, who agreed to review alternatives to the shuttle bus, including a possible exception for badged nearby residents.[72]

On September 8, at a press conference with Nelson and Guiton, Smith announced SR-402 would remain open to the public, except for sensitive space shuttle processing events such as a launch countdown. The new study had concluded that the shuttle was no longer vulnerable to certain weapons.[73] The SOBs invited Nelson to a celebration event, originally planned as a protest rally, to honor him for his work on the issue. *Today* credited Nelson with "working out the compromise" between NASA and the activists.[74]

As in prior contests, Nelson drew no Democratic opposition in his reelection primary. A lone Republican candidate, seventy-eight-year-old Stan Dowiat, filed to oppose him. Dowiat claimed that Nelson was too young to serve in Congress.[75] The district's voters didn't see it that way, returning Nelson to Washington by a 71 percent–29 percent margin, a greater victory than his 62 percent–38 percent triumph two years before.[76] Once again, Nelson stood alongside US-1 with his "Thank you!" sign the day after the election.[77]

With all its tiles permanently affixed, *Columbia* finally rolled out to Kennedy Space Center's Pad 39A on December 29, 1980. Bill Nelson was one of the few notables in attendance. Jimmy Carter had lost to Ronald Reagan the month before in the presidential election, so one administration was on its way out and another on its way in. Other than the space advocate community, most people were focused on their families and loved ones over the holidays. He told a *Today* reporter that he had already spoken with Reagan's incoming budget director, David Stockman, about the importance of protecting the government's science and technology spending. "We have to be careful in our zeal to cut the budget, and not cut off our nose to spite our face," Nelson said.[78]

Sensing an opportunity to promote Brevard County and the space program there, Nelson met with the county's commissioners to encourage them to help him promote the upcoming first space shuttle launch. "I'll try to get as many members of Congress here for the launching as possible. And I want to involve y'all and the mayors to explain the critical importance of the shuttle. I think we can do a good lick for the space program and the country by being good hosts for the congressmen when they come."[79] When launch day came, Nelson joined prominent politicians and celebrities at the VIP viewing site.[80]

Nelson flew out to California to attend *Columbia*'s landing. Upon his return to Orlando, Nelson held a press conference and said he had secured the full support of the House Science and Technology Committee members for an orbiting American space station by 1990. Echoing the raison d'être for Project Apollo in the 1960s, Nelson said, "We simply cannot afford for the Soviets to achieve and maintain space superiority." He also dismissed concerns about the shuttle's cost overruns; by his own calculations, the program went "only" 25 percent over budget.[81]

Labeled a "boll weevil" Democrat for his moderate to conservative positions on many issues, Nelson sometimes found himself voting with President Reagan instead of his own party's position.[82] Nelson supported Reagan's tax cut legislation, but after the nation entered a recession in the second half of 1981, Nelson concluded that the tax cuts had gone too far.[83] He remained politically popular in his district, whose borders had been redrawn after the decennial census. The *Today* editorial staff endorsed him in October 1982 for reelection. "His service in the last four years has been honorable, intelligent and closely representative of majority opinion on the Space Coast."[84] Even with different district borders, Nelson was reelected in November 1982 by the same 70 percent–30 percent margin as two years before.[85]

In the first year of his third term, Nelson mediated two space issues that made local headlines. One had to do with where to process the space shuttle's solid rocket boosters, while the other was the salvage of abandoned Project Apollo infrastructure.

Two minutes after a shuttle launched, its twin white solid rocket boosters separated from the external tank and fell back toward the Atlantic Ocean. Parachutes deployed from each depleted booster to slow its descent. After splashdown, the crews of two contract recovery vessels, *Freedom Star* and *Liberty Star*, retrieved the SRBs and transported them to Hangar AF at Cape Canaveral Air Force Station. Each SRB was broken down into its components, washed, then placed in a shipping container for rail transport to the Utah manufacturer, where they would be refurbished and reloaded with solid fuel.[86]

NASA contracted in 1976 with United Space Boosters, Inc. (USBI), a subsidiary of United Technologies, to assemble and retrieve the SRBs. USBI paid to build *Freedom Star* and *Liberty Star*, then leased them to NASA.[87]

By early 1982, USBI was looking to consolidate its SRB processing operations scattered across the Cape and KSC. The company negotiated the lease of thirteen acres of land, owned by the Canaveral Port Authority, to build a new processing plant. The land was located in north Merritt Island, near the barge canal that paralleled State Route 528. A trench would be dug to connect the canal to the plant.[88]

North Merritt Island residents weren't thrilled with the idea. The homeowners association circulated a newsletter accusing USBI of plotting to store hazardous waste on the property, which USBI denied.[89] The Merritt Island Executive Council, a coalition of local homeowner associations, sent a letter to KSC Director Smith lodging a protest. The letter alleged that an explosion could cause contamination and pollution in nearby residential areas.[90]

In late January 1983, NASA ruled in the residents' favor, informing the coalition by letter that the agency would no longer pursue the barge canal site. The letter said that NASA would consider "other appropriate sites."[91]

The vague implications of "other appropriate sites" became all too clear a few days later. Bill Nelson called a press conference on February 2 in Washington, DC, after he received a phone call from USBI employees who had learned that the company was thinking about moving SRB processing to Marshall Space Flight Center in Huntsville, Alabama, where USBI had its headquarters. "I am one upset congressman," Nelson told reporters. "I'm going to fight this move and I need the support of the Brevard community." Nelson said that the move would be "a human tragedy for 700 workers and their families."[92]

Nelson spoke with NASA Administrator James Beggs and Associate Administrator James Abrahamson, who oversaw space shuttle operations. According to Nelson, Beggs promised to consider the "human cost" of relocating 700 employees to Huntsville. Abrahamson said that existing Marshall facilities might be renovated more cheaply than building USBI a new facility on KSC property. He dismissed existing Cape hangars as "pretty old and would require a lot of work."

Despite lobbying from the Alabama congressional delegation, in early April NASA decided that SRB processing would remain in Brevard County, although where had yet to be determined. A brief NASA statement, released to the press by Nelson before the agency was ready, said there was "no sound basis" for moving to Huntsville. "I am one happy congressman," Nelson told a *Today* reporter.[93]

The facility problem was resolved in 1984 when USBI won a rebid of the SRB processing contract. Under a separate contract, USBI would build a new plant on KSC property. The $25 million forty-two-acre Assembly and Refurbishment Facility was constructed about a mile and a half south of the Vehicle Assembly Building.[94]

Around the time that Nelson intervened in the USBI dispute, another group had approached him with their own personal cause, this one being the salvation of an Apollo-era launch tower.

KSC on north Merritt Island was designed and built specifically to support the Saturn V rocket. Prior to KSC, most rockets, including its ancestral Saturn I and IB, were assembled at the Cape on the launch pad, minimally protected by a fixed service structure called a gantry. Components and payloads were mated and inspected at the pad. This approach would be a problem for the behemoth Saturn V, which, in its crewed lunar mission configuration, reached 363 feet. KSC was therefore designed with a revolutionary "mobile launch concept." The Saturn V stages and the payload would be integrated miles away from the pad in the Vehicle Assembly Building. The stack would be transported to the pad on a mobile launch platform with the service structure, called the Launch Umbilical Tower (LUT), affixed atop it. The platform was mounted to a tracked vehicle called a crawler-transporter, which transported the stack the three-plus miles to Pad 39A or 39B.[95]

Three mobile launchers were built for Project Apollo. The platforms were heavily modified for the Space Shuttle program. One of the most significant changes was to remove the LUT from each platform. Permanent fixed service structure gantries were erected at the two shuttle launch pads, salvaged from the top part of two LUTs.[96] The complete third LUT was to be scrapped, until a coalition of various preservation groups approached Nelson and other members of Congress asking that the remaining tower be preserved for display. Nelson and the others asked NASA to wait sixty days so that alternatives could be explored.[97] NASA agreed to consider a "preservation option" to determine the cost of saving and re-erecting the tower.[98]

In April 1983, just before the end of the sixty-day waiting period, Nelson's House Science and Technology Committee authorized spending $1.8 million to save the tower, as part of the mobile launcher's conversion for the Space Shuttle program. Nelson said he hoped the LUT could be resurrected at KSC's visitor center, Spaceport USA, but another $8 million would have to be found in a future budget to do it.[99] In September 1984, Nelson participated in a press conference where it was announced that the LUT would be reassembled east of Spaceport USA along with an exact-size replica of a Saturn V rocket. The preservation group borrowed $35,000 from the National Trust for Historic Preservation to begin their project, and hoped to solicit donations from the Apollo-era contractors. Nelson predicted the tower would be complete by 1986.[100]

The segments were deposited in a KSC industrial area vacant lot, behind the center's headquarters building. There they slowly corroded over the years from the salted ocean breeze. Congress never authorized the money to reassemble them, and the coalition members failed to raise any money for the project. Six years later, in July 1989, the *Orlando Sentinel*'s editorial page chastised the politicians and conservationists who had lobbied NASA to save the tower, only to fail to raise any money to do something with it. The editorial charged that Nelson had been motivated by "patriotic fervor and an illusion of public support."[101]

The segments remained until 2004, after the Environmental Protection Agency found that the gantry parts and their orange paint were leeching heavy metals and toxic substances into the soil. Despite impromptu efforts by a hastily organized preservation group, this time the segments were finally and irreversibly scrapped. NASA had to spend another $2 million to decontaminate and dispose of the remnants.[102]

* * * * *

In February 1982, then–NASA Administrator James Beggs requested that the NASA Advisory Council evaluate the practicality of flying private citizens on the space shuttle. The council appointed an ad hoc Informal Task Force for the

Study of Issues in Selecting Private Citizens for Space Shuttle Flight. In July 1982, they began their work to "provide an analysis of the factors that must be considered in any decision to fly private citizens as passengers on the Shuttle."[103]

The report was submitted to the council on June 16, 1983. In its cover letter, task force chairman John E. Naugle wrote, "We recommend that a modest program to fly private citizens as observers who would then communicate their experiences to the more general public be initiated as a first step, and we have suggested appropriate criteria to be met and selection procedures to be deployed." The report described flight participants as "observer/communicator" or "educator/communicator." The report warned:[104]

> It would be easy for people to misunderstand such a program as a self-serving public relations gimmick trivializing the space program, despite what is clear to us as NASA's well-meaning intentions.

To avoid that, the report recommended a number of "candidate suitability criteria." A "purpose-oriented approach," using an independent peer group selection process, was recommended. "The review process would have to be so open as to appear to be fair and not an 'old boy network.'"[105]

Out of the task force report came what was eventually known as the Space Flight Participant Program (SFPP). Alan Ladwig, a NASA civil service employee, was assigned in March 1984 to manage the SFPP. Its first program was Teacher in Space, which selected New Hampshire school teacher Christa McAuliffe to fly on *Challenger* with the STS-51L[106] mission in January 1986. The Council of Chief State School Officers was chosen by NASA to help develop the application process and coordinate candidate selection.[107]

In a 2017 NASA Oral History Project interview, Ladwig said that one reason Beggs created the program was that he "got tired of getting calls and letters from self-proclaimed VIPs that thought they should get to fly on the Shuttle."[108] Among those "self-proclaimed VIPs" were politicians, specifically members of Congress. Senator Jake Garn (R-UT), who since 1981 had been lobbying NASA to fly him on the shuttle, chaired the Senate appropriations subcommittee that oversaw NASA's budget. Morton Thiokol, the NASA contractor that manufactured the space shuttle's solid rocket booster motors, produced the SRBs in Garn's state.[109]

Another politician seeking a shuttle flight was Bill Nelson. After the task force report was released, Nelson sent a letter in early July 1983 to Administrator Beggs asking that he be considered for a shuttle flight. Since the report had recommended that priority be given to communicators, Nelson told the *Orlando Sentinel* that "he would communicate his experiences in talks to his colleagues or other audiences."[110]

Other members of Congress angled for a shuttle ride. The Lancaster, Pennsylvania, *Intelligencer Journal* reported in August 1983 that local Rep. Robert Walker, a Republican, had been assured by NASA that he was on an unofficial list of potential congressional passengers. He said he'd been told that Garn and Nelson were also on the list. Walker didn't believe that they'd fly any time soon; all shuttle crews had been assigned through the end of 1986, and that "the first non-working civilians will be journalists and artists."[111]

President Reagan announced the Teacher in Space program on August 27, 1984. Three months later, on November 7, Ladwig was about to appear on the NBC talk show *Late Night with David Letterman* to promote Teacher in Space. While waiting backstage, Ladwig received an urgent phone call from a colleague who alerted him that Administrator Beggs was about to make an announcement.[112]

To Ladwig's surprise, the first "non-working civilian" to go to space would not be a communicator such as an educator, a journalist, or an artist. Senator Garn, a politician, was selected for that honor. Beggs had bypassed the SFPP to independently appoint Garn. The task force warnings about the consequences of an "old boy network" selection had been ignored.

Ladwig recalled that representatives of the Council of Chief State School Officers were incensed. They were scheduled to appear the next day at NASA Headquarters for a Teacher in Space press conference. They felt "double-crossed" by the Garn announcement and almost canceled their participation in the event.[113]

Ladwig and the SFPP staff had never been informed in advance of the political flight plans. He was told by Jesse Moore, NASA's associate administrator for space flight, that the deal had been in the works for some time and was on a need-to-know basis. "I didn't have the need to know." Ladwig was assured that Garn would not fly before the Teacher in Space.[114]

The NASA press kit for Garn's shuttle flight described his presence as a "Congressional observer."[115] Beggs told *The Salt Lake Tribune* that it was "appropriate for those with congressional oversight to have flight opportunities to gain a personal awareness and familiarity" with the Space Shuttle program. Beggs also said that other members who served on NASA's authorization or appropriations committees would be considered if interested.[116]

After Garn's selection was made public, more members of Congress expressed their desire to fly on the space shuttle. Among the interested politicians identified in a December 1984 *Washington Post* article as having "submitted formal requests to NASA for a shuttle seat" were Rep. Edward Boland (D-MA), Eldon Rudd (R-AZ), Beverly Bron (D-MD), Larry Hopkins (R-KY), and Bill Nelson.[117]

Nelson won a fourth term in the November 1984 election, winning by about a 60 percent–40 percent margin over his Republican challenger.[118] When the Ninety-ninth session of Congress began in January 1985, Nelson was elected by fellow members of the House Space Science and Applications Subcommittee to be its chairman. The subcommittee had jurisdiction over NASA operations. Nelson told a *Today* reporter that he believed the selection improved his chances of being the next member of Congress to fly on the shuttle.[119]

To drive home that point, Nelson hosted Garn on his local televised talk show, *Dialogue with Bill Nelson*, which aired February 24, 1985, on Orlando's ABC affiliate WFTV Channel 9. The interview was taped while Nelson toured Johnson Space Center. Garn insisted that the first "private citizen" to fly on the shuttle would indeed be a teacher. "I am not a private citizen. I'm a public official with responsibility for NASA's budget." Garn implicitly passed the baton to Nelson by adding, "That's part of yours and my responsibility."[120]

Despite Jesse Moore's assuring Ladwig the prior November that a teacher would fly first, Garn flew first anyway, on STS-51D in April 1985. To make room for Garn, payload specialist Greg Jarvis, a civilian engineer for Hughes Aircraft, was moved to a later flight along with his microgravity fluid experiments.[121]

By early August 1985, media reports began to surface that NASA was looking to schedule Nelson for a flight. A NASA public affairs official told the *Orlando Sentinel* that the agency was "studying the flight schedule to identify the appropriate time." An unidentified source close to Nelson told the reporter that Nelson wanted the flight to occur before summer 1986 so it wouldn't conflict with his reelection campaign and be viewed as a publicity stunt.[122] *Florida Today* broke the news on August 31, citing "sources close to Nelson," that a formal invitation was imminent.[123]

The formal invitation was announced by Nelson on September 6. He told the assembled press that the experience "will allow me to be a better committee chairman by what I will learn from this opportunity."[124] When asked about criticisms that some might label the trip a junket, Nelson told the press, "Jake Garn has put most of that to rest. I think a lot of that criticism has died down."[125]

An overnight poll by Orlando station WFTV didn't quite see it that way. The unscientific phone-in poll showed that 1,246 viewers (50.3 percent) agreed Nelson should fly on the shuttle, while 1,232 (49.7 percent) said no.[126]

Members of Congress were also of differing opinions. The day after Nelson's formal announcement, US Senator Paula Hawkins (R-FL) told a group of Florida women that she supported shuttle flights by members of Congress and, "As a matter of fact, I've signed up myself." She felt that flying politicians on the shuttle was good for the program.[127] But Rep. Don Fuqua (D-FL), who chaired the House Committee on Science and Technology, believed that politicians had no business flying on the shuttle. "They get in the way," he told *The Miami Herald*. "Right now it should be just those who are trained—trained as astronauts or trained as scientists."[128]

Fulfilling his pledge to "communicate his experiences," Nelson wrote a periodic "space diary" during his training for *Florida Today* and other local newspapers. In his October 2, 1985, column, Nelson wrote that Administrator Beggs called him during a flight suit visor fitting "to discuss possible flight assignments."[129]

That flight was STS-61C. On October 4, a few days after Beggs's phone call, NASA announced that, once again, payload specialist Greg Jarvis would be reassigned in favor of a politician. According to media reports, Jarvis's employer Hughes Aircraft withdrew Jarvis and its Leasat 5 satellite from deployment on STS-61C due to technical problems, which opened up the seat for Nelson.[130] Jarvis and his fluid dynamics experiment were reassigned to STS-51L, the *Challenger* flight with Christa McAuliffe.[131] As did Garn, Nelson would also fly before the teacher.

In future years, some accused Nelson of singling out Jarvis's seat, forcing the Hughes employee onto *Challenger* and his doom. A March 19, 2021, article by David Axe in *The Daily Beast*, an American news website, is one example.[132]

> There's something of a black mark on Nelson in certain space circles. Some longtime veterans of the U.S. space program still loosely associate the former senator with one of NASA's greatest tragedies—the explosion of the space shuttle *Challenger* in 1986.

No public evidence exists to prove Nelson was responsible for moving Jarvis to the *Challenger*. The historical records support the news reports of the time. A 2015 essay by a former Hughes satellite employee recounts that Jarvis's flight was delayed from STS-61C *Columbia* to STS-51L *Challenger* because two Leasat satellites already in orbit were suffering technical problems. The concern was that Leasat 5 might have the same problem. Nelson was originally assigned to STS-61C, then to STS-51L, then back to STS-61C.[133] John M. Logsdon's *Ronald Reagan and the Space Frontier* also states that Jarvis was dropped from the STS-61C crew after Hughes removed its satellite payload.[134] In *The Daily Beast* article, Logsdon said it was unfair to blame Nelson for Jarvis's reassignment, because NASA reshuffled flight assignments all the time. Nelson "was just a passenger going along on a flight."

Nelson's version of the story matches the news reports. He wrote in 1988 that he was initially assigned to STS-51L "with Christa McAuliffe," then reassigned

Rep. Bill Nelson flies microgravity simulation training alongside STS-51L *Challenger* crew member Christa McAuliffe and other astronaut trainees on November 20, 1985. Image source: NASA.

to STS-61C after "Hughes Aircraft Company decided to pull its satellite off the *Columbia* mission" due to the failure of the two satellites. NASA Associate Administrator Jesse Moore called him on October 4 to notify him of the reassignment.[135]

After a record seven delays, STS-61C, with its crew including Bill Nelson, launched from KSC's Pad 39A on January 12, 1986. After three days of delays due to bad weather at KSC, *Columbia* finally landed at Edwards Air Force Base in California on January 18. Back on terra firma, Nelson said that his experience was "a story that I can take back to my colleagues in Congress, and I'm planning to do that, to share that with them at great length."[136]

The STS-61C crew held a post-mission press conference on January 23. Nelson said the flight left him "both fan and critic" of the space agency. He believed that the experience had given him insight to where several improvements were possible. "I am going to make a recommendation in several areas as to increased efficiencies in the operation of the whole system."[137]

Nelson told his story in *Mission: An American Congressman's Voyage to Space,* a book published in June 1988 by Harcourt Brace Jovanovich. He selected as coauthor Jamie Buckingham, a Melbourne pastor and prolific writer. The publisher's $25,000 advance was donated to Buckingham's charity that helped disadvantaged youths; any royalties would be split between the charity and the Bill Nelson Foundation, to fund space and science scholarships. The book was described in its dedication as: "A report to the American People and their representatives in Congress."

* * * * *

Nelson's STS-61C training and mission overlapped at times with the STS-51L mission. His 1988 memoir recalled his encounters with the *Challenger* crew. He flew weightless parabolas with Christa McAuliffe and Greg Jarvis on NASA's KC-135 "vomit comet." He shared a lunch with Judy Resnick. He watched Ron McNair's karate workouts in the KSC crew quarters gymnasium. He chatted with Ellison Onizuka. He noted Mike Smith's friendship with Jake Garn. Dick Scobee's wife June took aside his wife Grace at a party for the crew spouses.[138]

The morning of January 28, 1986—five days after the STS-61C postflight media event—Nelson was in his DC office watching the *Challenger* launch on television with his congressional staffers. His own liftoff replayed in his mind as he watched the shuttle attempt to defy gravity and achieve orbital velocity. Seventy-three seconds after it left the pad, he saw *Challenger* disintegrate just off the Space Coast, blown apart by a failed solid rocket booster rupturing the external tank. He thought of his lost colleagues, their families, and the children across America who had just witnessed a national tragedy on live television. The thought crossed his mind, "There, but for the grace of God, go I."[139]

Speaking on the House floor, Nelson extended his condolences to the surviving family members of "my personal friends."[140] He then joined his Science and Technology Committee, which was holding a press conference. When he arrived, the reporters requested that Nelson take the dais to answer questions. This was a scenario Administrator Beggs never foresaw when he envisioned flying members of Congress, but on that dark day Nelson was in a unique position to speak knowledgeably about the Space Shuttle program. He defended the agency when asked if NASA might have compromised safety in its pursuit of a more frequent launch schedule. "NASA does not launch until they feel that everything is right."[141]

President Reagan's independent commission members completed their investigation and released their report on June 6, 1986. They concluded that a flawed solid rocket booster design led to an O-ring failure in cold weather at liftoff. Contributing to the accident were communication failures within NASA management, and between NASA management and its contractor employees at

Morton Thiokol, the Utah-based company that designed, built, and refurbished the solid rocket motors.[142] The commission found that launch constraints due to O-ring limitations had been waived by the SRB project manager at Marshall Space Flight Center in Huntsville, Alabama. The constraints and waivers were not communicated to higher decision-makers. Thiokol engineers had raised concerns, but those concerns were not elevated to NASA upper management. The night before the launch, Thiokol executives felt pressured by NASA senior management at Marshall and Kennedy to prove it *wasn't* safe to launch, rather than it *was* safe. When the Thiokol executives held a caucus to discuss the weather's effect on the SRBs, a senior vice president told his engineering executive to "take off your engineering hat and put on your management hat."[143]

After the commission presented its findings, the House Committee on Science and Technology held several days of hearings. Nelson used much of his time to grill NASA and Thiokol executives about documents he had uncovered that suggested Thiokol had been required by their NASA contract to conduct an analysis of how the SRB would perform during extreme temperatures. NASA executives finally acknowledged that such an analysis had not been done, but that NASA nonetheless had certified the boosters for operational use. Nelson contended that the loss of the *Challenger* never would have happened if NASA and Thiokol had followed the temperature-testing requirements.[144]

In his 1988 memoir, Nelson saved the final chapter to comment on the *Challenger* tragedy, its causes and its consequences. He wrote, "There is no excuse for bad management and poor communications. It is those areas NASA is now struggling to change and improve."[145]

> A certain superiority and arrogance was to blame. NASA believed in itself. Why shouldn't it? It had had twenty-five years of success. And it expected others to believe in it, too. NASA expected Congress to rubber-stamp what it wanted; if NASA didn't get what it wanted, it would generally try to do what it wanted anyway. If NASA did not like something in a congressional spending bill, the agency had been known to wait until it was too late to change course and then explain to Congress that it was impossible to follow Congress's directive.

He had plenty of blame for his congressional colleagues too. Nelson believed, "A mindset developed within NASA: it was not accountable—to anyone nor to any institution." But he also blamed Congress for being a bit lax in enforcing its oversight responsibilities. His Space and Technology Committee "should have tried to hold NASA more accountable than it did . . . congressional oversight was no more than a matter for NASA to tolerate."[146]

* * * * *

Within less than thirty years after his father's passing, Bill Nelson went from Rock Point Ranch in Malabar to the "Rocket Ranch," as locals sometimes refer to KSC and CCAFS. He was raised in an affluent household as an only child by a doting mother, who was his constant companion as he became a shooting star in local social circles. He failed to fulfill the predictions that he would one day become Florida's governor, although he did try in 1990. Nelson was elected to the US Senate in 2000 and remained there until he was defeated for reelection in 2018. But President Joe Biden would nominate him for NASA administrator in 2021, an irony that his predecessor James Beggs might have appreciated if he were still alive.

When Beggs decided to fly members of Congress on the shuttle, he never could have imagined that he would be grooming his successor thirty-five years later.

Nor could have Beggs foreseen that he was grooming a congressman to ask informed and incisive questions that would expose NASA's failure to properly police its contractor.

In his February 1985 television interview with Bill Nelson, Senator Jake Garn accused the media of a double standard in its criticism of his "joyride."

> They can't have it both ways. They can't on the one hand say to you and I as members of Congress, when something happens in an agency, "Why didn't you know that? Why weren't you exercising your oversight responsibilities?" . . . That's part of yours and my responsibility. Not just to pass new laws, but to make certain the ones we've already passed work, to look at the funding that's already been approved, seeing that it's being spent as efficiently as possible.

Nelson's performance during the *Challenger* hearings validated Garn's perspective. It was no different from visiting a Veterans Administration hospital, or flying in an Air Force bomber, or riding in an Army tank. In Garn's view, it was part of his oversight responsibilities. That view was vindicated by Nelson exposing NASA's failure to properly supervise its solid rocket motor contractor.

As he had hoped, Nelson earned a reputation on Capitol Hill for expertise and insight with NASA human spaceflight operations. After he was elected to the US Senate in November 2000, Nelson rose to chair of the chamber's space subcommittee, where he could steer the panel's direction on policy issues. He would be in a position to determine the fate of President Barack Obama's revolutionary proposal in February 2010 to cancel NASA's Constellation program.

Should NASA continue with OldSpace legacy contractors who always delivered way late and way over budget? Or should NASA inject innovation and competition through a commercial approach that might rely on untested NewSpace companies while limiting the agency's financial exposure?

Should Bill Nelson be loyal to his party's president and fight for this radical change? Or should he protect Constellation, protect its contractors and their workforce at KSC, protect the agency's monopoly on human spaceflight?

Or could he craft a compromise?

5

Out of the Nest

The June 1989 issue of *Florida Trend* magazine ranked Florida counties for best business climate. Orange County—the home of Orlando, Walt Disney World, and other tourist destinations—ranked number one.[1]

Brevard County was relegated to a "second tier" of smaller markets. On that list, Brevard ranked number two. The author wrote that "Brevard is succeeding in spite of its bickering economic development agencies . . . In Titusville, the county seat, the permitting process for business can be disorganized and time-consuming, say those who have gone through it. And virtually no class A office space exists in the city."[2] The author detected what had been an obvious trend as Spaceport Florida took its first baby steps in the summer of 1989—pettiness and parochialism interfering with progress.

Despite the infighting, optimism was in the air. The George H.W. Bush administration continued Ronald Reagan's expansive policy of growing a vibrant commercial space sector. On July 20, 1989, the twentieth anniversary of the Apollo 11 landing, President Bush proposed what came to be known as the Space Exploration Initiative. NASA submitted to the National Space Council in November 1989 a "90-Day Study" that envisioned completing Space Station Freedom, followed by "a return to the Moon to stay, and a subsequent journey to Mars." The study said little about commercial space, preferring a NASA-centric approach.[3] Despite its lack of widespread political support—and therefore funding—Florida's state and local leaders nonetheless were go for launch. State and county government, the private sector, and educational institutions all saw potential in the proposal.

Events, national and local, outran the formation of Spaceport Florida. While the agency sought to hire its first executive director, various state, county, and university entities were pursuing their own agendas. No one entity coordinated them all. Seventeen different entities in Florida promoted the state's space interests.[4]

An October 1989 *Florida Today* article cited a lack of regional cooperation as one reason why the Space Coast had failed to establish local research and development clusters. Tom Keating, the vice president of economic affairs at the

Florida Chamber of Commerce, told the paper he was surprised to find four different economic chambers active in the county—the Space Coast Development Commission (SCDC), the Brevard Economic Development Council (BEDC), the Melbourne–Palm Bay Development Council, and the Cocoa Beach Area Chamber of Commerce. The BEDC had "created dissension among economic chambers," according to the article, by unilaterally declaring itself the leader in recruiting industry to the county.[5]

Two of these groups fought for space dominance in Tallahassee—the SCDC funded by the City of Titusville, and the BEDC funded by Brevard County. In January 1989, Brevard County commissioners selected the BEDC to coordinate the county's commercial space interests, including the future location of Spaceport Florida.[6] As noted earlier, the BEDC had told state executives to deal with them and not with the SCDC in coordinating local commercial space matters.

The BEDC went private in April 1989 and expanded its board to include two members from each of the other three boards.[7] Bob Allen, the SCDC executive director, told *Florida Today*, "It's a relief that we don't have to fight among ourselves anymore."[8] The organization changed its name to the Brevard Economic Development Corporation and hired a new chief executive. Weekly meetings began with other county economic development entities to encourage a more collegial atmosphere.[9]

As teased in Chapter 3, Florida politics and legislative malpractice led to the departure of Chris Shove from Florida space enterprise activism, and the downfall of the Space Research Foundation (SRF). Shove was arguably the most knowledgeable person in the state about Florida's commercial space legislation and the Martinez administration's plans for the Spaceport Florida Authority.

The tale begins in the days following the loss of *Challenger* and its crew on January 28, 1986. A Maitland, Florida, architect named Alan Helman led the organization of a nonprofit called the Astronauts Memorial Foundation (AMF) to raise money for a memorial to America's fallen astronauts. US Rep. Bill Nelson (D-FL), whose district included KSC, was recruited to head the campaign.[10] In June 1986, Governor Bob Graham signed a bill that created an alternate Florida license plate design honoring the seven *Challenger* crew members. (In Florida, license plates are also called "tags.") The tags would cost an extra $20, with $15 of that going to the AMF.[11] The memorial eventually was built at Spaceport USA.

Two years later, by mid-1988, the *Challenger* plates had generated far more revenue than was necessary to build and maintain the memorial. State legislators began to ask the question, What to do with all that money?

During this time, various commercial and university space initiatives were in their early stages—and looking for revenue sources. The solution, state legislators concluded, was to redirect some of the *Challenger* plate money from its intended purpose to other competing interests.

Brevard's state legislative delegation in February 1987 presented a bill that would create a commission promoting science research and development in Brevard County, called the Technological Research and Development Authority (TRDA). Rep. Bud Gardner (D-Titusville) said the authority would enable Florida Tech (FIT) to funnel state and federal funds, as well as private gifts, to the private university. No funding was attached to the bill, but the clear understanding was that TRDA was to benefit local businesses and industry, as well as FIT faculty and students.[12] TRDA passed both state houses and was signed by Governor Bob Martinez in June 1987.[13] In November, Martinez appointed five Melbourne-area businesspeople to TRDA's board of directors.[14]

In parallel to TRDA's creation, the Governor's Commission on Space was deliberating how Florida could compete with other states' commercial space initiatives. Stephen Morgan, who had pitched Martinez the commission idea, wanted to establish a space institute at FIT. In an October 2022 email, Morgan wrote that during the summer of 1987 he was focused on establishing the Space Research Institute (SRI) at FIT, which was unrelated to the parallel effort to establish TRDA. SRI and TRDA had no tacit agreement that TRDA would fund SRI, although both were aware of each other.[15]

FIT sought, and failed to win, a NASA designation that would have guaranteed SRI $1 million a year in space research. The institute's first budget proposal sought $8 million over the next three years, with $6 million coming from state funds. The rest was hoped to come from the aerospace industry or other grants.[16] But where to find the funding?

State Senator Tim Deratany (R-Indialantic), representing Florida Tech's interests, passed legislation in the Senate that would have reallocated half of the next fiscal year's *Challenger* plate revenue from the AMF directly to TRDA for dispersal to SRI. Rep. Gardner (D-Titusville), over in the House, represented AMF's interests. AMF objected to losing control over half their revenue. The AMF wanted the surplus to be used for a national educational program as a "living memorial" to the deceased astronauts.[17]

The compromise, reached on June 7, 1988, was to create a second research institute, called the Space Research Foundation. The SRF would be a subsidiary of the AMF. The AMF would retain half of the annual *Challenger* revenue for the memorial's construction and upkeep. The SRF would have to be funded out of whatever the AMF chose not to spend on the memorial. Another quarter would go to the state's Department of Education for a Challenger Astronauts Memorial Scholarship Trust. The remaining quarter went to TRDA for distribution to Florida colleges and universities, although Gardner told the *Orlando Sentinel*, "I suspect a lot of that will go to FIT."[18]

Why was a second space research institute needed? According to Chris Shove, the general assumption was that TRDA and SRI would direct most of their funding to FIT, as Rep. Gardner surmised. The SRF, however, could disperse research and development grants to any Florida university or college. Those decisions would not be made by a private university with its own self-interests, such as Florida Tech.

In a May 2024 phone interview with this author, former AMF president Ben Everidge said that the SRF was set up to be separate from TRDA "and for us to work on broad-based more commercial space policy, research policy." While the foundation was not comfortable with using *Challenger* tag money for projects other than its original purpose, "it was the state legislature's decision, it was their plate, it's their money." The SRF leadership assumed TRDA would focus on technological research.[19]

Edward Moran, the AMF's first president and CEO, resigned in August 1988 to run the Institute for Space Science and Technology.[20] The AMF's board of directors promoted Everidge, who was their government affairs director, to succeed Moran. A one-time legislative staffer for Bill Nelson in Washington, DC, Everidge had been recruited in August 1988 by Alan Helman to handle the foundation's government and public relations. To run the SRF, Everidge first hired David Webb, a space policy expert who had formed the Space Studies program at the University of North Dakota and served on President Reagan's US National Commission on Space. Webb's job was to organize the SRF and create a framework for operations.[21] Webb and Shove, at the time still the director of the state's Office of Space Programs, were among the 220 participants at a February 8 space research workshop in Orlando. *Florida Today* reporter Irene Klotz wrote, "Nearly 4,000 Florida researchers are involved in space-related programs or projects, but few realize how their work relates to that of their colleagues, conference participants said."[22]

Days after the Florida legislature passed the bill to create Spaceport Florida, Shove resigned from the Office of Space Programs, effective June 30.[23] Everidge approached Shove about taking over the SRF. Shove agreed, and by the end of June had joined the foundation as SRF's vice president. Everidge described him to *Florida Today* as "a godsend for us. Chris has credibility, experience and is already knowledgeable with many of the contacts we already have." The paper reported, "The foundation hopes to develop projects such as . . . [a] model moon base that can be used to support lunar research and attract tourists."[24]

It was around this time, on July 20, 1989, that President George H.W. Bush directed the National Space Council to develop a report for a permanent human presence on the moon in the twenty-first century, as a stepping stone to Mars.[25] Although the rhetoric was vague, the public seemed to take this as a proposal for a lunar base; *The New York Times*'s front-page headline the next day was "President Calls for Mars Mission and a Moon Base."[26]

Recalling the "analog moon base" idea from the UE&C report, Shove in August 1989 announced that the SRF would initiate a study to establish a lunar research facility on the Space Coast. "Moonbase 1" would be used to develop and demonstrate lunar settlement technologies.[27] Shove told the *Orlando Sentinel*, "Our top priority is the establishment of Moonbase 1. Moonbase 1 will attract both university and industry participants and will have national and international ramifications for Florida."[28] Vice President Dan Quayle, who chaired the National Space Council, wrote a letter supporting the initiative.[29]

Shove found a corporate backer, a Japanese construction company called Obayashi Corp. based in Tokyo. Sun Banks pitched in as the foundation's financial adviser. The SRF asked the Spaceport Florida Authority to help with a $100 million tax-exempt bond issue to finance the project. "It's going to be totally a commercially sponsored project, and that's the beauty of it," Shove told the *Orlando Sentinel*.[30]

Placing Moonbase 1 near Spaceport USA or elsewhere on KSC property would have meant a commercial operation trying to operate on federal land, a bureaucratic entanglement they wished to avoid. Obayashi and other potential investors were more interested in having the attraction closer to Orlando, the home of Walt Disney World and other amusement parks. According to Shove, Moonbase 1 was inspired in part by the hydroponics operation at EPCOT's The Land Pavilion. Guests ride a boat through greenhouses that display how horticulturists use innovative growing techniques and breed high-yielding crops, some of which are used to feed Disney guests.[31]

By October, the SRF had contracted to buy 300 acres for Moonbase 1 in a business park off State Road 528 east of Orlando International Airport. Ben Everidge told the *Orlando Sentinel* that an integral part of the project would be a visitor center that would help pay for lunar research.[32]

The project's demise began when the *Fort Lauderdale Sun-Sentinel*'s Tallahassee reporter John Kennedy became curious about how much money had been raised by *Challenger* license plate sales, and how the money was being spent.[33] On November 19, 1989, the *Sun-Sentinel* published a front-page investigative article concluding that the AMF was "a costly bureaucracy, paid for by taxpayers but not subject to any scrutiny."[34]

The article stated, "The foundation has extended its reach, with the state-financed organization now helping woo a Japanese corporation interested in developing a private, $100 million research park in Orlando." In particular, the paper called out Shove as "the foundation's top wage-earner." The report described Shove as "an intermediary to Japan's Obayashi Corp. and other firms looking to develop Moonbase No. 1 in Orlando."

At the AMF's annual meeting in Orlando on December 19, 1989, Shove submitted his resignation. According to the *Orlando Sentinel*, Shove told the board of directors that his work on Moonbase 1 had been undermined by foundation officials, and he could not effectively continue in the organization. Everidge told the paper that Shove and the foundation differed on the SRF's direction; the AMF board felt this would jeopardize the organization's nonprofit status.[35] Shove told *Florida Today*, "They don't want the moonbase. They felt it would be inappropriate for them to be involved."[36]

In a series of 2022 emails, Shove wrote that Everidge and the AMF board members were well aware of his activities; he brought it up with Everidge when recruited in June 1989. Everidge accompanied him to Tokyo to meet with Obayashi officials. But after the *Sun-Sentinel* article was published, an AMF board member and a staffer warned Shove that the board was having second thoughts about Moonbase 1. He had kept in contact with Orlando's mayor about the project, despite having been told not to do so, and this was seen as a justification for terminating his employment. There was also some speculation that partisan politics might be involved, as Everidge had once been a legislative aide to Bill Nelson. Shove also believed that the revenue for the SRF from the *Challenger* tags was "unsustainable," as the Florida legislature redirected funds to TRDA and the Department of Education, and began to offer more and more alternate license plates. He saw Moonbase 1 tourism dollars as a revenue stream for SRF to pay for facilities and research.[37]

In his May 2024 interview, Everidge said that *Sun-Sentinel* reporter John Kennedy's "work was top drawer," but at the same time the newspapers "were being fed an awful lot of false stories by politicians that were trying to do everything they could to undermine the money, take it back from everybody. It was a lot of money. They were trying hard to get that away" from the AMF.

His memory was that other papers in addition to the *Sun-Sentinel* were questioning the propriety of the AMF being involved with commercial enterprises. "Is AMF really the right partner to do this?" The board came to the decision that other parties were capable of doing the project, such as the new Spaceport Florida Authority. It was the right time for the AMF to bow out of the Moonbase project.

Everidge said he has "tremendous respect" for Shove. Noting today's commercial space enterprises at the Cape, he said that "none of this would have happened" without Bob Martinez, Jeb Bush, David Webb, and Chris Shove. From his experience, partisan politics had nothing to do with the events that led to the board dropping Moonbase 1 and Shove's departure. "It would have been crazy" for someone to have told Shove not to communicate with Orlando's mayor about the project. In his opinion, Shove would have been an excellent choice to be the first executive director of the Spaceport Florida Authority.

Shove recalled in one email the wisdom he was given by a senior federal space official. "The definition of a pioneer," Shove wrote, "is the skeleton on the side of the road with arrows in its back."

* * * * *

Even before the *Sun-Sentinel* article, questions were being raised about *Challenger* tag largesse, the lack of public scrutiny, and its use for projects other than the memorial.

The November 5, 1989, *Orlando Sentinel* editorial page opined that "the Legislature forced the foundation to veer from its goals of promoting space education." The implementing legislation lacked proper oversight for spending of the tag money. Florida Tech was the primary beneficiary, to the detriment of other state universities.[38]

Those under media scrutiny began pointing fingers. Everidge said he planned to ask the Internal Revenue Service to conduct an external audit of the foundation's finances. He also sent a letter to TRDA Executive Director Frank Kinney asking the authority to explain how it had awarded its grants. According to a January 4, 1990, *Florida Today* article, TRDA had "awarded $1.4 million to F.I.T.'s Space Research Institute while giving only $525,000 to space researchers at five other Florida colleges and universities." The article quoted Kinney as saying, "The grants were doled out without first taking proposals from competing schools." Everidge said, "We think they're discriminating against 11 of the universities in the state university system," and that the foundation supported new legislation redirecting the TRDA grants to the state's Department of Education. Kinney replied that TRDA had followed its directive to promote Brevard County as a regional center for high technology.[39]

A week later, TRDA sent a forty-five-page response that included a request to appoint an independent auditor to scrutinize the AMF. Kinney wrote, "While the TRDA understands that the AMF is requesting an audit by the Internal Revenue Service, we are concerned that this audit, if conducted, would not adequately answer the questions being raised concerning the AMF's use of Challenger license plates." Bud Gardner, who had been elected to the Florida Senate, told *Florida Today,* "We seem to be having what amounts to nothing more than an internal squabble. It's like a bunch of kids on a playground."[40] Gardner then introduced legislation that would redirect TRDA money to the Board of Regents of the state university system. Everidge told the *Orlando Sentinel* that the attacks stemmed from TRDA and other groups trying to get their hands on more tag money.[41] A February 4 *Orlando Sentinel* editorial called TRDA "a two-person outfit" that was "Sen. Tim Deratany's pet turkey."[42] *Florida Today* reported on February 14 that FIT had hired Deratany as a $35,000-a-year director of university relations while he still served in the Florida Senate.[43] After the revelation, Deratany resigned from FIT.[44] He told the *Sun-Sentinel,* "I think it's the Astronauts Memorial Foundation that has grossly abused and mismanaged the funds the public has placed with it . . . perhaps the whole thing should be done away with—the whole *Challenger* tag program."[45]

Looking back nearly thirty-five years later, Everidge commented in his May 2024 interview, "TRDA was a lot more political than any of us were." He left AMF soon after the events of early 1990 to start an academic consulting business.

Some members of the state legislature introduced bills to reform AMF, TRDA, or even disband them. In the end, legislators couldn't agree on a solution, so AMF and TRDA survived with the existing funding structure intact.[46] The AMF in April unveiled its plans for a Center for Space Education, which would rely on continued funding from the *Challenger* plates.[47] The center opened in 1994, funded by *Challenger* plate revenue, and remains active today.[48]

Bud Gardner introduced a bill that would have officially transferred the Space Research Foundation from the AMF to the Spaceport Florida Authority, but the bill failed. Brevard County land designated for the SRF was instead transferred to the SFA. The new authority also picked up some programs started by SRF.[49] The Moonbase 1 initiative vanished into history.

* * * * *

"This will be the year when the newly created Spaceport Florida Authority gets rolling," declared the *Florida Today* editorial page on January 1, 1990.[50]

Spaceport Florida stayed out of the *Challenger* tag controversy. The authority instead focused on standing up its own operations.

The SFA directors in November 1989 selected Edward O'Connor as the agency's first executive director. O'Connor had been a manager for Martin Marietta's Titan commercial launch program. The first commercial Titan was about to launch from Cape Canaveral. Martin Marietta made an attractive counteroffer, but O'Connor chose to join the authority.[51]

In his first interview with *Florida Today,* O'Connor predicted that the commercial space marketplace "will be in a deep economic struggle."[52] The uncertainty of the industry was the topic of a three-page special report in the October 23, 1989, *Florida Today.* Reporter Irene Klotz wrote that "the fledgling U.S. launch industry is finding its future hinged on politics." She found that satellite owners and rocket manufacturers were skeptical about the space authority. Klotz quoted Floyd Stuart, a vice president with satellite manufacturer Hughes Communications. "It's hard to say a commercial spaceport will succeed. I think today our needs are met. It doesn't seem to be a strong requirement."[53]

The article reported that much of the current demand for launch services was due to a backlog caused by the space shuttle's delays and the *Challenger* accident. Some companies in the global commercial satellite industry had already signed launch contracts with the European launch company Arianespace. Other nations, such as the Soviet Union and China, were also entering the market. Satellite owners worried that the major American commercial launchers—Martin Marietta, General Dynamics, and McDonnell Douglas—already had guaranteed government customers with NASA and the US Air Force, so commercial customers wouldn't be a priority. And once the shuttle backlog was gone, the future of an American commercial launch industry was uncertain.

Despite the uncertainties, in 1990 the SFA began developing a detailed business plan for submission to the Florida legislature. Five Florida universities offered their assistance, including FIT.[54] The report, presented in April, predicted six suborbital launches by Florida universities from Cape San Blas without charge later in the year. The authority's first commercial launch from the Cape was projected for April 1991.[55]

Titusville state legislators Senator Bud Gardner and Representative Charlie Roberts (who had succeeded Gardner in the House) introduced in their chambers legislation to increase the SFA's development bond cap from $210 million to $500 million through 1995. The bills also lifted restrictions on the types of launches the authority could support in Brevard County; its 1989 enabling legislation had forbidden the SFA from supporting launches at the Cape of payloads more than 150 pounds, and from making structural changes to any launch pads.[56] The authority began planning bonds for improvements to launch pads and processing areas at the Cape, and a program to lure high-technology firms to Florida.[57] The legislation passed both houses and was signed by Governor Martinez in June.

One of the projects the SFA picked up from the late Space Research Foundation was assisting universities in flying small research experiments aboard the space shuttle. Melbourne's FIT and Orlando's University of Central Florida announced in May they had joined a consortium of universities hoping to fly a space object tracking experiment, funded by the Pentagon. Just as FIT had the Space Research Institute, UCF formed the Space Education and Research Center.[58] SFA also brokered a $7 million agreement for UCF to use SPACEHAB[59] module lockers for laser research in the shuttle's payload bay. SFA's O'Connor estimated that the authority had saved the university $1.2 million.[60]

The first refurbished launch pad at the Cape was not by Spaceport Florida, but by the DOD. Launch Complex 20, last used in the mid-1960s for the Titan II ICBM, began renovation in 1988 under a contract issued by the Pentagon's Strategic Defense Initiative Organization to test a suborbital rocket called Starbird. A university consortium, and the authority, also expressed interest in the pad.[61]

As the Cold War came to an end, the Soviets sought to enter the global commercial space business, in partnership with American companies. Just as the United States began converting military missiles into civilian payload launchers, so did the Soviets begin to market their launch vehicles. United Space Boosters, Inc., the contractor that assembled and retrieved the space shuttle's solid rocket boosters, sought US government approval to operate a commercial spaceport at Cape York, Australia, on behalf of a private consortium. The Cape York Space Agency planned on buying launch vehicles from the Soviets. USBI would operate the complex and launch the vehicles. If approved, USBI would learn directly from the Soviets how to operate Russian Zenit rockets. American launch companies raised concerns about USBI helping a foreign competitor that could squelch the fledgling US commercial space industry before it was ready to compete.[62] Foreign launch companies enjoyed government subsidies that American companies did not, offering lower prices US companies couldn't match.[63] The Bush administration allowed USBI to proceed with restrictions, but in the end the project failed due to objections by Australia's First Nations people and the lack of capital investment.[64]

Unforeseen was a request by Soviet space officials to launch their Proton rockets from Cape Canaveral. The SFA was approached by the Space Commerce Corp. of Houston, a marketing company representing Soviet interests, about using the Cape for Proton commercial launches. The Soviet Union of course had its own launch sites, but, from an orbital mechanics perspective, it made sense because Cape Canaveral was much closer to the equator. The idea might also circumvent American laws prohibiting the overseas transfer of advanced technology, which had kept the Soviets from winning contracts to launch American commercial satellites. The SFA's O'Connor, questioning if it were even legal for the Soviets to launch from American military launch pads, forwarded the request to the National Space Council.[65] After that, the idea apparently went nowhere.

The National Space Policy Directive (NSPD-2) issued by the White House on September 5, 1990, directed that "U.S. government satellites will be launched on U.S.-manufactured launch vehicles unless specifically exempted by the President." It also required government agencies to "actively consider commercial space launch needs and factor them into their decisions on improvements in launch infrastructure and launch vehicles aimed at reducing cost, and increasing responsiveness and reliability of space launch vehicles." The press secretary's release declared, "The United States seeks a free and fair international commercial space launch market to further the use of outer space for the betterment of mankind."[66] O'Connor told the *Orlando Sentinel*, "I think the new space policy deals with the commercial market for the next decade, showing that it's truly going to be a multi-national marketplace and the recognition of the United States that support to the development of new systems will be necessary."[67]

Spaceport Florida's first launch, ironically, was "international" but not in the way one might expect. SFA launched a 12-foot (3.5-meter) tall sounding rocket called Viper on behalf of an FIT research program that wanted to study an upcoming eclipse. Because the eclipse would reach totality in Mexico, not Florida, the agency decided to launch Viper from a Mexico village on the Pacific coast. SFA and TRDA provided grants to fund the mission. Viper successfully launched on July 11, 1991, but the payload transmitted no data.[68]

TRDA and Spaceport Florida got along much better than the TRDA-AMF rivalry. The two agencies began 1991 planning several cooperative ventures.[69] Bob Allen, the executive director of Titusville's SCDC and an early promoter of creating the authority, was appointed by Governor Chiles to TRDA's governing board.[70] TRDA awarded a $200,000 one-year grant in June to several bidders, including one for a space law and commerce research institute. TRDA required that the building be located in Titusville, shared by an authority program to develop an advanced launch processing system for commercial rockets.[71]

The SFA staff were well aware that they faced several obstacles, one of which was the obsolescence of Cape launch facilities. A USAF study obtained in January 1991 by *Florida Today* concluded that much of the base should be rebuilt. At the time, many facilities and infrastructure were thirty to forty years old. Investment in Cape facilities had declined starting in the mid-1970s, when federal policy dictated that the nation's future primary launch vehicle would be the space shuttle, at KSC and Vandenberg AFB.[72]

Launch operations needed to change too. In the old days, a launch vehicle's components were assembled and inspected on the pad, a process that could take weeks or months. The modern approach was to assemble and integrate the payload in a hangar before transport to the pad. Eastern Range officials feared that, without a significant investment in new infrastructure, the commercial launch industry would launch with overseas competitors. The study estimated that a $2.4 billion investment would be required over seven years to upgrade launch vehicle assembly and integration facilities.[73]

Recognizing the problem, the SFA planned to issue bonds to construct new launch operations buildings at the Cape that could be leased to commercial launch companies. The authority sought a $25 million grant from the US Department of Transportation (USDOT) to build a joint-use launch control center complex, but federal funding was unavailable. SFA could fund a study, though, and use its bonding power to pay for construction. The problem, though, was that any SFA bond proposal had to go through public hearings in the state legislature, be approved by the governor's cabinet, and also be authorized by the state treasury department.[74] Senator Gardner and Representative Roberts sponsored a bill, passed in April 1991, that gave the SFA the power to sell bonds without approval by the legislature, as well as the power to condemn property already under SFA control.[75]

Another complication was a recessionary economy. The American economy entered a recession in July 1990, ending in March 1991. The recession triggered a period of decreased employment that lasted into the early 1990s.[76] The commercial space industry was still heavily dependent on federal government spending, and future congressional commitment to space spending was uncertain.[77]

Spaceport Florida sought another $25 million grant from USDOT to fund launch complex improvements, but USDOT lacked the funds for such programs. Rep. Jim Bacchus (D-Orlando), who succeeded Bill Nelson in the House, tried to insert $25 million for the project into the proposed FY 1993 budget for USDOT's AST, because the Air Force didn't have the money either.[78] When the FY 1993 budget passed, Congress authorized $10 million out of the DOD budget, because these were military pads, but the grant program would be disbursed by AST, which had the legal authority to oversee the nation's commercial space enterprises.[79]

California's Western Commercial Space Center also sought to apply for a grant, $4 million to build a small commercial launch pad at Vandenberg Air Force Base. WCSC administrators concluded that DOD bureaucrats weren't interested in the program. Executives at both the Spaceport Florida Authority and WCSC wanted the program transferred to USDOT because they felt the Pentagon was hostile to their plans and imposed too many rules. Congress, however, didn't seem inclined to amend the legislation to move the funds to AST's budget.[80] The grants were never appropriated.

Concluding that strength lie in numbers, O'Connor and SFA officials began to organize spacefaring states into one coalition, called the Aerospace States Association. Florida was joined by Alabama, California, Colorado, Hawaii, and Virginia. As one, they would lobby Congress to support commercial space projects, upgrade launch facilities, and compete with other spacefaring nations. The group first met in Miami, with Florida the host, in late July 1990. O'Connor told the *Orlando Sentinel*, "There's a vested interest that cuts across traditional, competitive state lines."[81] By January 1992, twenty-four states had joined the ASA, including those without existing infrastructure such as Montana, Wisconsin, and Arkansas, hoping to attract investment from private industry or NASA.[82]

"Spaceport Florida May Fold," read a front-page headline published in the January 5, 1992, *Orlando Sentinel*. "The 3-year old agency that was supposed to bring Spaceport Florida to life may run out of money and be forced out of business by mid-summer," the article predicted. Florida House Speaker T. K. Wetherell told the paper, "They're going to have to make it" without any new state money. "That was part of their commitment." O'Connor, according to the reporter, had become little more than "a state-paid lobbyist for the commercial space industry." The predicted commercial market had not yet emerged. The authority, and commercial launch companies, had encountered environmental and bureaucratic obstacles.[83]

The budget shortfall was resolved with a $600,000 interest-free loan from the Florida Department of Transportation (FDOT). The loan was approved by the state legislature in late June 1992 and issued in October 1992. A second $600,000 loan from FDOT was issued in October 1993. Both loans were payable once the SFA received a grant from USDOT, in four semiannual installments of $150,000 each. The loans required the SFA to apply each year for the grant. Authority executives assumed they'd be able to repay the loans from the transportation grant programs inserted into the federal budget by Rep. Bacchus, but those programs never transpired. With no funding source, the loans payable were carried for years on the books as "non-current liabilities."[84]

The loans apparently were never repaid. This will be discussed in Chapter 6.

On August 22, 1992, Spaceport Florida launched its first rocket from Cape San Blas. A ten-foot-tall, ninety-pound rocket named Microstar lifted off at 10:00 a.m. Eastern Time. Its payload, built by Florida State University, was designed to take ozone measurements about twenty miles above the launch site. The authority, finally, had a tangible result.[85]

* * * * *

With a successful launch on the board, the SFA began its Rockets For Schools program. In partnership with the federal AST, twenty high school students were selected from across the nation to spend four days in Brevard County learning all phases of payload preparation and rocket launch, including the launch of a small sounding rocket from the Cape. The authority also hosted at its Titusville laboratories middle and high school teachers from across Florida who would be trained to operate weather satellite receiving stations in their classrooms, in a program funded by TRDA and the National Oceanic and Atmospheric Administration.[86] Titusville High School science students flew brine shrimp on the space shuttle to determine how microgravity affects them, an experiment sponsored by Spaceport Florida. Florida Education Commissioner Betty Castor said, "Space research is no longer limited to colleges and universities. Florida's public schools are taking advantage of unique opportunities through Spaceport Florida."[87]

By the end of 1993, Spaceport Florida's future was much brighter. The authority had sponsored six suborbital launches, including that first launch from Mexico in July 1991. The fifth and sixth, from Cape San Blas on December 11, were in partnership with Florida State University.[88] The foreseen commercial launch boom had yet to emerge, but at least the authority demonstrated that its educational activities were helping students across the state, from middle and high schools to colleges and universities.

Although the DOD-USDOT grant program sponsored by Rep. Bacchus fell victim to bureaucratic intransigence, another Pentagon program emerged that finally provided a revenue source to upgrade Cape launch facilities.

As the Cold War came to an end, elected officials began to discuss how to reap a "peace dividend" by redistributing part of the federal military budget. Florida Governor Lawton Chiles on May 22, 1992, announced the formation of a state Defense Reinvestment Task Force to obtain federal peace dividend dollars, which would be used for retraining defense workers and helping companies apply their defense technologies to commercial uses. Chiles appointed Bacchus to chair the state task force. Bacchus said, "I see my role primarily as getting all the federal dollars I can find."[89]

Along with US Senator Bob Graham (D-FL), Bacchus introduced legislation on June 3 that would create a direct grant program for states to respond to their unique defense investment needs.[90] The FY 1993 appropriations bill passed in October 1992 by Congress provided $1.7 billion for transition programs. The state task force prepared a plan to pursue their fair share.[91]

Bill Clinton was elected president on November 3, 1992. An idea came out of his campaign to use the Pentagon for making venture capital–like investments in the defense industry's commercial sector. This "dual use" approach suggested that it was in the Pentagon's interest for the industry to also find commercial markets. Defense contractors could remain in business servicing the American military, while also finding civilian customers to keep their companies afloat during a time of government budget cutbacks.[92]

Standing at a Westinghouse plant in Linthicum, Maryland, on March 11, 1993, Clinton proposed $20 billion be allocated for his Defense Reinvestment and Conversion Initiative. The proposal would be amended to the federal FY 1994 budget draft already working its way through the Senate. The proposal included $4.7 billion in additional funding for "dual use technology and commercial military integration programs established under the FY 1993 defense authorization of appropriations bills," according to the *Congressional Record.*[93] On April 14, Bacchus and Graham reintroduced their defense reinvestment bill from the last session, hoping that the Clinton initiative would include their provision for sending some of the money directly to the states for use as they saw fit.[94]

Within days, Spaceport Florida was advertising for technical writers and "industrial partners" to help prepare and submit applications "for federal grants for dual use space launch facilities to support Department of Defense (DOD) and commercial space launch requirements. These grants have been formally approved and will be administered by the United States Air Force."[95]

Florida and its defense contractors rushed to submit their proposals in competition with those from other states. Separate from the SFA, in 1992 the state legislature passed a Governor Chiles proposal creating a public-private nonprofit called Enterprise Florida, an economic development partnership of the state Department of Commerce and the Florida Chamber of Commerce.[96] Defense conversion proposals were submitted both to the state's Department of Commerce and Enterprise Florida. Brevard-based space industry companies such as Martin Marietta were among those submitting proposals.[97]

The Chiles administration in March 1993 proposed Florida's own version of the defense reinvestment program. The proposal would have allocated $10 million a year in tax credits over the next ten years to any retooling Florida defense contractor. The bill died in the House in April, lacking statewide support as potential job losses were viewed to impact only a few places, such as Brevard County.[98]

Another bill that would have helped Spaceport Florida also died in the legislature. The bill would have directed to the authority half of the sales tax revenue generated by the business it generated. The bill was introduced in February 1993 in the Florida House by Rep. Charlie Roberts (D-Titusville), and in the Florida Senate by Senator Patsy Kurth (D-Palm Bay). The legislation died in the Senate in April.[99]

One legislation that did pass was an amendment to the SFA statutes. Section 331.360 allowed FDOT to "enter into a joint project agreement with, or otherwise assist, the Spaceport Florida Authority . . . and may allocate funds for such purposes in its five-year work program." The amendment made it clear that the funds could not be used for "administrative or operational costs of the authority."[100] The space launch industry was now viewed by FDOT as just another transportation mode to be included in its long-range transportation plans. FDOT's airport matching grant funds could be used by SFA to attract possible federal matching grant dollars for launch infrastructure development.[101]

During 1993, the SFA began to convert a Cape launch site for commercial use. Spaceport Florida submitted a proposal to the USAF to modify Launch Complex 46, a former Trident missile test site, for small commercial rockets.[102] The USAF awarded the SFA $2.15 million to convert LC-46 to dual use. The authority would have to provide $1.3 million in matching funds.[103] The SFA had plans for an eighty-foot launch tower and upgraded flame trench at the pad; industry partners reportedly invested another $1 million in the project in addition to the funds awarded by the USAF.[104]

Despite the mix of success and failure in 1993, SFA's future prospects certainly had improved entering 1994. The Air Force had decided to invest $1.1 billion through 2002 to upgrade the Cape's range safety and tracking equipment, as well as the base infrastructure. The USAF intended to reduce the time between major launches from forty-eight hours to about one day, which would make the Cape more competitive with foreign launch services.[105] That could only benefit Spaceport Florida, which sought less money ($480,000) from the state for FY 1995 than the year before ($600,000), even though their staff had grown from seven to ten.[106]

The authority won a new ally with the Brevard Economic Development Corporation. The BEDC's chairman, Dan Johnson, pledged to help market Cape launch facilities to potential commercial customers. "We haven't done a good job of nurturing commercial space," Johnson told *Florida Today*. "We haven't been partners before." The March 24, 1994, article reported that USAF cooperation with customers had improved. Security measures had been streamlined, and base officers had opened a "one-stop" office for customers.[107] A workshop convened by BEDC heard from speakers who warned that the US share of the global commercial launch market could shrink to zero if government and industry didn't join together to compete with other nations.[108]

That message didn't resonate in Tallahassee. Both Representative Roberts and Senator Kurth tried once more to convince the legislature to guarantee the SFA a funding source by redirecting to the authority a portion of the sales tax it generated, but again the Senate let it die. The final state budget gave Spaceport Florida only $300,000 to continue operations for the next fiscal year.[109]

That April, the USAF opened another round of grant solicitations under the dual use program. In October, the authority won another $2.74 million for the modification of LC-46.[110] At one point, the SFA had joined the Western

Commercial Space Center and the Alaska Aerospace Development Agency to develop a combined proposal, but the SFA and AADC dropped out to go their own ways. The WCSC won $4.7 million in seed money for Vandenberg's commercial space industry, while the AADC won $750,000 for design and study of a commercial satellite launch pad. The State of New Mexico received $750,000, and Practical Solutions of Virginia received $1.3 million.[111] Despite going separate ways on their grant proposals, the SFA and WCSC collaborated on designing their commercial launch sites to assure compatibility for commercial customers.[112]

The authority slowly was racking up more wins than losses, but critics raised two complaints—the agency wasn't self-sufficient, nor had it launched a rocket from the Cape. In 1989, Governor Martinez had challenged the SFA to launch its first orbital flight from Cape Canaveral by 1992. Supporters argued that the SFA never received the $10 million Martinez had originally requested, and funding each year typically had been just enough to keep the agency afloat. For FY 1996, state legislators included $440,000 in the budget for the SFA, but future funding was dubious at best. State Rep. Harry Goode (D-Melbourne) told *Florida Today* that legislators "pretty much agree that, 'It's one more shot, boys.'" After that, if the authority could not sustain itself, the legislature might not provide any more funding.[113]

Despite their best efforts, the Spaceport Florida leadership couldn't outrun the pace of national and global events. A January 1995 *Orlando Sentinel* tally found that nearly 3,000 jobs had been lost from civilian employers at KSC and CCAFS over the last two years due to industry consolidations, a reduction of 16 percent in the workforce.[114] Republicans in January 1995 took control of both houses of Congress, promising to balance the federal budget as part of their Contract with America.[115] NASA was not exempt; in May 1995 the agency announced it planned to eliminate over 3,000 more jobs from KSC over the next five years.[116] KSC Director Jay Honeycutt feared a "brain drain" as experienced space workers accepted buyout offers with a bonus of up to $25,000 to retire early.[117]

Another complication was a general trend by three successive White House administrations—Ronald Reagan, George H.W. Bush, and Bill Clinton—to allow onetime adversarial nations to bid for US commercial satellite launch contracts. After the *Challenger* accident, and the late emergence of US commercial launch companies, satellite companies were desperate for delivery systems to fulfill customer orders. The administrations were also concerned that former Soviet republics desperate for revenue, such as Russia and Ukraine, might sell their launch technologies to rogue regimes such as Iran and North

Korea. Those three administrations entered into a series of trade agreements that placated US satellite manufacturers, but alarmed long-established launch interests on the Space Coast.[118]

Agreements signed by the Clinton administration in 1995–1996 modified quotas inherited from its predecessors limiting how many satellite launches China, Russia, and Ukraine potentially could win each year through 2000, and how much they could charge. From the Clinton administration's perspective, the quotas protected US launch companies, limiting predatory pricing by these nations. The quotas were also a foreign policy tool. A January 1996 agreement raised Russia's quota from sixteen to twenty as a reward for its help limiting launch technology transfer to rogue nations. This decision continued a precedent established by Clinton's predecessors, granting foreign access to the US commercial satellite industry as a reward for converting old ICBMs into peaceful payload launchers—the metaphorical conversion of swords into plowshares.

American companies sought partnerships with these foreign launch startups. Hughes Aircraft in 1988 supported an export license application for its satellites to launch on a Chinese Long March booster. The Reagan administration approved the request, reasoning that approval was an incentive for China not to sell its Silkworm missiles to Iran.[119] Lockheed Martin partnered with two Russian companies in 1992 to market sales of the former Soviet Proton launcher; in 1995, those companies formed a new company called International Launch Services (ILS) to market the Proton and the American Atlas.[120] Boeing partnered in 1995 with interests in Russia, Ukraine, and Norway to start Sea Launch, a project that would launch Ukrainian Zenit boosters from a marine launch platform based in Long Beach, California.[121]

Those who championed a commercial space boom for Brevard County had long warned of the consequences if the state and federal governments failed to adequately support their initiatives.

The time of reckoning had arrived.

National assets had gone to waste. America's rivals were ascendant. The Spaceport Florida Authority seemed caught in a conundrum. Successes were a justification for reducing legislative support. Failures were a justification for reducing legislative support. The SFA had been founded with the promise that it would soon fund itself, but that had yet to happen. Despite the best efforts of the authority's leaders, launches and therefore jobs were going overseas. And while American politicians hesitated to fund commercial space initiatives, other spacefaring nations were more than happy to subsidize their space industries, unbothered by ideological purity.

* * * * *

Under pressure from Congress to reduce his agency's budget, NASA Administrator Dan Goldin formed an independent review team led by Apollo-era Director of Flight Operations Christopher Kraft to "review the Space Shuttle Program and propose a new management system that could significantly reduce operating costs." The February 1995 report recommended that NASA separate itself from daily operations of the space shuttle to reduce costs by consolidating operations "under a single-business entity."[122]

To compete for the contract, in August 1995 Lockheed Martin and Rockwell International partnered to create a new company called United Space Alliance. Later known by its acronym, USA planned to bid against several competitors for the NASA contract to operate the space shuttle. Boeing, which already held the single prime contract for the space station project, was considered by many to be the forerunner.[123] But in November, NASA scrapped the planned competition and decided to award the prime contract outright to USA.[124] Boeing acquired Rockwell's aerospace and defense businesses in August 1996, including Rockwell's half of USA.[125] A January 8, 1996, *Orlando Sentinel* article estimated that 1,150 civil service and 2,000 contractor jobs would be lost at KSC due to the new contract, although some might wind up with USA or the new commercial enterprises projected to arrive later in the decade.[126]

In its annual report issued at the end of 1995, Spaceport Florida reported, "US launch companies currently control only 30 percent of the world's commercial market, due mainly to competition from Europe's *Arianespace* government-supported consortium . . . emerging competition from Japan, China, Russia and Ukraine threaten to drastically reduce Florida's number of commercial launches." The report noted that China, Russia, and Ukraine "highly subsidized" their launch systems, to the point that their launch prices "cannot be matched by market-based competitors." The federal government had imposed regulations limiting the number of US commercial satellites these three nations could launch, but the allotted totals "could soon exceed the total market available to U.S. launch companies."[127]

Governor Chiles sent a letter to President Clinton on January 26, 1996, urging that the administration "adopt a pro-active policy that encourages foreign launch systems to use U.S. launch sites . . . A clear statement in the final trade agreement with Russia that U.S. launch sites are available for use by Russian launch vehicles would be an important step toward accomplishing this objective." Chiles also asked Clinton to "establish Cape Canaveral as an international spaceport." The authority hoped to develop Florida launch sites for Russian, Ukrainian, and Japanese rockets, and that these sites would be assigned to the state agency for management and operation. The SFA suggested Launch Complexes 34, 37, and 41, with a preference for 37. Authority executives speculated that Russia might even use the Cape to launch International Space Station components. The "international spaceport" designation, however, wasn't included in the agreements.[128] Federal regulations still required American government payloads to launch on American rockets, guaranteeing some business for the Space Coast.[129]

In October 1996, memoranda of understanding were signed between Florida and Ukraine, hoping to create "favorable conditions" for business, including space. One MOU, between the SFA and the National Space Agency of Ukraine, sought to "develop cooperation in the field of commercial space launch." A second MOU, between the State of Florida and the Autonomous Republic of Crimea (Ukraine), sought to "develop cooperation in the tourism, recreation branches of economy, agricultural production, education and culture."[130]

Rep. Jim Bacchus chose not to run for a third term, so in January 1995 the Democrat was succeeded by a Republican, David Weldon, a Melbourne physician. As did his Democratic predecessors, Weldon sought to protect Brevard County's aerospace interests in Congress.[131] With the Republicans now in charge of the House, Weldon was named vice chair of the House Subcommittee on Space and Aeronautics. Just as Governor Chiles hoped the Russians might launch Proton rockets from Cape Canaveral, Rep. Weldon hoped the Ukrainians might launch their Zenit rockets from SFA's new commercial launch site.[132] Neither ever launched from the Cape, preferring to remain at the long-established spaceport near Baikonur, Kazakhstan.

Representatives from the commercial space authorities in Florida, California, and Alaska testified on June 12, 1996, before the House Subcommittee on Space and Aeronautics. They complained that the Clinton administration had given America's rivals an advantage in the emerging commercial launch industry.[133] The House eventually passed the Space Commercialization Promotion Act of 1996, which stated, "The Congress declares that a priority goal of constructing the International Space Station is the economic development of Earth orbital space." But the bill did not address the launch trade quotas, and the Senate failed to take up the bill.[134]

Although Congress did not act, the Florida state legislature finally seemed to get the message, approving $780,000 for Spaceport Florida in its FY 1997 budget.[135] Entering the summer of 1996, the SFA focused on completing Launch Complex 46. The pad would be designed to support any small rocket. Authority leaders believed a new small satellite launch market was about to emerge. NASA was also building a number of science probes that required only small launchers. To support their missions, NASA in August awarded the SFA a $500,000 grant to build an enclosure on the LC-46 service tower and for a modern support center for commercial customers. The Alaska and California authorities also received $500,000 each. The SFA received an additional $281,250 NASA grant from the same fund in September. O'Connor estimated that LC-46 could support up to twelve launches a year, generating $3 million to $25 million per launch for the Space Coast.[136]

Larger commercial satellite vendors, however, continued to head overseas. In November 1996, Hughes Electronics announced that the company had hired the Japanese consortium Rocket Systems Corp. to launch ten of its satellites between 2000 and 2005. The $10 billion deal was another lost opportunity for US launch companies and the Cape.[137]

The American commercial launch industry meanwhile shrank to just two liquid-fueled rocket vendors. After acquiring Rockwell in August, Boeing in December agreed to buy McDonnell Douglas for $13.3 billion. The deal made Boeing the largest US commercial launch company, bigger than Lockheed Martin.[138] Fewer domestic launch vendors meant less competition and fewer choices for customers. The cheaper alternatives were overseas.

Days later, the USAF selected McDonnell (soon to be part of Boeing) and Lockmart as finalists for a $1.4 billion contract to develop new rockets for the military. These rockets could also be used to launch nongovernmental customers, so the winner arguably would enjoy an implicit government subsidy.[139]

Lockmart's bid would use a Russian engine in development called the RD-180. The engine, developed by NPO Energomash, was the latest in a long line of kerosene-fueled Soviet engines that traced back to the 1940s. RP-1 kerosene was a cheaper and safer fuel than the liquid hydrogen and solid fuels popular with American launch companies in the latter twentieth century. Pratt & Whitney, based in Palm Beach County about 125 miles south of the Cape, partnered with Energomash to develop and market the engine in the United States. Lockmart selected the RD-180 to power its future Atlas boosters. If Lockmart won the contract, it could result in more jobs not only at the Cape, but also in Palm Beach.[140] Although the use of the Russian engine made economic sense at the time, in future years it would create a national security headache that threatened to ground America's launch capabilities.

As discussed in Chapter 2, during this time Lockmart had won a NASA contract to develop an experimental single-stage-to-orbit spaceplane designated the X-33. Lockmart intended to evolve the X-33 into a commercial vehicle called VentureStar. In 1996, NASA awarded a separate contract to Orbital Sciences (today, part of Northrop Grumman) to develop an advanced technology demonstration spaceplane designed the X-34. The vehicle was to be carried into flight by a Lockheed L-1011, released to conduct its tests, and land at a runway. The X-34 was intended to eventually conduct a powered test flight using a new rocket motor called Fastrac developed by NASA's Marshall Space Flight Center in Huntsville, Alabama. After three unpowered test flights, the program was canceled in 2001 due to a lack of funding.[141]

In early 1997, all that was in the near future for Spaceport Florida. In an April session with the *Florida Today* editorial board, incoming KSC Director Roy Bridges stated his intent to attract future X-34 landings to the center's shuttle runway. He envisioned KSC becoming the project's "home base" one day, and noted that Spaceport Florida was working the state legislature to secure funding for a hangar.[142]

The authority succeeded in winning $4 million in Florida's FY 1998 budget to build its "space hangar" near the runway. The hope was that the hangar would lure not only the X-34, but perhaps VentureStar and any other reusable space vehicles that might succeed the space shuttle. The SFA also received $752,844 for operating expenses in its next fiscal year. Florida Tech's TRDA received $750,000 to support the Center for Space Education.[143]

On May 29, 1997, Spaceport Florida opened Launch Complex 46 for commercial use. Rep. Weldon delivered its operating license from the FAA. SFA executives estimated they would charge customers about $300,000 per launch.[144] But in future years, only two missions launched from LC-46: the NASA Lunar Prospector atop an Athena II booster in January 1998 and a Taiwan science satellite atop an Athena I in January 1999. The Athena line was a solid-fueled booster offered by Lockmart and ATK (a descendant of Thiokol). No other vendors used LC-46 for orbital flights until after renovations that began in 2015.[145]

SFA began planning a second commercial site, at Launch Complex 20, which was built in the late 1950s to support the original generation Titan missiles and unused since the military's Starbird tests in the early 1990s. While LC-46 was designed for more powerful rockets, LC-20 would be much simpler, designed for smaller rockets that launched suborbital or orbital payloads. Among the customers the agency targeted was an Israeli company hoping to find a US partner for launching small satellites into orbit with its Shavit rocket. But LC-20 was only used for three small suborbital student launches in the year 2000, then it went dormant.[146]

The authority tried to attract rocket manufacturing to Florida, but that effort failed as well. By the late 1990s, Boeing and Lockmart were developing next-generation liquid-fueled rockets. The SFA tried to attract the companies to build them in Florida, but Boeing chose to manufacture the Delta IV in Alabama, and Lockmart chose to manufacture the Atlas V in Colorado. Both vehicles, however, would launch from the Cape. The Delta IV would use the refurbished LC-37, while the Atlas V would use the renovated LC-41. The vendors would pitch both vehicles to the military for the $1.4 billion contract to launch the next generation of national security payloads.[147]

"Florida's Dominance Is Long Gone," read the headline of a *Florida Today* article in a special section titled "Florida's Future on Space," published on December 14, 1997. Written by space reporter Todd Halvorson, the article found that "a dramatically changing market for launching rockets and fierce national and global competition could drive the state's coastal spaceports right out of business."[148] The article concluded that reusable launch vehicles (RLVs) such as the X-33 and VentureStar were the future. Coastal launch sites until this point in time were preferred because expendable boosters fell into the ocean. The single-stage-to-orbit reusable design meant no part would be expended. An RLV could launch from anywhere.

Local political leaders took the *Florida Today* conclusions seriously and called a summit meeting on January 12, 1998, at KSC. About seventy-five state and local political, government, and businesspeople attended. The group decided to form a task force to lure VentureStar to the Space Coast. Rep. Dave Weldon said he had verbal commitments from the SFA and from Enterprise Florida to pay for the "Florida VentureStar Capture Team" that would lobby Lockmart to bring their RLV program to Brevard County.[149]

Documents obtained by *Florida Today* in March 1998 showed that Florida faced several obstacles in their pursuit of VentureStar. Lockmart forecasted forty launches a year, and worried about delays caused by the Cape's frequent summer thunderstorms. A multibillion-dollar upgrade of Cape facilities by the USAF was three years behind schedule. NASA officials identified three potential VentureStar launch sites, but all three would impact existing operations, either at KSC or at CCAFS.[150]

Governor Chiles proposed $1.6 million in the state's proposed FY 1999 to help attract VentureStar.[151] Not only did that request pass both houses, but the state legislature also added another $1.7 million for a space operations center that Spaceport Florida would operate near Port Canaveral.[152]

By the time Lockmart executives toured the Cape in June 1998, the list of launch site candidates had grown to eighteen. A Lockmart executive told *Florida Today* the company would select the cheapest and "best site of combination of sites." A final decision was expected by fall 1999.[153]

That decision never came.

By the year 2000, VentureStar's viability had devolved from bleary-eyed optimism to optical illusion.

"Rocket Ship of Future Can't Fly" declared the front-page headline of the March 19, 2000, *Orlando Sentinel*. The article reported, "A new technology once billed as the key to the ambitious program is on the verge of being abandoned." States bidding to host the program were "hedging their bets. What began as campaigns to land VentureStar have broadened into more generic efforts to bring home any next-generation vehicle."[154]

A year later, in February 2001, NASA canceled the X-33 as well as the X-34. When filled with hydrogen chilled to -427°F to keep that fuel liquid, the graphite composite materials used to manufacture the X-33 aerospike engine's fuel tanks cracked and corroded. A subsequent GAO investigation concluded that, "NASA did not develop realistic cost estimates, timely acquisition and risk management plans, and adequate and realistic performance goals." NASA chose to scrap both programs in favor of a new approach called the Space Launch Initiative. Without its partner providing a critical capital investment, Lockmart chose not to continue with VentureStar. The commercial satellite market, which Lockmart had foreseen as being VentureStar's primary customer, had gone dormant.[155]

The X-33 was too far ahead of its time to be technically viable. X-33 and VentureStar became just another abandoned paper exercise. Launch, land, and relaunch didn't become practical until SpaceX perfected the Falcon 9 in the late 2010s. That's a story for Chapter 10.

* * * * *

It is difficult to make predictions, especially about the future.

Attributed to Danish Proverb[156]

SFA Executive Director Ed O'Connor announced in September 2000 his intention to retire early the next year.[157] He and his staff had done their best over the last decade to grow the SFA from the commercial spaceport idea midwifed by the Florida Governor's Commission on Space in the late 1980s. They had largely avoided the internecine squabbles of Brevard County politics that had sabotaged other initiatives. At the time of his announcement, the SFA's annual revenue had grown to $2.7 million, along with $3.9 million in state grants. The authority's balance sheet listed among its assets the commercial space pad at LC-46, a second under construction at LC-20, the Reusable Launch Vehicle Hangar (RLVH) built for the X-34, and a Solid Rocket Motor Upgrade facility used by Lockmart at Camp Blanding.[158]

The SFA had provided funding to upgrade LC-37 for the Boeing Delta IV, financing a horizontal integration building. The authority also helped Lockmart finance an upgrade of LC-41 for the Atlas V. Plans were in the works for a space life sciences building to support ISS research, and a space commerce park. A "Space Operations Control Center" was completed in the authority's Titusville headquarters.[159]

With the X-33 and X-34 canceled, the authority leased the RLVH to USA, which intended to use it to store space shuttle equipment. USA's two-year lease called for monthly payments of $10,000 rent.[160] No one could have foreseen that, less than two years later, the hangar would hold the remains of the orbiter *Columbia* as the accident board conducted its investigation.[161]

Not all had gone well. Hurricane Opal in October 1995 washed out the Cape San Blas launch site; it was never rebuilt.[162] The SFA never convinced the state legislature to guarantee the agency a reliable revenue source from the sales taxes it generated. Congress never passed legislation that would have exempted SFA bonds from federal taxes. The federal legislative branch did, however, pass the Commercial Space Act of 1998, which required the federal government to acquire "space transportation services from United States commercial providers whenever such services are required in the course of its activities." It also required the NASA administrator to "prepare for an orderly transition" to commercial space transportation services, "including human, cargo, and mixed payloads." The act required a number of studies both by NASA and the DOD aimed at encouraging those agencies to reduce red tape and transition their facilities to the private sector. SFA executives viewed the act as positioning the Cape to compete for what was believed to be an imminent RLV industry.[163]

Florida remained heavily dependent on federal funding for technological research and development. A 1997 study found that the DOD provided 72 percent of the state's federal research and development funding. NASA was second at 16 percent.[164] Experts believed that Florida lagged behind California and other nations in training high school students to participate in a workforce required to master physics, chemistry, and calculus. One expert ranked California, if it were a nation, number four globally in R&D because of the state's superb university system that "adequately trains students for science in the real world." The expert ranked Florida number twenty.[165]

Similar inadequacies were perceived on the Space Coast. An October 1996 *Florida Today* investigative report concluded that Brevard County advanced education fell short compared to other regions that support aerospace. The article speculated that residents might start new careers elsewhere if they couldn't find locally the education they needed.[166]

In June 1996, the SFA partnered with Brevard Community College, the University of Central Florida, and FIT to form a consortium called the Florida Space Institute (FSI). The idea was to team up in soliciting federal and state grant money, as well as participate in luring businesses to Brevard and to maintain the local workforce.[167] In January 1997, the FSI won its first contract, $500,000 from the DOD to design a small satellite for testing laser communications technology.[168] In October 1997, NASA donated a used satellite antenna dish to the FSI for a space-research ground station.[169] The FSI to this day remains based at the University of Central Florida.[170]

In the spring of 1999, the state legislature created the Florida Space Research Institute (FSRI), which sounds similar to the FSI but was a separate entity. The FSRI was a consortium of the SFA, Enterprise Florida, Florida Aviation Aerospace Alliance, Florida Space Business Roundtable, and three KSC contractors—Boeing, Bionetics, and Command & Control Technologies.[171] The FSRI teamed with the NASA-sponsored Florida Space Grant Consortium to provide grants to Florida universities and colleges.[172] The state's FY 2001 budget included $800,000 for FSRI.[173]

The budget also included $10 million for the Space Experiment Research and Processing Laboratory (SERPL), known today as the Space Life Sciences Lab (SLSL). Construction began in February 2001. Florida was expected to pay $30 million of the cost, with NASA funding the remaining $17 million.[174]

O'Connor left his successor a vibrant growing organization that finally had the support of the Florida state legislature, and the Florida delegation in Congress. But the unanimity came too late. Most commercial satellite launches had gone overseas, and satellite demand had dwindled by the end of the decade. Spaceport Florida had built their commercial pad, but the launch industry seemed indifferent. Without commercial space enterprise to generate a revenue stream, all the SFA had left were federal government projects such as the X-33 and X-34. Their cancelations were the first of many federal initiatives that built hopes, only to be dashed by the whims of Congress or the next administration in the White House.

The commercial sector certainly wasn't any more reliable. Many entrepreneurial initiatives failed to achieve their promises. Overseas commercial launchers from Russia and Ukraine had toured the Cape but shown no further interest.

The Danish proverb reminds us, "It is difficult to make predictions, especially about the future." Even if county, state, and federal politicians and bureaucrats had united to support American commercial space initiatives, one could argue it might not have mattered.

Congress chose not to pursue President Bush's Space Exploration Initiative and its long-term $500 billion price tag. Florida could have backed the SRF's Moonbase 1 project, believing it would pay off once NASA returned in earnest to the moon, only to be left with an attraction that was no more than of passing interest to tourists visiting Walt Disney World and other Orlando tourist destinations.

Ronald Reagan, George H.W. Bush, and Bill Clinton all would have had to change their geopolitical foreign policy strategies for the Cape and other states to have a chance at competing with Europe, Japan, Russia, and Ukraine. What would have been the consequence if they had decided to ban American-made satellites from launching on foreign rockets? A protectionist policy might have led to Cold War–era enemy weapons being sold to rogue nations, or at the very least Russian and Ukrainian engineers going to work for hostiles.

Even so, launch sites are often dictated more by orbital mechanics than the lowest cost. The closer a launch site to the equator, the greater the boost from Earth's rotation—but only if the customer wants its payload placed in a posigrade orbit, that is, moving in the same direction as Earth's rotation. Launching a rocket over land is a hazard, so it's unlikely that a posigrade (west-to-east) launch would occur from sites such as California or New Mexico. Cape Canaveral might be more bureaucratic and expensive, but it's safer.

By the end of the twentieth century, the SFA had become the agency envisioned by the Florida Governor's Commission on Space back in the late 1980s, but it's difficult to make predictions, especially about the future. For the commercial launch industry, Florida had built it, but they'd yet to come.

6

What's in a Name?

When Chris Shove helped the state legislature craft the statute that created the Spaceport Florida Authority in 1989, he based it on language that created the Port Authority of New York and New Jersey in the early 1920s. Following the advice of legal counsel, he also based the bill on language that created the Reedy Creek Improvement District for the Walt Disney World Company in 1967.[1]

The Reedy Creek bill created a special taxing district with the same authority as a county government. Landowners within the district—the sole landowner being Disney—would be responsible for paying the costs of municipal services. The district would be responsible for overseeing land use and environmental protections within its borders.[2]

Florida law empowers the state legislature to create special districts and authorities with broad powers for special purpose government operations. Shove designed the SFA to extend a government agency's sovereign immunity to commercial space enterprises using the authority's facilities. His intent was to limit their liability, thereby lowering insurance costs.[3]

In the early 1990s, the state legislature created several new authorities that would at times overlap their interests with the SFA.

In January 1992, as the authority was struggling to survive, Governor Lawton Chiles (a Democrat) formally proposed the creation of another new entity called Enterprise Florida. The idea came from studies by the Florida Chamber of Commerce, which was looking for ways that Florida could be more competitive with other states and nations. Chiles wanted to downsize the state government, and saw Enterprise Florida as a way to give the governor and business executives more control over state commerce. Enterprise Florida and local development groups would replace the state's Department of Commerce within three years. The bill passed the legislature in late March.[4]

Three years later, Governor Chiles sought approval to abolish the Department of Commerce as planned, transferring its remaining responsibilities to Enterprise Florida and other public-private partnerships. He believed that the proposal would save the taxpayers money, because the industry benefiting from the agency's activities would fund much of it.[5] The proposal passed the

Democratic-majority House, but the Republican-led Senate wouldn't take it up, so the proposal died in May 1995.[6] Not helping were two audits that found a lack of oversight in other state public-private partnerships, resulting in wasteful spending and shoddy work.[7] In November, Florida's Auditor General released a finding that the State of Florida needed better oversight of such partnerships.[8] A revised version of Chiles's proposal, requiring performance measures, passed in May 1996. It also "leased" to Enterprise Florida any Department of Commerce employees wanting to join the private sector so as to protect their pensions vested with the State of Florida.[9]

Chiles also wanted to convert the state's Division of Tourism into a public-private partnership tentatively called Travel Florida. The change would give the industry more control over how revenue raised from rental car surcharges was used to promote Florida tourism.[10]

The proposal died in the 1995 legislative session, but was revived in the 1996 session and approved. The Division of Tourism was replaced by the Florida Tourism Commission, which was to perform a function similar to Enterprise Florida. The commission created a not-for-profit agency called the Florida Tourism Industry Marketing Corporation. That unwieldy term was replaced in August 1997 by a new name—Visit Florida.[11]

Both Enterprise Florida and Visit Florida had interests that overlapped at times with Spaceport Florida. The SFA had a mutual interest with Enterprise Florida in promoting new and growing high-tech industries in the state. In April 1995, for example, Enterprise Florida and Lockheed Martin partnered in a venture creating a high-tech small business incubator in central Florida. Lockmart expected their investment to create commercial opportunities for its defense technologies.[12] The SFA also had a mutual interest with Visit Florida in promoting tourism. Spaceport USA attracted tourists from around the world, although attendance had slacked from three million in 1990 to two million in 1994. The Florida Tourism Commission responded by placing ads in newspapers promoting KSC's visitor center.[13]

As the various ideas for privatizing government services circulated around the state legislature in 1995, the proposed legislation included a new entity that would oversee all the partnerships, as well as rural and minority business programs. It would be called the Governor's Office of Tourism, Trade and Economic Development (OTTED).[14] The office returned in the proposed 1996 legislation, intended to oversee spending and measure how many jobs were created by the partnerships.[15] The final version, passed in May 1996, created OTTED to oversee Enterprise Florida and the Florida Tourism Commission. An inspector general position was created within OTTED to ensure proper oversight.[16] Tax dollars for the two partnerships would flow through OTTED. Concerns were raised that OTTED added another layer of bureaucracy to the government, but those concerns were dismissed by state Senator John McKay (R-Bradenton), who sponsored the bill. He said that OTTED would have a small staff and simply "pass through" money to the agencies under its purview. State Rep. Fred Lippman (D-Fort Lauderdale) described OTTED as no more than "a contracting agent."[17]

The language of the final legislation was interpreted as giving OTTED some oversight of the SFA. The OTTED was directed to "monitor the activities of public-private partnerships and state agencies in order to avoid duplication and promote coordinated and consistent implementation of programs." The bill authorized OTTED to enter into contracts with the SFA "in connection with the fulfillment of its duties."[18]

According to multiple individuals working in state government at the time, this resulted in the SFA having to submit its annual budgets, reports, and financial statements to the OTTED, which could now control how state money was spent by the authority. The agency's board of supervisors still governed its activities, and the use of any revenues outside of state tax dollars remained within the authority's purview. The OTTED was also now in a position to participate, as a representative of the governor, in national space decisions made by the federal government.

OTTED soon proved the legislature's oversight concerns valid.

Florida's film marketing agency had once been part of the state's Department of Commerce, but those functions were privatized after Commerce was disbanded. In February 1997, an audit of the Florida Entertainment Commission found sloppy bookkeeping and bloated claims of its success. The commission was replaced with the Florida Entertainment Industry Council, but the same abuses persisted. In January 1998, the OTTED suspended the council's $200,000 in state financing for failing to meet its contractual requirements. A new not-for-profit called Film Florida was created in August 1999 to promote the state's film industry. Film Florida would work with an OTTED-appointed state film commissioner.[19]

In May 1999, the *Orlando Sentinel* obtained a copy of a state Auditor General report that concluded Enterprise Florida had failed to oversee its finances. "The auditor found that, 'basic contract-procurement best practices were not used by the president of Enterprise Florida on contracts for which he was solely responsible for the selection and negotiation process.'" Enterprise Florida had procedures in place, but failed to follow them, the auditor concluded. "'This promotes an atmosphere lacking accountability,'" the *Sentinel* quoted. The president had claimed inappropriate reimbursements for taking his wife on trips and had claimed her auto insurance premiums as well.[20] By August, the inspector general found that "EFI appears to be taking corrective action."[21]

Jeb Bush (a Republican), the state's commerce secretary under Governor Bob Martinez in 1987–1988, was elected governor, and took office on January 5, 1999. Bush began to rework Enterprise Florida and other public-private partnerships into his own vision. On January 29, 1999, Bush proposed combining Enterprise Florida with the SFA, Visit Florida, and three other agencies into one group for state financing.[22] He appointed Antonio Villamil, a US undersecretary of commerce for economic affairs under President George H.W. Bush, to be OTTED's director and oversee the consolidation.[23]

After the political sausage was ground, in May 1999 the Florida legislature granted Bush a diluted version of his proposal. Although the legislature didn't approve combining the six agencies, it did approve pooling into one account 20 percent of state funding for Enterprise Florida, Visit Florida, and the SFA. By creating the pool, it gave Villamil the authority to shift funding priorities.[24]

When the *Orlando Sentinel* reported the findings of the Enterprise Florida audit, Villamil told the reporter, "We want to make sure the right financial controls are in place that would allow for the growth of the partnership." He said his office was looking as well at its other public-private partnerships. Villamil told the *Miami Herald* in September 1999 that tougher standards were now being applied to several partnerships, including the SFA. Although his office didn't want to "micro-manage" these partnerships, his intent was to establish specific performance goals and measurements. The 20 percent withheld from these agencies' budgets would only be released if they met their goals. He also expected these agencies to do a better job of working together.[25]

Enjoying a robust national economy, the state legislature in May 1999 passed an FY 2000 budget that gave the SFA a healthy budget. The SFA received $720,000 for its operating budget, as well as $1.41 million to develop a five-year transportation infrastructure plan that included $1.2 million for the planned X-34/VentureStar hangar, and $500,000 for a planned industrial space park. The legislation also created the Florida Space Research Institute discussed in Chapter 5, and budgeted $1.5 million for a new Florida Commercial Space Finance Corporation[26] that would loan money to launch companies and their customers doing business in Florida. The legislature ultimately dropped an idea to redirect sales tax revenue from the Kennedy Space Center Visitor Complex (KSCVC, formerly named Spaceport USA) guests to fund space research and development managed by the University of Florida.[27]

In March 1999, State Rep. Charles Bronson (R-Satellite Beach) introduced SB 2540, the Florida Space Transportation Planning Act, which made the SFA eligible for project funds available from the same pot as airports, seaports, and highways. The bill was passed by the legislature in May 1999.[28] The statutes up to that point had allowed FDOT to "enter into a joint project agreement with, or otherwise assist," the SFA and allocate funds to the Authority as part of FDOT's five-year work program, but the funds could not be used for the SFA "administrative or operational costs." SB 2540 added new language to Section 331.360 of the SFA statutes, stating it was now the "duty, function, and responsibility" of FDOT to "promote the further development and improvement of aerospace transportation facilities." Space was now another mode of transportation. The SFA was required to "develop a spaceport master plan for expansion and modernization of space transportation facilities" for submission to FDOT, which could incorporate the plan into its five-year work program. SB 2540 also required the creation of a Spaceport Management Council "to provide coordination and recommendations on projects and activities" at Florida's space launch facilities.[29] In November, the Spaceport Management Council began work on the five-year transportation infrastructure plan. Representatives from the SFA, OTTED, Enterprise Florida, FDOT, KSC, and the Air Force 45th Space Wing at Cape Canaveral were part of the council.[30]

This might explain why it appears that the FDOT loans discussed earlier were never repaid. You may recall that, in the early 1990s, the state legislature was reluctant to provide the SFA with project money. The compromise was to loan the authority $600,000 in 1992, and another $600,000 in 1993. The loans were carried on the SFA books for years as "non-current liabilities."

This author attempted to locate SFA financial statements for each year through 2006, when the agency folded. Some were unavailable; state policy requires that an agency retain records for only ten years, with certain exceptions. The SFA's 2000 financial statement, released in November 2000, reported that one of the $600,000 loans would begin payments in March 2001; $200,000 would be repaid each year in 2001, 2002, and 2003. It appears the payments were never made. A footnote in the 2004 financial statement read, "This debt has not been repaid per the amortization schedule; however, OTTED has not requested such payments be made."

As we'll discuss later in this chapter, the SFA was replaced in 2006 by another authority, Space Florida. That agency's first financial statement made no mention of the loans; the consensus of those willing to talk about the matter is that they believe the loans were "forgiven" or "forgotten."

Some sources suggested that, if the loans were "forgiven," it was because of the 1999 legislation amending Section 331.360. The new language gave the SFA access to FDOT's capital project funds. Projects described in the new master plan would be eligible for state transportation funding. Had the SFA repaid the loans, the payments may have returned to the SFA anyway for these future projects. So why repay the loans if the money comes back to the authority?! A rhetorical question, for which no answer could be found.

An example of the closer relationship between SFA and FDOT is the $4 million "Fast Track" program grant awarded by FDOT to SFA in November 1999 to build what is now known as Space Commerce Way. Fast Track was a Governor Bush program that transferred money from a planned high-speed rail project to fund statewide transportation needs. The authority submitted a request for funding to build a road that would provide public access to a planned Space Commerce Park and its anchor tenant, the Space Experiment Research & Processing Laboratory (SERPL), now known as the Space Life Sciences Lab (SLSL). Space Commerce Way was completed in 2003. The original commerce park was never built, for reasons discussed later in this chapter, but today's Exploration Park broke ground nearby in 2010. The $4 million was to be matched by $4 million from NASA; KSCVC ultimately paid NASA's share, as the road improved access to the park. In 2018, KSCVC opened a new entrance plaza from Space Commerce Way.[31]

* * * * *

In April 1999, Governor Bush visited Cape Canaveral and declared that Florida lacked a "coherent vision . . . There are a lot of great ideas, but they're kind of scattered now." Bush announced the appointment of Lieutenant Governor Frank Brogan (a Republican) as the lead state executive for space business. Bush and Brogan directed OTTED to develop recommendations for how Florida could improve its commercial space competitiveness. OTTED retained a USDOT think tank called the Volpe Center to review Florida space activities and recommend future actions. The selection of the Volpe Center, along with the Florida Space Transportation Planning Act, demonstrated an evolution in state leadership thinking—commercial space was just another form of transportation, albeit a unique one.

The ninety-three-page final report was released in December 1999. It refuted those who believed commercial space primarily benefited only the Space Coast. The report found 356 commercial space firms in Florida, half of which were outside Brevard County. Over $61 million in sales tax revenue were generated by the companies; that revenue was distributed statewide. The space industry had a higher annual average income for its employees ($42,792) than any other significant category, such as service/retail ($19,281) and manufacturing ($30,743).[32]

The Volpe Center declared that Florida was "home to the most sophisticated and wide-ranging launch infrastructure in the world."[33] The report identified three significant challenges for Florida to attract customers and provide competitive services:[34]

1. Cape launches were based on traditional government and military requirements rather than the private sector.
2. Facilities and technologies were antiquated. A modernization program was installing already obsolete technologies based on 1980s computers and systems.
3. Supporting a new generation of space technologies with unforeseen requirements.

The final section, "A Course of Action," recommended goals and objectives in steps gathered into two broad categories—Leadership in Space Transportation and Diversification of the State's Commercial Space Economy.[35]

The report encouraged more independence for the authority, but at the same time the SFA should do a better job communicating its activities to the state legislature. The authority should also develop "an institutional capability within the state to conduct ongoing market and technology assessments, and make this information available to the public."

The Volpe Center asserted that the Cape already had many competitive advantages over its rivals, so it was unnecessary to "over react" to real or perceived competitive threats. The SFA should act as an advocate for its commercial customers, as a "one-stop" shop for business partners.

The report implicitly endorsed the actions taken earlier in the year by the Jeb Bush administration, calling on the governor's office to "take the lead in addressing national policies and institutionalized federal operating procedures that impede greater commercial use of the launch facilities." The SFA should be given "clear guidelines and authority to offer economic development and financing packages to attract new launch service providers to the Cape and retain current ones."

Volpe also encouraged Florida to diversify its space industry beyond just launch services into other segments of the space market. The authors wrote that "the greatest opportunities for diversification exist in the end-user services markets," such as telecommunications, data transfer, and broadcast services. They noted, however, the volatility of those markets, evidenced by a number of bankruptcies earlier that year. Florida, they concluded, "lacks a comprehensive strategic assessment of its opportunities as well as a plan to pursue them."

A Volpe Center economic analyst testified about the report before the state's Senate Commerce and Economic Opportunities Committee. "Florida basically has everything you need to be a leader," she told the panel. "It's just a matter of getting everyone together on the same page."[36]

A December 26, 1999, *Florida Today* editorial column urged the state to "Get busy. Get organized. Get *moving*." (*Italics* emphasis in the original.):[37]

Message received.

On January 14, 2000, a 2.5-hour "space summit" was held at KSCVC. In attendance were NASA Administrator Dan Goldin, Florida Governor Jeb Bush, US Senator Bob Graham (D-FL) and Rep. Dave Weldon (R-Palm Bay). Also there were SFA Executive Director Edward O'Connor, KSC Director Roy Bridges, and USAF Brigadier General Donald Pettit, who commanded the Cape's 45th Space Wing. Executives from Boeing and Lockmart also attended, as did many others. The event, open to the public, was organized and hosted by Graham and Weldon.[38]

Many of the challenges noted in the Volpe Center report arguably were to be addressed by the proposed SERPL and adjacent Space Commerce Park. According to a March 2000 State of Florida business prospectus, the projects would be "needed to meet the anticipated increase in scientific and technological ground support associated with the [ISS]." SERPL would be "a world-class laboratory with the capability and systems necessary to host ISS experiment processing as well as life sciences and microgravity-related research." It would be a "magnet facility" for the Space Commerce Park, "an ideal location for businesses and research groups with a need for close proximity to the Space Center's launch and landing facilities and capabilities."[39]

SERPL design work had begun, but the state legislature had yet to approve the $30 million to build it. KSC Director Bridges said NASA would pay to lease the building, but the state's universities had to provide the scientists willing to do the research. NASA Administrator Goldin noted that, unlike NASA's other space centers, KSC didn't have close ties to nearby research universities, therefore Florida lost on many research grant opportunities. No Florida university ranked in the top thirty schools receiving NASA research grants.[40]

Graham and Weldon returned to Washington intent on advancing the Florida commercial space agenda. Graham introduced in the US Senate the Commercial Space Partnership Act, which would have allowed NASA to lease its property to "any person or entity," including state or local government. The bill went nowhere.[41] Graham and Weldon hoped to advance the Spaceport Investment Act they'd introduced in their chambers last June, once again pursuing federal tax-exempt status for spaceport bonds, but that bill went nowhere too.[42]

Space Coast advocates had better luck in Tallahassee. The state legislature's FY 2001 budget, adopted on May 5, 2000, included $10 million to begin work on the SERPL facility, and the $4 million in matching funds to build Space Commerce Way. The budget also included $800,000 for the FSRI.[43]

On September 11, 2000, the authority dedicated a new seven-story, 100,000 square foot Horizontal Integration Facility at Launch Complex 37. The HIF had been built for Boeing's Delta IV rocket with $24 million of SFA financing. The event was a major milestone for the authority. For the first time, the SFA had resurrected and opened an orbital-class launch complex for a specific commercial launch company with a credible client, the US Air Force. The SFA financed and constructed the facility, then leased it to Boeing, which was investing $250 million in Delta IV facilities at the Cape.[44]

Smaller commercial companies were also looking at the Cape. The SFA courted Texas-based Beal Aerospace, which was planning a 224-foot tall three-stage rocket that might use the old Titan-era pad at Launch Complex 15. The authority offered Beal a $10 million incentive package to build a manufacturing and launch facility at the Cape, but Beal folded in November 2000. Beal blamed potential competition from future NASA launch vehicle programs.[45] A Titusville-based company called E'Prime Aerospace hoped to launch rockets, based on retired USAF Peacekeeper ICBMs, from KSC but couldn't find significant investors and had already racked up $15 million in debt by 2000.[46]

As the twenty-first century began, the US economy slowed and in March 2001 entered into a mild recession. The September 11, 2001, terrorist attacks also impacted the American economy, in particular travel-dependent industries. Consumers lost confidence in the economy, leading to a decline in retail sales.[47]

The US commercial launch industry already was in decline, due to a mismatch between launch vehicle capacity, satellite manufacturing capabilities, and launch market demand. *Florida Today* reported in January 2001, "Commercial satellite launches will decline this year," and that the launch industry was about to go through "a tough period." Ed Gormel, who had just replaced Ed O'Connor as SFA's executive director, predicted attrition in the industry workforce. Gormel cited several causes for the downturn: the slowing national economy, bankruptcies in the satellite industry, fiber-optic cables replacing satellite transmissions, and a longer lifespan for existing satellites.[48]

Governor Jeb Bush's brother became president on January 20, 2001. Some may have assumed that the relationship would uniquely benefit Florida but, at a time of mild recession and the September 11 terrorist attacks, Florida space was far from a priority for President George W. Bush. The war on terrorism affected the Space Coast as well. Enemies who had already aimed at American symbols such as the Pentagon and New York City's World Trade Center might find KSC a tempting target. Security measures were tightened at the Cape. Space exploration didn't have the attention of Congress so much as waging a war against an elusive enemy. If money were to be spent on space, it would be to protect the homeland.[49]

* * * * *

"Allegations of Theft Jolt Spaceport," read the front-page headline of the March 22, 2001, *Florida Today.* The article reported that a Spaceport Florida Authority bookkeeper had been placed on leave, accused of embezzlement. An independent audit had turned up evidence of impropriety. The authority's finance manager was placed under arrest on March 30 and released on $4,500 bail. She was accused of having taken up to $100,000 from SFA payroll accounts, which she transferred electronically to her personal bank account. The finance manager was charged with grand theft, as well as two counts of forgery and uttering a forged instrument. She eventually pleaded nolo contendre to the charges. The judge sentenced her to eighteen months in prison, to be followed by fifteen years' probation, and $137,000 in restitution.[50]

Executive Director Ed Gormel told *Florida Today* that the SFA would improve its accounting methods, and tighten its security passwords. Authority credit cards for almost all employees were rescinded to tighten spending; agency standard practice had been to issue employees credit cards for business expenses. One former employee was found to have charged $2,700 in personal purchases to an SFA charge card; the debt was repaid, so no charges were filed.[51]

The allegation was bad timing for a $16 million request by Jeb Bush to help fund completion of the SERPL. Some state legislators were already asking if NASA could be a reliable partner. The State of Florida had funded the RLVH for the X-34, only to have NASA cancel that project and leave the state with an empty hangar. Why fund completion of the SERPL when the same might happen? The embezzlement allegation left some legislators even more skeptical of the authority's credibility. Rep. Rob Allen (R-Merritt Island) told *Florida Today,* "I'm not about to promote to my colleagues the Florida Spaceport Authority until I know the problem has been fixed. I'm very angry that Spaceport Florida has jeopardized our space industry with their methods." Others were forgiving, viewing the incident more as an aberration than a pattern.[52]

The authority countered with an April 2 live audio connection between the ISS and state leadership gathered on the Tallahassee House floor. NASA astronaut Jim Voss thanked them for funding SERPL so far. "I think you'll be able to do great work there," Voss said. "You're going to reap some benefits not only for the state but for all of mankind." Lieutenant Governor Brogan joked, "We've just been lobbied from outer space."[53]

The next day, the SFA announced an agreement with Boeing to base an experimental spaceplane project at the Cape called the X-37. The craft had been a joint project of NASA and the USAF since 1999.[54] Authority officials thought the project might be based at Launch Complex 20 by 2003.[55] That never happened, but the spaceplane project evolved into the Boeing X-37B; NASA transferred the project in 2004 to the Defense Advanced Research Projects Agency (DARPA). At that point, it became a classified project. Two X-37B planes were built; they're based today in two former space shuttle hangars next to KSC's Vehicle Assembly Building. The X-37B flies long-duration missions for the Space Force, conducting largely classified experiments and testing reusability of the space vehicle.[56] As of this writing, it has launched atop the ULA Atlas V and the SpaceX Falcon 9 and Falcon Heavy.

Despite the bookkeeper scandal, the overall track record of success won the SFA the $16 million it had requested to complete SERPL. The authority's operating budget was reduced by $110,000 to $700,000.[57]

Around the time that the state legislature approved the SFA's FY 2002 budget, the USDOT's Office of Commercial Space Transportation (AST) issued a gloomy report that forecasted a significant downtown in commercial launches worldwide through 2010, averaging only thirty-two launches annually through 2010.[58] The report was issued four months before the September 11, 2001, terror attacks. The 2002 report made no mention of the attacks, other than a passing observation that, "Terrorist activity during 2001 has, however, increased government interest in potential relationships with commercial providers."[59] The 2002 forecast issued next May was gloomier, forecasting an average of only twenty-seven annual commercial launches from 2002 to 2011.[60]

* * * * *

State politicians and officials began using the phrase *Cape Canaveral Spaceport* as a marketing term for space operations at the Cape. One of the earliest uses was by Governor Bush in May 1999, as he planned a domestic trade mission to California. By August, *Florida Today* was using the term, reporting it was now used by the SFA, NASA, and the USAF. The *Tampa Bay Times* reported in November a proposal for rail service from the Tampa Bay area to the "Cape Canaveral Spaceport." The SFA's annual report, published in November, described "Cape Canaveral Spaceport" as "Florida's 'anchor' space facility." Brigadier General Donald Pettit of the 45th Space Wing at Patrick Air Force Base used the governor's January 2000 Space Summit to call for NASA and the USAF to create a "united" Cape Canaveral Spaceport.[61] The state legislature amended the SFA's statutes in 2002 to allow the term to be used to describe any authority property within the boundaries of KSC, CCAFS, or Patrick Air Force Base.[62]

In November 2001, the Spaceport Florida Authority (SFA) changed its name to the Florida Space Authority (FSA). Ken Haiko, chair of the authority's board of supervisors, told *Florida Today*, "We need to reach out more to the entire state of Florida. When they think of the spaceport, they think of Brevard County.

This is more all-encompassing." The authority's logo changed from showing a rocket launch to one representing space enterprises.[63] The name change was formally ratified by the state legislature amending the statutes in 2002.

On February 5, 2002, newly appointed NASA Administrator Sean O'Keefe appeared in Tallahassee alongside Governor Bush to announce that he would consolidate space shuttle orbiter maintenance operations at KSC.[64] From the Space Shuttle program's inception, orbiters were transported back to their manufacturing factory in Palmdale, California for significant maintenance and overhaul. O'Keefe decided to move those operations to KSC, which posed a numbers problem. NASA had four orbiters, but only three orbiter hangars at KSC.

The proposed solution was to refurbish the hangar built for the X-34. The RLVH was already leased to USA for storage, but would require significant upgrades to process an orbiter. Certain legislators, despite sharing a partisan affiliation with Bush, saw this as an opportunity to redirect control of the authority away from the governor's office to NASA's space contractors. Rep. Bob Allen and Rep. Carlos Lacasa (R-Miami) proposed reducing the hangar's lease to $1 per year. They wanted the FSA to finance the hangar's $24 million retrofit, adding work stands, bridges, an overhead crane, and hydraulics. They introduced legislation to expand the FSA's governing board from seven to eight seats, and to mandate that four of those eight seats go to NASA contractors. OTTED Director Pam Dana told *Florida Today* that such appointments would create a conflict of interest should the board vote on contracts affecting their companies. Lacasa claimed that the unused RLVH was an example of what happens when "you don't have the right people on the board . . . The reason things like this happen is you have a board that doesn't necessarily respond to industry needs."[65] Lacasa's rhetoric overlooked the very recent historical evidence that the RLVH was built specifically to address the industry's need for a reusable space vehicle hangar, which NASA had projected only a few years before. The hangar had gone unused only because NASA and industry canceled the X-33/VentureStar and X-34 programs.

California's congressional delegation, upset that jobs would be lost at Palmdale, saw Florida's internal bickering as an opening to reverse O'Keefe's decision. Past efforts to consolidate operations at KSC had been blocked by the California delegation; now that the Florida governor's brother was president, some believed that nepotism was behind the decision. Rep. Dave Weldon (R-Palm Bay) warned all Florida parties that the bickering could create the perception that the authority wasn't ready to take on the responsibility of orbiter maintenance, providing an excuse to extend Palmdale operations. The feuding ended when all parties met in the governor's office to compromise. The final decision was to have all eight board members appointed by the governor, but four of them must "represent space industry." "Labor interests" and "minority interests" were also guaranteed one seat each. The remaining seats were at-large appointments. The lieutenant governor could cast any tie-breaking votes.[66]

While state politicians ground the sausage, the authority focused on completing the master plan mandated by SB 2540 passed in May 1999. It would be the Cape's first master plan in the half century that rockets had launched from there. Most construction projects, especially in the early years, had been to service ad hoc military R&D programs. A launch complex was built for a specific test project, with no thought for what might come later; when a project was complete, the site was often abandoned in place. The FSA now intended to issue a master plan looking fifty years into the future. Ed Gormel compared their challenge to the Wright Brothers trying to imagine the Boeing 747.[67]

The 690-page plan was released on August 28, 2002.[68] At the media event, Rep. Weldon said the idea for the plan came from the X-33 competition; local space officials realized they needed to plan for future launch sites. Gormel remarked, "We have to make sure that this document . . . does not become a doorstop. It's our responsibility to reference it, to use it."[69]

The plan was produced by a collaboration of the FSA, NASA, the USAF 45th Space Wing, and their prime consultant ZHA Incorporated. It had five chapters and eleven appendixes. The document was intended to be "a strategic blueprint for meeting and exceeding customer expectations in the 21st Century." The authors sought to develop "opportunities to respond to and lead changing market conditions over a 50-year planning horizon and beyond."[70]

The plan looked at studies attempting to forecast launch demand, both worldwide and specific to Florida, during the next ten years as well as to 2025. The general conclusion was, "The Florida launch level is likely to increase slightly over the current historical average of 20 launches per year . . . This assumes that launch prices and government priorities will remain stable over the forecast period. It also assumes there will be no emerging markets within the forecast period to change the types or priorities of launches."[71]

It's always hard to foresee unanticipated leaps of technology. The plan studied multiple scenarios for future reusable launch vehicles, including what it called multiple stage to orbit (MSTO) with a controlled or parachute return.[72] None of the concepts mentioned in this study ever emerged, but today's SpaceX Falcon boosters would fit into that category. The Blue Origin New Glenn lands at sea on a recovery ship. The ULA Vulcan may one day jettison its engines for recovery by parachute.

The 2002 study, lacking a crystal ball, created three "planning horizon" scenarios to forecast future launch demand. The first assumed 30 launches per year. The second assumed 44. The third assumed an astonishing 251 launches per year, with half of those horizontal launch vehicles.[73] The horizontal launch/landing industry has not yet emerged to the extent envisioned in 2002, although the Boeing X-37B flies classified military missions, and Sierra Space has a contract to launch its Dream Chaser orbital spaceplane to the ISS. Both launch vertically on a booster but land horizontally. Stratolaunch, essentially two 747 fuselages joined together by a common wing, was originally intended to launch and land at the KSC runway but now offers hypersonic flight test services in the Mojave Desert.[74] Swiss Space Systems intended to use an Airbus A300 to launch a reusable suborbital space plane from the KSC runway to deploy small satellites, but the company went bankrupt in January 2017 amid allegations of bank fraud.[75] Virgin Orbit, a modified 747 with a small rocket launcher, had already launched tiny "cubesat" payloads for NASA from the Mojave Desert and could have used KSC's runway in the future, but the company filed for bankruptcy in April 2023 after laying off 85 percent of its workforce and accumulating a debt of $1 billion. The 747 was acquired in May 2023 by Stratolaunch.[76]

To meet its forecasted horizontal launch demands, the plan recommended a new launch site north and west of KSC launch pads 39A and 39B. "This site produced the greatest and most flexible opportunity to accommodate supporting development associated with horizontal launch complex systems." For vertical launches, the plan recommended the development or renovation of launch sites along the coastline. "This arrangement creates the potential opportunity for commercial launch users to operate out of shared launch support facilities and shared launch complexes, thus reducing their cost basis through the joint amortization of fixed capital investments." KSC's Launch Complex 39 was "anticipated to transition to shared-use facilities as technological developments and operations begin to overlap one another and co-exist within facilities formerly designated for single programs."[77]

Anticipated tourism demand would be centered around an "expanded and improved" KSCVC. The Apollo Saturn V Center, just completed in 1996, would be closed with its assets relocated to KSCVC; the old site would revert to "conservation use." The USAF space museum, based at Launch Complex 26, would close and relocate adjacent to KSCVC. A parcel near the visitor hub would be allocated for a hotel and conference center.[78]

With the notion in mind of viewing both KSC and DOD operations as one spaceport, the plan proposed consolidating Cape Canaveral administrative operations to KSC's Industrial Area. The base's civil engineering, planning, and data processing offices would also move to KSC.[79]

The plan was ambitious in its scope. It shifted the focus from leasing disparate launch sites to offering services and capabilities. Looking back through the prism of history, though, few of its ambitions have been achieved to date. It intended to be a spaceport version of the master plans used by municipalities across the nation, which are typically enforced by a local zoning ordinance. Those plans are implemented by professional staffers, working for a public agency, guided by a representative body, chosen by a local electorate. Staff recommends action to the representative body, which approves, denies, or modifies the request. The spaceport plan had been mandated by the state legislature for integration with FDOT capital project requests, but otherwise had no body representing an electorate, only the advisory Spaceport Management Council. There was no local zoning ordinance to enforce the spaceport plan. To this day, the Cape's future is directed by the whims of members of Congress, the Florida state legislature, the Florida governor's office, the spaceport authority, and the NASA and DOD bureaucracies.

So how to implement its recommendations? The report dedicated only three pages to its denouement. Among their recommendations, the authors rather boldly suggested that NASA and the USAF surrender control over their properties—"the management structure that is developed should have the ability to at least partially remove the Spaceport from the federal budgeting process and provide access to private capital markets."[80] A congressional report, issued by the Commission on the Future of the United States Aerospace Industry in November 2002, had made a similar recommendation. That commission report recommended that "NASA should also consider turning over day-to-day management responsibilities for its field centers to the respective state governments, universities, or companies." (The commission did not recommend that the DOD do the same.)[81]

Implicit in all these recommendations was that the FSA, or a similar entity, have dominant and independent control over Cape operations—much as Chris Shove had intended when he crafted the Spaceport Florida Authority in 1989 based on the Port Authority of New York and New Jersey. But exempting the spaceport from "the federal budgeting process" was easier said than done; the master plan authors were asking Congress, the NASA and DOD bureaucracies, and the Florida legislature to surrender their political interests at the Cape. It was a nice idea in theory, but almost impossible to achieve. The idea has been explored many times, most recently as 2020, but without any significant progress.[82]

* * * * *

As the year 2002 ended, *Florida Today* space editor John Kelly was cautiously optimistic that the launch industry might be about to emerge from its multiyear funk. He wrote, "The commercial and military satellite business seems to have stopped dipping and started coming-out of a steep slip in demand." In his view NASA, under new Administrator Sean O'Keefe (a former director at OMB), was "getting its financial house in order." Kelly believed that the space shuttle's future was "sound for at least another decade and assures the next-generation manned spaceship will be launched from Brevard."[83] That "next-generation" ship was the Orbital Space Plane (OSP); with the X-33 program canceled, the OSP was the latest attempt to design a successor to the space shuttle. The OSP was intended to be "a multipurpose spacecraft that can perform crew rescue vehicle and crew transfer vehicle missions for the space station," according to an April 2003 Boeing press release. Boeing, Lockmart, and Orbital Sciences all had OSP contracts. Today's X-37B evolved out of OSP technology.[84] The space shuttle, meanwhile, would remain NASA's primary launch vehicle. Kelly wrote, "Now NASA plans to keep the intrepid shuttles flying through 2020 and possibly longer."

And then *Columbia.*

The Columbia Accident Investigation Board (CAIB) released its report on August 26, 2003. The authors pulled no punches:[85]

> Although an engineering marvel that enables a wide-variety of on-orbit operations, including the assembly of the International Space Station, the Shuttle has few of the mission capabilities that NASA originally promised. It cannot be launched on demand, does not recoup its costs, no longer carries national security payloads, and is not cost-effective enough, nor allowed by law, to carry commercial satellites. Despite efforts to improve its safety, the Shuttle remains a complex and risky system that remains central to U.S. ambitions in space. Columbia's failure to return home is a harsh reminder that the Space Shuttle is a developmental vehicle that operates not in routine flight but in the realm of dangerous exploration.

The report cited several causes for the accident, but notably called out NASA's institutional culture as resistant to change. It found that "the NASA organizational culture had as much to do with the accident as the foam." The authors believed that unless the culture changed, "we have no confidence that other 'corrective actions' will improve the safety of Shuttle operations. The changes we recommend will be difficult to accomplish—and will be internally resisted."[86]

The Space Life Sciences Lab in March 2004. Image source: NASA.

The board members found that the same resistance occurred after the *Challenger* accident. Management maintained "the internal belief that NASA was still a 'perfect place,' alone in its ability to execute a program of human space flight."[87]

In the days after the accident, the FSA sought to measure the impact on local and state business. Working with the Economic Development Commission of Florida's Space Coast (known as EDC), the FSA emailed about 200 Florida businesses to ask how the tragedy would affect them. The gathered data could be helpful if the federal government issued emergency loans and grants, even though the incident was not a natural disaster. Ed Gormel believed the impact would be less than from the *Challenger* accident, because the Florida space economy had diversified since January 1986 and was less dependent on government space programs. The FSA provided the state Agency for Workforce Innovation with a vendor list of 288 contractors, 111 of which were in Brevard County, that might be impacted if NASA were to ground the shuttle fleet. The agency prepared to offer assistance to any employees who might be laid off as a result.[88]

OTTED Director Pam Dana told the FSA to "slow down the horses," warning that any impacts were as yet unknown. "To create angst where you don't need to is bad practice," she said.[89] Dana's instincts were proven correct a month later when NASA Administrator O'Keefe testified before Congress in late February that no layoffs were anticipated due to *Columbia*.[90]

* * * * *

Ed Gormel announced in March 2003 his intention to retire.[91] He was succeeded by Winston Scott, a Miami native and retired NASA astronaut who was an associate dean at Florida A&M University/Florida State University College of Engineering. Lieutenant Governor Toni Jennings said, "Winston brings to us an instant credibility in the space community. Astronauts are the stars of the space program. He can give us an insider's point of view of how things go."[92]

Another launch site was added to the authority's portfolio in November 2003. The DOD signed over Launch Complex 47 to the FSA. LC-47 was to perform the same function as the authority's Cape San Blas site once had in the early 1990s. The simple pad had been used for decades to launch weather sounding rockets. Now it could be used by the State of Florida for educational projects as well.[93]

That same month, the new space station laboratory building was dedicated off the recently completed Space Commerce Way. The facility was no longer called the Space Experiment Research & Processing Laboratory (SERPL); now it was the Space Life Sciences Lab (SLSL). The SLSL was already being used by NASA contractors for early research into how astronauts might grow food in microgravity one day aboard the ISS. NASA and the FSA hoped the SLSL would become the anchor for an International Space Research Park.[94]

This assumed that NASA would continue to prioritize microgravity experiments aboard the space shuttle while the transportation system was also used to complete construction of the ISS. But *Columbia* had been lost. Because *Columbia* was heavier than the other orbiters and had an internal airlock, it wasn't cost-effective for ISS missions. *Columbia*, therefore, typically flew science missions, the ESA's Spacelab, classified DOD missions, and satellite deployments. A SPACEHAB commercial microgravity research module was lost on STS-107.[95] The remaining three orbiters would be needed to complete the ISS. Science missions, for the time being, were no longer a priority.

President George W. Bush delivered his Vision for Space Exploration (VSE) speech on January 14, 2004. He said that NASA's first goal would be "to complete the International Space Station by 2010 . . . We will meet our obligations to our 15 international partners on this project." Once the ISS was completed, "In 2010, the Space Shuttle—after nearly 30 years of duty—will be retired from service."[96]

When President Ronald Reagan proposed Space Station Freedom during his 1984 State of the Union address, he said that it would "permit quantum leaps in our research in science, communications, in metals, and in lifesaving medicines which could be manufactured only in space." He also promised that NASA would "invite other countries to participate" that shared American goals.[97]

George W. Bush changed the station's primary mission. NASA's second goal, he said, would be to develop a "Crew Exploration Vehicle" (CEV) that would be used initially to service the ISS after the space shuttle's retirement, but its main purpose "will be to carry astronauts beyond our orbit to other worlds." His third goal was to return crew to the moon by 2020, "as the launching point for missions beyond." The ISS would no longer emphasize "quantum leaps" in science as Reagan had envisioned. Bush said the station would be used to "help us better understand and overcome the obstacles that limit exploration," such as the long-term effects of space travel on human biology.

A budget chart released by the White House, which came to be known as the "Vision Sand Chart," showed that ISS would end in FY 2016, with its funding shifted into "exploration missions."[98] The station would be completed by 2010, only to be decommissioned five years later.

NASA's third goal, Bush said, was "to return to the moon by 2020, as the launching point for missions beyond." He projected that "extended human missions to the moon" would begin "as early as 2015."

The VSE pivoted NASA's focus away from ISS life sciences research to building hardware for what would soon be called Project Constellation. By April 2006, the SLSL had lost about one-third of the 150 scientists conducting microgravity research experiments, some of which were to be payloads sent to the ISS. Across the agency, about 240 university researchers had received project termination letters from NASA within the past year. It appeared that NASA had abandoned much of its microgravity research in fields such as cancer cells and protein crystal growth. Charles Quincy, chief of KSC's biology division who oversaw the SLSL, told *Florida Today,* "We can't afford to leave the building empty and do nothing."[99]

The National Research Council in 2006 found that NASA's new plans to prioritize ISS for crewed deep-space exploration research "no longer include the fundamental biological and physical research that had been a major focus of ISS planning since its inception." The loss of this microgravity research, intended for terrestrial benefits, "is likely to limit or impede the development of such technologies."[100]

The plans for Space Commerce Park fell apart in September 2005, as NASA and the FSA negotiated a lease agreement for the federal land on NASA property. NASA wanted final say on developer agreements and subleases, a fifty-fifty split of any profits generated by the park, and the power to expel tenants without compensation. FSA officials felt the restrictions put the state's investments at risk and created government obstacles for what was supposed to be a commercial park. Florida had already invested $30 million in building the SLSL and Space Commerce Way. From the authority's perspective, there wasn't enough interest from prospective developers or tenants to justify spending another $90 million on the project. Any significant demand for the SLSL's facilities was now years in the future.[101]

* * * * *

The seeds for the FSA's demise were planted in early 2005, when a Florida Senate staff report concluded that too many state agencies had "space" in their title or mission statement. A bill was drafted to create a Commission on the Future of Space in Florida, which would recommend to the state legislature reforms to help the state compete for the commercial space industry. SB 1026 died after the Senate and House versions could not be reconciled, but Governor Bush issued an executive order in June 2005 to create the panel anyway. The commission was to be chaired by Lieutenant Governor Toni Jennings.[102]

The commission held its first meeting on July 13, 2005, at the FSA in Cape Canaveral.[103] Three speakers were on the agenda, one of whom was "Internet entrepreneur—and now rocket builder—Elon Musk," as *Florida Today* described him. Musk told the commissioners that Florida needed more aerospace engineers and technicians to compete for commercial space ventures. SpaceX was headquartered at the time in an El Segundo, California warehouse, just south of Los Angeles International Airport. Musk said he chose to locate SpaceX in the Southern California labor market because it was the "nexus" of aerospace talent. Michael Griffin, who had become NASA administrator the month before, also spoke to the commission.[104] Griffin had accompanied Musk and others to Russia in 2002 in an ill-fated effort to purchase an old Soviet ICBM that Musk could use to send a small greenhouse to Mars, a symbol of Musk's vision to terraform the Red Planet. Their failure led Musk to conclude they could build a rocket themselves, which led to the creation of SpaceX.[105] The third speaker was astrophysicist and media personality Neil deGrasse Tyson.

The commission's report was released on January 18, 2006.[106] In summary, the report offered a "vision" of eighteen recommendations grouped into four objectives:[107]

- A skilled workforce, world-class education system, and leading-edge research capability.
- A focused business attraction, retention, and creation strategy.
- A world-leading infrastructure and launch environment.
- A new organization to provide statewide leadership, advocacy, and coordination for space-related policy.

Perhaps the commission's use of the word *vision* was an homage, if not outright linkage, to NASA's Vision for Space Exploration. The VSE was in response to the CAIB finding a "lack of an agreed national vision for human space flight." Executive Order 05-120 issued by Governor Bush credited the president with having "established a bold new vision" and declared that "Florida must ensure that the state is optimally positioned for NASA's new Vision for Space Exploration."[108] The commission report's executive summary refers to the VSE, then on the same page repeatedly uses the word *vision* in referring to the commission's own proposals for national leadership in "new commercial space opportunities and the integration of space, aeronautics, and aviation technologies. If realized, this vision will position Florida for sustained economic growth and prosperity for decades to come. Attaining this vision will require a strong public and private commitment to a world-class space and aeronautics industry."[109]

The report recommended that the FSA be merged with the FSRI and the Florida Aerospace Finance Corporation into a new organization. Its name? Space Florida. "This restructuring is intended to increase the overall visibility of space and aeronautics" in the state. "Space Florida would serve as the central organization for coordinating and communicating all space-related matters within the State." The new agency would be "structured as a public/private partnership similar to Enterprise Florida or Visit Florida." It would "have a contractual relationship" with OTTED, similar to its predecessor.[110] Linking Space Florida's success to NASA's VSE meant the agency's success or failure would rise or fall with the VSE. It also linked the political space successes of the Bush brothers.

The commissioners urged the state to pursue the Crew Exploration Vehicle. "The State should commit to partner with NASA and industry on necessary infrastructure and facility upgrades to enable the Cape to accommodate the CEV."[111]

The authors also saw commercial space tourism as "in its infancy, but it is not unreasonable to foresee a time when its status will be comparable to that of commercial aviation." The report cited an October 2002 Futron Corporation study, which concluded that "suborbital space travel is a promising market—Futron's forecast for suborbital space travel projects that by 2021, over 15,000 passengers could be flying annually, representing revenues in excess of US$700 million."[112] The report concluded, "The space tourism industry represents a major opportunity for Florida," relying on the Cape's history, the state's "world-renowned hospitality" and the belief that in the future suborbital technologies could be used for "high-speed movement of goods in suborbital trajectories."[113]

The report was released in Tallahassee in coordination with the governor's office releasing its proposed fiscal year 2007 budget. Much of Jeb Bush's space budget proposal was targeted at luring CEV assembly and maintenance to KSC. Of the $55 million the governor requested, $35 million would be spent to assist in refurbishment of a KSC facility for the CEV. Another $11 million would be for the creation of Space Florida.[114]

House Speaker Pro Tempore Leslie Waters (R-Seminole) introduced HB 1489, the Space Florida Act, on March 6. The final version of the bill passed the state legislature in early May, and was signed by Governor Bush on May 30. The legislation gave the governor and the commission much of what they wanted. "Space Florida will be the single point of contact for state aerospace-related activities with federal agencies, the military, state agencies, businesses, and the private sector."[115]

Among the bill's more significant directives for Space Florida:

- Complete a business plan by March 1, 2007.
- Enter into various agreements with other state agencies, including the Department of Education, the Department of Transportation, Enterprise Florida, Workforce Florida, and the Florida Commission on Tourism.
- Seek federal support and develop partnerships "to renew and upgrade the infrastructure and technologies" at KSC, CCAFS, and the Eastern Range.
- Pursue "the development of commercial spaceports in the state."
- Contract for SLSL operations.

The FSA's governance structure was replaced. Space Florida would be governed by a board of directors, led by Florida's elected governor or designee, who would be the chair. The board would have seventeen members appointed by the governor, including twelve from the private sector, one of whom would represent "organized labor." The appointees should "reflect the racial, ethnic, and gender diversity, as well as the geographic distribution, of the state."[116]

The old executive director position was eliminated. Space Florida would now be led by a president, appointed by the board of directors. (Similar state agencies, such as Enterprise Florida and Workforce Florida, were also led by a president.) The bill required the board to hire the first president by September 1, 2006.[117]

HB 1489 appropriated for fiscal year 2006–2007 $35 million from the state's General Fund "to be used for infrastructure needs related to the development of" the CEV. The bill also provided $3 million for Space Florida operations and $4 million for "implementation of innovative education programs and financing assistance for aerospace business-development projects."[118]

The $35 million for CEV infrastructure drew the attention of Lockmart executives. Company representatives announced on February 22, 2006, that CEV final assembly work would be at KSC—if they won the contract. (The other bidders were Boeing and Northrop Grumman.) KSC's Operations and Checkout Building, known as "the O&C," was originally used for processing Gemini and Apollo spacecraft in the 1960s. If the company won the contract, it would use the O&C for CEV assembly, servicing, and maintenance. Lockmart representatives estimated that the contract would create about 300 to 400 jobs for the Space Coast, with perhaps more to come. They specifically cited Governor Bush's incentive proposal as one reason for their choice.[119] NASA announced at the end of August that Lockmart had won the contract to build the spacecraft, now named Orion.[120] Lockmart took over the O&C in January 2007.[121]

With his days numbered as executive director, Winston Scott submitted his resignation in July.[122] He was succeeded on September 1 by Steve Kohler, the CEO of Winner Global Defense, a Pennsylvania-based manufacturer of security and anti-terrorism equipment.[123] Kohler from 1996 to 2001 had been executive director of an economic development task force called the Governor's Action Team for Pennsylvania Governor Tom Ridge.[124] Governor Ridge went on to serve as President Bush's first Secretary of Homeland Security. Jeb Bush, the president's brother, personally selected Kohler for the Space Florida job. Kohler lacked aerospace experience, a deficiency the Space Florida board acknowledged. "What was sought was a completely fresh outlook," Kohler told the *Tampa Bay Times*. "I'm coming with a wide open aperture."[125]

Space Florida's new governing body met for the first time on October 20, 2006. Steve Kohler represented the agency as its new president. Governor Bush presided over the meeting as chair of the board of directors.[126] Bush left office a little more than two months later, his second term completed. He was succeeded by Charlie Crist, a Republican serving as the state's attorney general.

* * * * *

Florida Today published two editorials at year's end advising Governor-elect Crist and the incoming Brevard County state legislature delegation how to represent the interests of the Space Coast in 2007.

The recommendations were unsurprising. The two columns were titled, "Pull Out the Stops" and "Fighting for Brevard." The newspaper's editors urged Crist to support Space Florida, and for Steve Kohler to work closely with the governor. The authors blamed "years of inaction from Tallahassee that's left Florida behind other states in attracting new space industries." They cited in particular the space tourism industry, claiming that "New Mexico and California are light-years ahead in attracting start-up space tourism firms."[127]

The columns noted that the clock was running out on the Space Shuttle program, with 5,000 to 8,000 jobs lost by 2010. Space tourism, however, turned out not to be the panacea, nor was the hangar built to lure spaceplanes such as the X-33 and X-34, nor was the space commerce park with the state's life sciences laboratory as an anchor tenant.

State legislators had every right at this time to be skeptical of NASA as a jobs generator. The X-33, the X-34, the OSP, the ISS, and Project Constellation were all major NASA programs they hoped might create jobs to replace those lost at the shuttle's retirement. None of them did, at least to any significance.

When the SFA was formed by the Florida legislature in 1989, it was with the expectation that the agency would sustain itself with revenue from the private sector it was formed to serve. That didn't happen. In the early 1990s, desperate to stay afloat, the SFA borrowed money from the state Department of Transportation, money that apparently was never repaid. As the economy improved, and services came online, the authority found some revenue streams but it remained largely dependent on state budget money and grants, wherever they might be found.

Over the eight years of Governor Jeb Bush's administration, in exchange for more stable and reliable state revenue sources, the agency gradually found itself losing its independence and its original name. It became accountable to the OTTED, and by statute was financially intertwined with FDOT, Enterprise Florida, and other state agencies. As Bush left office, his last significant space legislation created Space Florida, with a new directing body led by the governor himself. Bush successfully convinced the legislature to invest $35 million for upgrades at the Cape to lure the CEV, renamed Orion. But Orion would not launch until the end of 2014, and its second flight would not be until the end of 2022. As of this writing, it still has not flown with a crew, now projected for no earlier than April 2026.

The problem was far more fundamental than described by the journalists at *Florida Today*. Seventeen years had passed since the birth of the SFA, but the state's space enterprise was still beholden to the whims of federal agencies that took marching orders from the White House and Congress. Commercial entrepreneurs had proven to be an unreliable lot, unwilling to significantly risk capital without government subsidies to sustain their business model.

One man was about to change all that.

7

The End of the Beginning

When *Florida Today* subscribers opened their newspapers the morning of December 13, 2002, they saw the front-page headline, "Company to Launch Small Rockets for Less Cost."[1]

> A dot.com tycoon has picked Cape Canaveral as a launch site for a new breed of rockets that could lift half-ton satellites for about half the current market price, state government officials said Thursday.
>
> Space Exploration Technologies Corp., or SpaceX, plans to launch its proposed Falcon rocket from a state-owned pad at a Space Launch Complex 46, Gov. Jeb Bush and Florida Space Authority officials said.

In his 2015 book *Elon Musk,* Ashlee Vance detailed the circumstances that led to the creation of SpaceX in June 2002. The FSA agreement six months later was not mentioned; Musk's focus at the time was to stand up a company capable of building a small rocket called Falcon (no appended number indicating the number of engines) he could evolve into the "Southwest Airlines of Space." SpaceX would build a better and cheaper engine than its putative competitors, propelling a rocket launched from a mobile vehicle at any site a customer so desired.[2]

By December 2002, SpaceX had little more than an old warehouse in El Segundo, California, but LC-46 made sense for both the FSA and the space startup.

From the authority's perspective, LC-46 would save SpaceX the $20 million to $30 million it might otherwise cost the company to refurbish an existing launch pad.[3] The FSA had already invested $8 million in the mid-1990s modifying LC-46 to support multiple launch systems.[4] But since its opening in May 1997, the pad had supported only two launches—an Athena II in 1998, and an Athena I in 1999. The FSA would welcome SpaceX or any other customer for LC-46.

By April 2003, Musk was telling the media he had his first customer for the Falcon, "a government agency he won't identify" that could launch as early as December that year from Vandenberg Air Force Base in California, which would serve as an initial launch test site before moving to Cape Canaveral.

The customer was later revealed in October to be the DOD Office of Force Transformation's TacSat-1 data communications microsatellite.[5]

This was one of the earliest examples of "Elon Time," a term that became popular in social media around the mid-2010s. It describes Musk's overly optimistic projections for when a product will be available.

This author interviewed an individual who said he was in the first Cape meeting Musk had with FSA and USAF personnel. His recollection was that Musk complained about the Air Force having too many regulations, and claimed SpaceX would launch with far fewer personnel than Boeing or Lockheed Martin. The individual said that the USAF staff "were willing to work with him but were ticked off at his obnoxious attitude."

The Wallops, Virginia, Space Flight Center had also courted SpaceX. As had the FSA, Wallops failed to fulfill its own visions for a robust commercial launch market. When Musk selected Vandenberg in April as its initial launch test site, Wallops wrote off SpaceX as another missed opportunity.[6] Musk told the *Lompoc Record* that Vandenberg personnel had "gone the extra mile" to assist his fledgling firm. SpaceX selected Vandenberg's Space Launch Complex 3 West, an old Atlas launch pad, to test and launch its "mostly reusable vehicle." Musk told the *Record,* "The first stage is recovered. It's a parachute-to-water landing in the same way that the shuttle boosters are recovered."[7]

After technical delays and environmental reviews, SpaceX finally conducted a static test fire of its Falcon 1 (one engine) at SLC-3 West on May 27, 2005. By then, Musk had become frustrated with his military landlord's bureaucratic delays. The USAF would not allow the Falcon 1 to launch until after a Titan IV with a classified military reconnaissance satellite payload launched from an adjacent pad. Air Force officials feared that a Falcon failure could somehow damage the Titan and its payload. Musk told the *Lompoc Record* that it was "a very tiny risk, but they're unwilling to take even a tiny risk." Loath to wait any longer, Musk decided to relocate his launch test to the Kwajalein Atoll in the Marshall Islands.[8]

Vance released a sequel in 2023, *When the Heavens Went on Sale.* He wrote that SpaceX at this time did have some supporters within the Pentagon. USAF General Simon "Pete" Worden was an Air Force iconoclast. Worden believed the military should develop a "responsive space" capability, quickly deploying small satellites in response to a perceived threat. Musk pitched his Falcon 1 to Worden, but "wanted to know if we would use it." Worden connected Musk to DARPA, urging them to give SpaceX a research contract.[9] In November 2003, SpaceX was one of nine companies awarded contracts between $350,000 and $540,000 to develop conceptual designs for an "affordable and responsive space lift capabilities" program coincidentally named FALCON (Force Application and Launch from CONtinental United States).[10] Adding SpaceX to their ranks gave the company some credibility; a DARPA website credits SpaceX with helping to develop the FALCON technology, and that the program helped to certify the company for launching DOD payloads.[11]

Launching from Cape Canaveral was viewed by the company as an inevitable progression, but not an immediate one. Tim Buzza, who was the fifth SpaceX employee hired by Musk, said in a 2013 NASA Oral History interview that the company chose to go to Kwajalein because the Falcon 1 was "an initial test program" that "wasn't well suited to go to Cape Canaveral."[12]

Musk spoke at the National Space Club Florida Committee's monthly luncheon in Cocoa Beach on July 8, 2003. He told the assembled that his Falcon 1 could launch from Cape Canaveral as soon as 2004 (Elon Time). And possibly within the next 3.5 years, SpaceX might have a more powerful Falcon, one capable of launching humans someday.[13]

That larger rocket may have been the Falcon 5 (five engines). On December 4, 2003, Musk staged a promotional event at the National Air and Space Museum in Washington, DC. A police escort led a prototype Falcon 1 to a parking place across from the museum, in front of the FAA. Standing at a podium placed at the nose end of the mockup, Musk announced the "follow-on" to the Falcon 1, "called the Falcon 5." He predicted its first launch would be in 2005[14] (Elon Time).

SpaceX announced in May 2004 its first commercial customer for the Falcon 5. Bigelow Aerospace was a Las Vegas space startup founded by Robert Bigelow, owner of the Budget Suites of America hotel chain. Bigelow had licensed NASA's canceled TransHab expandable habitat technology, intending to deploy commercial space habitat platforms in low Earth orbit. His prototype payload, Genesis Pathfinder, had a Falcon 5 launch date at Vandenberg for November 2005. A Bigelow representative told Space.com, "Money has changed hands."[15]

Elon Time cost SpaceX the Genesis contract. By March 2005, Bigelow had contracted with a Russia-Ukraine-Kazakhstan company, ISC Kosmotras, to launch his Genesis I and II prototypes on decommissioned Soviet Dnepr ICBMs. Genesis I launched in July 2006. Genesis II launched in June 2007.[16] Falcon 5 never launched.

Musk's plans for a more powerful rocket led SpaceX to walk away from LC-46 in favor of Launch Complex 36, which since the early 1960s had supported the Atlas-Centaur. LC-36 was about to be decommissioned; according to a January 30, 2005, *Florida Today* article, SpaceX intended to be operational at the site by the end of 2006.[17]

But by the fall of 2005, SpaceX was planning an even more powerful booster, the Falcon 9 (nine engines).[18] The upgrade to nine engines was driven by NASA opening its Commercial Crew/Cargo Project Office (C3PO). The November 7, 2005, NASA press release said that the agency would soon "issue a draft solicitation requesting commercial service demonstrations for space station crew and cargo delivery and return."[19] In his 2013 NASA Oral History interview, Tim Buzza recalled that the Falcon 5 would not have adequate thrust for sending the SpaceX cargo Dragon spacecraft to the ISS. Five engines wouldn't do the trick, but nine engines would.

This also drove SpaceX to look yet again for another pad at the Cape. LC-36 at that time wasn't designed to handle a booster with the thrust projected from a Falcon 9. The company turned its attention to another recently deactivated pad, Launch Complex 40, most recently occupied by the Lockheed Martin Titan IV. A December 28, 2005, *Florida Today* article quoted Gwynne Shotwell, at the time the company's vice president of business development, as saying the company expected to launch commercial cargo payloads from the Cape if it won a contract. LC-40 was a possibility, although NASA was also looking at the pad for the CEV program, which evolved into Constellation's Orion.[20]

The demise of the FSA coincided with the award of a commercial cargo contract to SpaceX. An August 23, 2006, *Florida Today* editorial reminded Governor Bush that the incoming Space Florida leadership will "have to hit the ground running." Regarding the commercial cargo awards, "there's no guarantee the rockets would be launched from the Cape on their missions." The new agency would need to "create a more business-friendly climate that would make it easier for such firms to use the Cape, and not be hog-tied by bureaucratic red tape."[21]

In April 2007, the Air Force Space Command announced it had granted SpaceX a five-year license to use LC-40 for the Falcon 9. (The license was nonexclusive; technically, the USAF could order SpaceX to share the pad with other licensees.)[22] The SpaceX press release announcing the Falcon 9 had said that "SpaceX still plans to make Falcon 5 available in late 2007." By the end of February 2007, though, the Falcon 5 had disappeared from the SpaceX website, replaced by the "Falcon 9 Heavy."[23] The Falcon 5 had become an evolutionary dead end.

* * * * *

SpaceX broke ground on its Launch Complex 40 renovation on November 1, 2007. That same day, the year-old Space Florida hosted a summit conference of commercial space industry leaders. Among the speakers were Elon Musk, XPRIZE Foundation founder Peter Diamandis, XCOR Aerospace CEO Jeff

SpaceX CEO Elon Musk speaks at the Space Florida groundbreaking ceremony for SpaceX at Launch Complex 40 on November 1, 2007. Image source: NASA/George Shelton.

Greason, and Bigelow Aerospace corporate counsel Michael Gold. The keynote speaker was retired astronaut Susan Helms, who was commanding the 45th Space Wing.[24]

The event coincided with the opening day of the World Space Expo, staged November 1–4 at KSCVC. That event was also hosted by Space Florida, along with Florida's Space Coast Office of Tourism. The idea was to boost tourism while also promoting NASA's upcoming fiftieth anniversary.[25]

The year 2007 began with the fledgling agency still learning how to spread its wings, never mind how to fly.

Newly hired Space Florida president Steve Kohler had been selected for the job by Republican Governor Jeb Bush, but Bush had left office on January 2, 2007, succeeded by another Republican, Charlie Crist. As was the tradition with his predecessors, Crist withdrew the names of pending appointees named by Bush, including the twelve private sector members of the Space Florida board of directors. The new governor didn't appoint his first four representatives to that board until June, finally creating a quorum so the board could act on pending agenda matters. The remaining eight were not appointed until August.[26]

A June *Florida Today* editorial called out Crist's "foot-dragging," not just for the delay in naming new board members, but also for his veto of an aerospace investment fund championed by Rep. Bob Allen (R-Merritt Island) that had passed the legislature the month before.

During the 2006 legislative session, Allen had promoted the idea of an "Aerospace Workforce Challenge Fund" to protect and attract Florida aerospace jobs. He foresaw a $500 million pot that would expand and upgrade Florida aerospace facilities: $250 million would come from the state legislature, while labor unions said they would provide a matching $250 million. The proposal died in the House, which was contemplating the Space Florida Act (HB 1489) discussed in Chapter 6.[27] Allen revived the idea in the 2007 legislative session, this time calling it the "Florida Energy, Aerospace & Technology Fund," or FEAT. Rather than targeting just the aerospace industry, Allen broadened its focus to include the clean energy and high-tech fields; the language was included in bill HB 7123. With a new governor in Tallahassee, a fellow Republican, Allen thought his proposal might have better luck. "Now we have enlightenment," he told a reporter.[28]

Crist turned out not to be as "enlightened" as Allen had hoped. Although it passed the legislature, Governor Crist vetoed HB 7123 for reasons that had nothing to do with its aerospace provisions.[29]

"Crist's foot-dragging on naming new board members and whacking the job-creation fund are the latest examples of why Florida is choking on the competition's dust," opined a June 27, 2007, *Florida Today* editorial.[30]

Space Florida officials, meanwhile, had been meeting with NASA officials to plan for the end of the Space Shuttle program. Project Constellation was designing two boosters. The initial Ares I would be a single-stick design evolved out of a shuttle solid rocket booster. A more powerful Ares V would follow sometime later; it would have a center core evolved out of the shuttle's external tank, with two evolved SRBs strapped to either side in a shuttle-like configuration. Atop either booster would be the Orion crew vehicle, conical in shape, an evolved version of the Apollo capsule. The Ares V could also be used for launching cargo. NASA hoped to have Orion operational on an Ares I booster by the end of 2014, assuming no technical delays or congressional budget cuts.[31] That meant a minimum four-year gap without crewed spaceflight at KSC. *Florida Today* reported on May 12, 2007, that an estimated 5,000 to 8,000 KSC jobs would be lost at the end of the Space Shuttle program; including indirect employment within the county, the job loss could be 12,000 to 19,000.[32]

The business plan required by Space Florida's 2006 enabling legislation was released to the public in March 2007. It called for the State of Florida to obtain a horizontal launch license from the FAA, still believing that commercial space ventures such as the Virgin Galactic spaceplane were the future.[33]

Sir Richard Branson had teased that Cape Canaveral was the next logical destination for Virgin Galactic after New Mexico's Spaceport America. During a February 2006 visit to KSC, Branson told *Florida Today*, "At some stage, we're going to need an East Coast base. If you have to set up on the East Coast somewhere, I can't think of anywhere better than Cape Canaveral."[34] *The Governor's Commission Report* released one month earlier had urged the state to "move forward with efforts to develop a **commercial spaceport** targeted initially at horizontal launches and located separately from the Federal lands at the Cape. Structured as an operating authority similar to an airport, such a facility could serve space tourism and other commercial operators."[35] (The **bold** emphasis was in the original.)

To help promote its vision of horizontal launch experiences, Space Florida cosponsored a flight by astrophysicist Stephen Hawking aboard a Zero-G aircraft in April 2007. The Boeing 727 flew parabolic sorties; during each nosedive, passengers briefly experienced weightlessness. Freed from his wheelchair, Hawking was able to soar in microgravity.[36]

Space Florida and Zero Gravity Corporation announced on March 19 the creation of the Florida Microgravity Education and Research Center to provide microgravity educational opportunities to Florida teachers and students.[37] Space Florida signed two contracts with Zero-G worth $1 million to create the center, but the contract was terminated in 2009. According to a February 15, 2009, *Orlando Sentinel* article, the two entities disagreed on the program's training curriculum. A former Space Florida executive told the *Sentinel* that the agency lacked the staffing and capability to implement the program.[38]

Seeing little progress made by Space Florida, the *Florida Today* editorial page scolded the agency. A September 7, 2007, column complained that the agency didn't attend a Brevard County Commission session strategizing a response to the pending shuttle job losses. Steve Kohler was targeted in particular, claiming, "Neither he nor the agency has anything of substance to show for their work." The editorial alleged that Space Florida had a "poor" record of coordinating with local government and economic development entities. It also blamed "no substantive coordination between local officials, key Tallahassee policymakers and Florida's congressional delegation"—a common complaint heard since the creation of the first Spaceport Florida Authority in 1989.[39]

Another editorial, published October 2, 2007, charged that state leadership was ineffective. "That puts more of the onus on [Brevard County] commissioners to press ideas for moon-Mars jobs, including state incentives to lure new commercial space companies here."[40]

Despite the criticisms, a nonscientific online poll conducted on the *Florida Today* website found that 64 percent of 133 respondents believed that Space Florida was "working hard enough to create needed jobs for the post-shuttle era."[41]

On October 9, 2007, two Space Coast politicians traveled to Tallahassee to meet with Governor Crist. US Rep. Tom Feeney (R-Orlando) and State Rep. Thad Altman (R-Melbourne) told the governor that he needed to provide retraining and business incentives for those in the workforce who would lose their jobs once the Space Shuttle program ended in 2010. "The Brevard County folks are really attentive to this," Feeney said. "I don't want to say panicked, but they are very concerned."

"It's on the precipice of panic," Crist replied, according to *Florida Today*. With the 2008 presidential primaries approaching, Crist assured them he would press the candidates about the importance of space to the state economy.[42]

* * * * *

By the fall of 2007, Republican and Democratic presidential candidates were already holding televised debates. Eight Republican candidates participated in an October 21 Fox News televised debate held in Orlando. Space, government or commercial, did not come up as a topic.[43]

The candidates spent much of that night criticizing Senator Hillary Clinton (D-NY), their presumptive opponent. But other Democrats sought their nomination, including first-term Senator Barack Obama (D-IL), a onetime community organizer seen by many as a rising star within the party. Clinton had the benefit of a battle-tested political machine inherited from her husband Bill, the forty-second US president, but had also inherited the political baggage that came with the Clinton surname. Obama, who had turned forty-six on August 4, appealed to a younger generation looking for new leadership without a tarnished track record.

National polls favored Clinton. During the 2007 Democratic candidate debates, Clinton's mastery of the issues was evident, in contrast to Obama's inexperience and lack of detail.[44] A *Washington Post*–ABC News poll released October 1 showed that Clinton was favored by 53 percent of Democratic-leaning respondents, while Obama was favored by only 20 percent. Respondents judged Clinton "the strongest leader" by 61 percent to 20 percent, and a better candidate to handle the economy by 56 percent to 17 percent.[45] A Gallup poll released December 18, 2007, found that Clinton was favored by 45 percent of Democratic or Democratic-leaning respondents. Obama was second with 27 percent.[46]

Presidential campaigns are expected to issue policy papers detailing the candidate's position on a topic. Clinton's campaign attracted Lori Garver, a onetime executive director of the National Space Society who left that post in 1996 to become a special assistant to NASA Administrator Dan Goldin, and was eventually appointed as associate administrator for the NASA Office of Policy and Plans. After attending campaign events for Clinton and Obama, Garver chose to volunteer for Hillary's campaign in May 2007. Garver wrote in her 2022 biography *Escaping Gravity* that Clinton gave her "a more fulsome answer and discussion" than did Obama to her question about the candidate's future for NASA. She developed space policy documents for the campaign and represented her candidate at space policy events.[47]

Clinton delivered a May 31 speech to the Silicon Valley Leadership Group in which she proposed a twenty-first-century "innovation economy." One of her ideas was "competitive prizes to encourage innovation," which would have sounded familiar to any NewSpace believers in the crowd. Clinton warned of a coming "skill shortage" as the Apollo-era engineering generation reached retirement age.[48] On October 4, 2007, the fiftieth anniversary of the first Sputnik launch, the Clinton campaign released a scientific innovation policy statement that promised to "enhance American leadership in space." A Hillary Clinton administration would pursue "an ambitious 21st century Space Exploration Program, by implementing a balanced strategy of robust human spaceflight, expanded robotic spaceflight, and enhanced space science activities." The policy statement also promised to fully fund NASA's earth sciences program, establish a space-based climate change initiative, and reverse funding cuts to NASA's aeronautics research.[49]

A subsequent statement provided to *SpaceNews* magazine by Clinton campaign staff pledged "a successful and speedy transition" from shuttle to "a next-generation space transportation system that can take us back to the Moon and beyond."[50] The pledge did not specifically mention that Constellation would be the system chosen to achieve that goal.

The Obama campaign, meanwhile, had no Lori Garver. It had no space expert at all, until an inadvertent gaffe forced the interim appointment of Steve Robinson, a legislative assistant in the senator's office who had a doctorate in molecular biology and had taught high school science classes in Oregon. Robinson had joined Obama's office through an Albert Einstein Distinguished Education Fellowship designed to help science teachers become involved in public policy. After the fellowship ended, Robinson remained in Obama's office to work on education issues. During his off-duty hours, Robinson worked on Obama's presidential campaign. His primary responsibility was developing education policy.[51]

In late November 2007, the Obama campaign released an education policy paper titled, "Barack Obama's Plan for Lifetime Success Through Education."

The fifteen-page paper concluded with a final paragraph titled, "A Commitment to Fiscal Responsibility." Within that paragraph was a sentence that, within space punditry circles, was considered a major gaffe. It would haunt both his campaign and his administration. The sentence began:[52]

> The early education plan will be paid for by delaying the NASA Constellation Program for five years.

In an April 2023 interview with this author, Robinson said he hadn't been involved with the funding side of the education policy. It was only after the notoriety of the Constellation recommendation that he was asked to act as interim space policy adviser until a permanent appointee could be found. According to another source in the Obama campaign, the idea to include Constellation funding probably came from a list of "pay fors" compiled from various think tanks and congressional committee staffers. Candidate Obama most likely didn't specifically suggest delaying Constellation funding but would have been generally aware "at a high level" of what government programs made the "pay for" list.

The space pundit community reacted quickly, and not at all kindly.

"Obama Pits Human Space Exploration Against Education," read the headline on a November 21, 2007, *Wired* magazine website article. "A five year delay of the Constellation program would leave the United States government without its own human launch capability for nearly ten years."[53] The *Space Politics* blog received 112 comments, some supportive, others critical.[54] In a column titled, "Clinton Favors Future Human Spaceflight," *Washington Post* staff writer Marc Kaufman opined that Obama "made clear that he is not enamored with NASA's effort to build a new spacecraft to take astronauts to the moon and beyond." Kaufman quoted a Clinton campaign spokesperson: "Senator Clinton does not support delaying the Constellation program and intends to maintain American leadership in space exploration."[55]

On January 6, 2008, the Obama campaign produced a campaign position paper titled, "Barack Obama's Plan for American Leadership in Space." (A PDF copy was obtained by this author from an Obama campaign staffer.) The document stated that Obama would support development of "the Orion Crew Exploration Vehicle (CEV) carried by the Ares I Launch Vehicle . . . The CEV will be the backbone of future missions, and is being designed with technology that is already proven and available." The statement also supported completion of the ISS, robotic probes, "space platforms" to monitor climate change, science research, and surveillance satellites "to strengthen national security."

Steve Robinson recalls that he did not write the January 6 policy paper, nor does he know who did. His records show him making edits on subsequent versions through May 2008. He said that it was typical for the policy papers to pass "through many hands, many edits, many pens." The Obama policy team had the philosophy that many participants providing input, some with "out of the box" thinking, might produce fresh ideas.

Unlike Clinton's Lori Garver, the Obama campaign did not have a one-stop space policy expert. This led some to believe that Obama was flip-flopping, when in fact the policy had gone through many hands responding to inquiries and pressure from the space advocacy community, rather than producing a proactive comprehensive policy managed by an expert.

The Florida presidential primaries were scheduled for January 29, 2008. The Republican-majority state legislature had voted in 2007 to advance their primaries from early March to late January, hoping Florida would be more prominent in the nomination process. The national Democratic and Republican parties didn't approve. The national Democratic party stripped Florida of its delegates to their convention, while the national Republican party chose to cut Florida's delegation by half.[56]

The major Democratic candidates, as required by their national party, boycotted Florida's primary season, although surrogates campaigned on their behalf.[57] The national Republican party did not forbid its candidates to campaign in Florida. Some Republican candidates visited Brevard County, but only former New York City Mayor Rudy Giuliani offered any space specifics. An opinion column with Giuliani's byline was published January 26 by *Florida Today.* Giuliani endorsed sending Americans back to the moon and then to Mars, and the commercial cargo program. "We will expand private-sector access to Cape Canaveral launch pads," he wrote. "To help prepare astronauts for longer stays in space, we will fund the Space Life Sciences Lab," clearly an appeal to the locals who were desperate to find tenants for the SLSL. He concluded, "From day one, a Giuliani administration will strengthen America's leadership in space."[58] Rep. Thad Altman, Giuliani's Florida campaign cochair, told *Florida Today,* "That's going to help us bridge the gap."[59]

Space policy didn't seem to be a significant issue, either to the candidates or to the voters statewide. A January 23, 2008, *Florida Today* article commented that Clinton was the only major candidate to offer a detailed space policy until Obama and Giuliani had in recent days. "Space exploration isn't a top priority for the candidates, partly because it does not swing many voters and also because of the crowded primary cycle," the article stated.[60] A space policy did not appear on the Senator John McCain (R-AZ) campaign website until just before the primary's election day; it was a single paragraph that claimed McCain had "sponsored legislation authorizing funding consistent with the President's vision for the space program." It stated that McCain had been "a staunch advocate for ensuring that NASA funding is accompanied by proper management and oversight to ensure that the taxpayers receive the maximum return on their investment."[61]

Hillary Clinton won Florida's Democratic primary with 49.8 percent of the statewide vote. Barack Obama finished second with 32.9 percent. No other candidate had above 15 percent.[62] In Brevard County, Clinton finished ahead of Obama, 50.6 percent to 29.4 percent.[63]

John McCain won the Florida Republican primary with 36.0 percent of the votes, followed by former Massachusetts Governor Mitt Romney at 31.0 percent. Giuliani finished third at 14.7 percent.[64] In Brevard County, McCain won 35.9 percent, Romney 31.0 percent, and Giuliani 14.0 percent.[65]

John McCain clinched the Republican nomination after March 4 primary victories.[66] The Democratic race at that time was far from decided. By the end of March, however, Obama had taken the delegate lead among Democratic candidates, 1,623 to 1,499, with 2,024 needed to win. Clinton was resisting growing pressure from within the party to bow out and cede the nomination to Obama.[67]

Sometime in March, the policy team sent Obama an internal campaign memo responding to the candidate's request "to change your position on delaying the NASA Constellation program," and what might be the consequences. (A campaign staffer provided this author brief relevant excerpts from the memo, although not the memo itself.) This confirms that Obama was certainly aware of the campaign gaffe and was thinking about how to fix it.

If he was, the GAO gave him reason not to do so.

On April 3, the GAO released a report as part of a staffer's testimony that day to the House Subcommittee on Space and Aeronautics. The findings cast pessimism across the program's future:[68]

> There are considerable unknowns as to whether NASA's plans for the Ares I and Orion vehicles can be executed within schedule goals, as well as what these efforts will ultimately cost. In fact, we do not know yet whether the architecture and design solutions selected by NASA will work as intended.

The report noted that NASA was still "working through significant technical risks," such as oscillation by the solid-fueled Ares I first stage that could "cause unacceptable structural vibrations."

In late May, the National Space Society held its annual International Space Development Conference, at the Capital Hilton Hotel in Washington, DC. On May 30, CNN journalist Miles O'Brien moderated a space policy panel with advisers representing McCain, Obama, and Clinton. McCain was represented by Floyd DesChamps, a mechanical engineer who was a senior staffer on the Senate Committee on Commerce, Science and Transportation. Steve Robinson represented Obama. Lori Garver represented Clinton.[69]

Their views largely coincided, with no more than nuances in their responses. DesChamps commented that McCain would monitor NASA spending, noting that the senator had supported a spending cap on ISS construction. Garver emphasized Clinton's early detailed release of a space policy and lifelong interest in human spaceflight.

O'Brien asked Robinson about the early proposal to delay Constellation funding. Robinson's reply described a more holistic approach by his candidate, noting that the proposal was in the education policy paper, not the space policy paper. "This is within the context of my boss, Senator Obama, has come out and said that overall he believes that science funding nationally needs to double, and that includes an increase in NASA funding over the years."

Later in the event, O'Brien suggested to Robinson that delaying Constellation would cost a generation of schoolchildren the inspiration of watching humans go to Mars, the way the Apollo program inspired O'Brien's generation. Robinson, a high school science teacher, said that what inspired the Apollo generation might not inspire the current generation, for whom human spaceflight was no novelty. Robinson reminded him that NASA was already on Mars with robotic craft.

"We shouldn't limit what inspires us to just exploration by humans," Robinson said.

"There aren't any high schools named after robots, are there," O'Brien noted.

"No, but they are at high schools ***building*** robots," Robinson replied.

A week after the forum, Hillary Clinton conceded the race on June 7. She urged her supporters to "do all we can to help elect Barack Obama as the next president of the United States."[70]

* * * * *

Despite the "widely held view" that he would endorse Hillary Clinton, Florida Senator Bill Nelson (D-FL) withheld his early endorsement of any Democratic candidate. "Clearly the nominee is going to need Florida in November, so I'm going to wait and see how they treat us in the meantime," Nelson told the *St. Petersburg Times* in early January.[71]

His neutrality didn't last long. The evening of the January 29 Florida primary, Nelson appeared on stage with Clinton at the Signature Grand Ballroom in Davie, a Fort Lauderdale suburb, to endorse her.[72]

As Clinton fell behind in the Democratic delegate count, Nelson sought to find a solution that would allow Florida's delegates to be counted. At first, he suggested a redo, a vote-by-mail Democratic primary that would comply with the national party's rules, but found little support.[73] His next proposal, typical of his tendency to seek compromise, was the Republicans' solution. That compromise would give Clinton only half her delegates' votes, but half was better than nothing. Nelson told the *Orlando Sentinel* that he had approached both Clinton and Obama on the Senate floor with the compromise; he said both candidates at least listened.[74]

That was ultimately the solution the national party adopted. At a May 31, 2008, Democratic National Committee Rules and Bylaws Committee meeting in Washington, DC, the Florida state delegation presented its case to have all its delegates seated at the convention. Arguing on his state's behalf, Nelson said he would prefer that all of Florida's delegate votes count but would accept the 50 percent compromise.[75]

Despite endorsing Clinton, Nelson sought to influence the space policies of whomever the eventual nominees would be of both major parties. On February 20, he told a local nonpartisan political club, "It's going to be up to us to educate the two candidates about space."[76] At a Brevard County–sponsored workshop on April 28, Nelson urged those present to pressure the remaining presidential candidates. "These candidates are going to be here and we need to work 'em over," Nelson said. "We need to tell them what's been wrong thus far and how to change it." Former Rep. Robert Walker (R-PA), now a lobbyist for Brevard County, said that all three candidates were beginning to question if Constellation had been a mistake.[77] Perhaps this was due to the critical GAO report released earlier in the month.

By late May, as the two Democratic candidates raced to clinch the nomination, both Obama and Clinton campaigned in Florida. Obama held a town-hall style rally May 21 in Kissimmee, an Orlando suburb. *Florida Today* reporter Susanne Cervenka wrote that Obama "promised to work with NASA officials to develop a focused mission for the future of the space program . . . Obama said he would fund a strengthened space program, including the Orion program."[78]

That was the closest either came to Brevard County that month. In a May 26 article, *Florida Today*'s Patrick Peterson lamented that "Brevard County doesn't offer what presidential candidates want: big venues and a concentrated population." Describing the county as "white, older, Republican and unlikely to change," the Space Coast offered little to attract the Democratic candidates during the primary season's denouement.[79]

Hoping to draw more attention to the imminent loss of space shuttle jobs in Florida, Nelson scheduled for June 23 a field hearing of the US Senate Committee on Commerce, Science, and Transportation. The hearing was held at the Canaveral Port Authority.

Starting on June 18, advertisements appeared in *Florida Today* promoting a rally in front of the Canaveral Port Authority before the hearing. The ads were placed by a group called "Link to Launch," describing itself as "a movement started by people of Florida's Space Coast to raise the awareness of the nation and our policy makers that space needs to be a priority for America." The ad stated, "*Florida Today* proudly supports this community effort."[80]

In a series of June 2023 emails, space activist Dale Ketcham confirmed that he was one of the organizers of Link to Launch. At the time, Ketcham was director of the Spaceport Research and Technology Institute at the University of Central Florida. The core organizers were people who also organized the annual Space Day every year at the Florida state capital in Tallahassee, an event that promoted and advocated for Florida's space economy.[81]

According to *Florida Today*, Link to Launch hoped for 6,000 people to attend the rally. Only 1,500 "worried space workers" showed up, according to the newspaper; many employees couldn't get leave from work. Reporter Jessica Raynor wrote that Bill Nelson "gave them what they wanted—a rallying cry that echoed their concerns about their uncertain future."[82]

Link to Launch disappeared as quickly as it appeared. The Internet Archive shows the group's website was never updated after the event.[83] According to Dale Ketcham, the rally's goal was to remind NASA's administrators participating in the field hearing that Florida's workforce was just as important as those in other states about to be impacted by the transition from Shuttle to Constellation.

Florida Today watchdog reporter Matt Reed wrote in his column that he'd spoken with Nelson after the event. Nelson said that he'd been discussing Constellation with Senator Obama. "To Barack's credit, he has a very supple mind," Nelson said. He'd reminded Obama that children might embrace science if "you get them hooked on spaceflight."[84]

The event was held in the early months of what was becoming known as the "Great Recession." The US economy had entered a recession in December 2007. Home prices were in free fall, leading to more mortgage defaults and greater losses for those holding such securities. American gross domestic product was in decline, while unemployment was on the rise.[85]

The EDC's Lynda Weatherman submitted testimony at the hearing that included an analysis which forecast Brevard County would lose almost 13,000 jobs in the next three years due to the transition from the Space Shuttle program to the Constellation program—coming to be known locally as "the gap"—including 6,400 KSC jobs, with a cumulative lost income of $650 million. Space Florida's Steve Kohler predicted that the gap after the space shuttle "will lead to very difficult times in Florida, which could be especially difficult considering the current economic troubles our nation faces."[86]

The Great Recession coincided with the end of the Space Shuttle program, a complication unforeseen when the George W. Bush presidential administration in January 2004 announced Space Shuttle program's denouement and the subsequent gap. The county's unemployment rate on January 1, 2008, was 4.9 percent. On July 1, it was 6.5 percent. At year's end, the rate was 8.6 percent. During 2009, the unemployment rate would peak at 11.5 percent by New Year's Day 2010.[87]

For a region well familiar with the Saffir-Simpson hurricane wind scale, the next few years would be an economic Category 5.

* * * * *

While the presidential election campaign unfolded, Space Florida received more editorial criticism from *Florida Today,* Steve Kohler in particular.

The newspaper ran an investigative series in November 2007 titled, "After the Shuttle: Ensuring Brevard's Future in Space." In a November 18 editorial, the paper declared that, "Florida has squandered too much time meeting this challenge. Now there's no more time to waste." The paper blamed Governor Crist, state lawmakers, Florida's members in Congress, and local Space Coast leaders. "That includes Space Florida, the state's space-recruiting agency, which has been anemic in producing results."[88]

In its end-of-2007 review, another editorial concluded that Kohler so far had been a failure. "The director of Space Florida, the state's space industry recruiting arm, turned in a lackluster performance with no tangible results toward the agency's main goal: Using innovation and incentives to lure 21st Century space industries and entrepreneurs to Cape Canaveral and Florida."[89]

Another opportunity was missed in February 2008 when NASA chose Orbital Sciences Corp. of Dulles, Virginia, for the second commercial cargo contract, joining SpaceX. Orbital chose to launch from the NASA Wallops Flight Center in Virginia instead of the Cape because the facility was relatively close to their corporate headquarters. According to *Florida Today,* four of the losing bidders had been working with Space Florida to choose a Cape launch site, but none of them were selected.[90]

Governor Crist requested $8.5 million for Space Florida's fiscal year 2009 operating budget, but in May 2008 the state legislature cut that request by 43 percent to $4 million. The cuts were part of an overall reduction of the state's budget by $5 billion from FY 2008, accepting the dark reality that the Great Recession would slash state revenue. The legislature did approve $1.25 million for workforce retention and training programs, and a partial tax refund for aerospace companies that retained workers.[91] The approved budget also included $14.5 million for improving launch infrastructure at the Cape.[92]

With that improvement money in the bank, Space Florida announced an agreement on August 7, 2008, with the USAF to use the $14.5 million for refurbishing LC-36, the former Atlas Centaur site SpaceX had briefly considered in 2005. The USAF granted Space Florida a five-year license for the complex. Space Florida's officials envisioned LC-36 as a "Commercial Launch Zone" where customers could operate free of taxes and tariffs, similar to a duty-free Foreign Trade Zone.[93] Critical until now, the *Florida Today* editorial page praised the deal, but cautioned, "Now comes a bigger challenge—making it work."[94]

Space Florida held a press conference in Pensacola on December 4, 2008, to announce a $500,000 medical and training program for suborbital space tourists, called Project Odyssey. Half the funding was to come from Space Florida, with the other half from OTTED. Project Odyssey was to be in partnership with the Andrews Institute, a sports medicine facility based in the Florida Panhandle. Lieutenant Governor Jeff Kottkamp was at the event to announce the agreement. Steve Kohler told a *Pensacola News Journal* reporter that space-based tourism was a growing commercial enterprise, with several thousand potential customers, perhaps launching from Florida.[95]

Project Odyssey was the beginning of the end for Steve Kohler. On January 24, 2009, the *Orlando Sentinel* published the first in a series of investigative articles about questionable contracts linked to Space Florida. The article questioned the propriety of OTTED's half of the funding coming from a grant program intended to target "maintenance and expansion of military missions in Florida." The article also noted that Embry-Riddle Aeronautical University in Daytona Beach had complained to OTTED that Kohler had stolen the idea from them; Kohler told the *Sentinel* that he'd come up with the idea himself after touring the Andrews Institute.[96]

The *Sentinel* also reported that, as the Project Odyssey deal came to a close, an OTTED staff employee involved in the negotiations had left for the Andrews Institute to run the project, creating a potential conflict of interest. The *Pensacola News Journal* reported on January 27 that Governor Crist had asked the state's inspector general to look into the deal. The article quoted an Andrews Institute marketing manager as saying that the OTTED employee had been recommended by OTTED and Space Florida, and that he'd signed a contract with Andrews after resigning from OTTED.[97]

The inspector general's investigation was released on April 10, 2009. It found that the OTTED employee had helped design the program and participated in determining its director's $150,000 compensation, the position he resigned to fill. Space Florida announced that the agency had frozen $200,000 of the $250,000 it was to give Project Odyssey, and would reevaluate its future participation in the space tourist training program. The former OTTED employee resigned from his Project Odyssey director position on April 14. Governor Crist decided not to file an ethics complaint against him, but the Florida Ethics Commission decided on March 3, 2010, to file charges anyway. An administrative law judge dismissed the charges on May 3, 2011, finding that "The evidence did not demonstrate that Respondent had any substantial or significant input into the development of Space Florida's ideas or 'near-term unfunded opportunities.'" The employee's job was to coordinate the various aspects of the project, but he did not play a significant role in negotiating the deal or the Andrews Institute director's compensation. "Put simply, Respondent was a lower level employee of OTTED who did not have authority or control over decisions made in regards to Project Odyssey." Although the *Pensacola News Journal* reported his exoneration the next day, the *Orlando Sentinel* did not do so until June 23, 2011. Unlike the earlier sensational reports that had been published on the front page, this article was buried on page B9.[98]

In parallel to Project Odyssey was another controversy involving Space Florida's upgrade of LC-36. On January 29, 2009, the *Sentinel* reported that state auditors with the Office of Program Policy Analysis and Governmental Accountability (OPPAGA) had faulted the agency for not developing a master plan for the launch site.[99] Kohler responded with a guest column in the February 8 *Florida Today*. He wrote, "We think Florida citizens should know that we believe the report in its entirety validates the actions Space Florida has taken in developing an updated and integrated master plan."[100] Kohler also sent a letter to state legislators claiming that news reports of the OPPAGA findings were "inaccurate and lack critical facts."[101]

A copy of the OPPAGA memo obtained by this author shows that it begins, "We concluded that Space Florida has met most of its statutory responsibilities for creating partnerships with other state and local entities to recruit and retain aerospace businesses in Florida." The writers believed that completing a master plan and improving the business plan "would help Space Florida communicate its goals, objectives, and performance to the Legislature and other stakeholders." OPPAGA recommended that the legislature amend state law to require Space Florida to complete its master plan by a specific date, and to prohibit Space Florida from spending the balance of its spaceport development funds until the plan was completed.[102]

The OPPAGA memo also raised the concern that a multiuser facility at LC-36 was impractical. Although a NASA official told OPPAGA it was feasible, several launch company representatives said "each launch vehicle has unique specifications. One of these representatives contended that this may result in Space Florida investing state funds to improve a launch facility that may benefit only a small number of aerospace companies."[103]

"Critics Blast Space Florida as $50M Waste" declared the *Sentinel*'s front-page headline in its February 15 edition. The paper's reporters claimed to have reviewed "thousands of documents and emails" regarding the agency. They quoted SpaceX founder Elon Musk as one of the agency's critics, saying, "I don't see any point to the organization." In addition to the issues discussed earlier in this chapter, the article reported that Starfighters Inc., which operated a fleet of F-104 jets out of the KSC runway, had accused Space Florida of not paying the company $250,000 it believed it was owed for flying the agency's logo on the fuselage of its jets at air shows.[104]

Kohler responded with yet another guest column, this one in the February 19 *Sentinel*. He called the February 15 article "one-sided" and complained that the *Sentinel* reporter had ignored "volumes of documents" provided by Space Florida. "The article failed the readers and the aerospace industry when it neglected to mention any of the achievements made in ensuring Florida remains a vital space leader for the future." Implicitly responding to Elon Musk, Kohler wrote that Space Florida had "advocated" for the USAF to lease LC-40 to SpaceX, and "provided more than $2 million in cash, facilities and in-kind support."[105]

In light of the OPPAGA recommendations and the inflammatory media reports, the Florida Senate Commerce Committee sought to stop the LC-36 upgrade. The committee chairman wanted to freeze $10 million of the $14.5 million the legislature had provided Space Florida for LC-36, and told the agency to drop its plans to request another $43 million over the next three years. Kohler agreed to go along with the freeze.[106]

The *Sentinel* published yet another investigative article on April 28, 2009, this time questioning the money Space Florida was spending to upgrade LC-36 to a multiuser commercial launch pad. The article reported that Brevard County's congressional representatives had declined to support a $5 million Space Florida request for federal dollars to help renovate the complex, citing the agency's "controversial reputation," a reputation the *Sentinel* helped create. The article also quoted Elon Musk, describing him as a "fierce critic of the agency" because he saw LC-36 as competition for his Falcon 9 plans at LC-40. "There's just something really wrong with that picture," Musk said, "especially when you consider they really don't have anyone for sure who is going to use it. They are just hoping someone is going to use it. It just seems like a huge waste of Florida taxpayer money." Space Florida's chief financial officer was quoted as acknowledging that the Great Recession made it hard to find banks or private capital interested in financing the project.[107]

The next day, on April 29, the *Sentinel* reported that Space Florida had issued a no-bid lobbying contract to Blank Rome, a Pennsylvania-based law firm with "close personal ties" to Steve Kohler. A Space Florida executive alerted the OTTED that the contract was improper, skirting agency rules to issue the contract without approval by its board of directors.[108]

Kohler was linked to another contracting scandal involving a former NASA chief of staff. Courtney Stadd was accused in March 2009 by the US Attorney's Office in Washington, DC, of steering NASA funds to one of his consulting clients.[109] On May 2, the *Sentinel* reported that Stadd's consulting company had also received a $25,000 no-bid contract in February 2008 from Space Florida to study the creation of a commercial launch pad at Cape Canaveral. Stadd had been appointed by Governor Jeb Bush in July 2006 to Space Florida's first board of directors, and chaired the search committee that chose Kohler as the agency's first president. Charlie Crist, after becoming governor in January 2007, released Stadd from the board because he was not a Florida resident. Although Stadd was considered qualified to perform the study, his no-bid selection by someone he'd once hired created yet another perceived conflict of interest.[110]

That was the end for Steve Kohler. On May 7, he submitted his resignation, blaming media coverage, which he wrote had "created considerable distraction, compromising needed forward momentum for the agency." The *Sentinel* reported that Lieutenant Governor Kottkamp had met in Tallahassee with Brevard County officials and large aerospace companies; they had urged him to dismiss Kohler.[111]

The next day, Senator Bill Nelson announced he would request a $14 million earmark in the federal budget "for a commercial launchpad and other space projects" in Florida: $5 million would go to convert LC-36, $5 million would go to a thermal vacuum facility, and another $4 million for "a space-outreach program."[112] Nelson told the *Sentinel,* "It's time to clean house at Space Florida and move ahead with an aggressive schedule for a commercial space industry in Central Florida."[113]

On May 19, the Space Florida board voted unanimously to hire Frank DiBello as an interim successor. DiBello had finished second to Kohler in the 2006 selection process. A new search led to offering the job to Shana Dale, NASA's deputy administrator during the George W. Bush administration, but critics questioned her lack of economic development experience. After Dale turned down the job, the committee selected DiBello, who was the favorite of the aerospace industry and local political leaders. DiBello remained in the job until he retired on June 30, 2023.[114]

* * * * *

Having won their parties' presidential nominations, Barack Obama and John McCain spent the summer and fall of 2008 campaigning to win the nation's November 4 general election.

Obama went on to win the election and the electoral college. Out of 538 total electoral college votes, 270 were needed to win. Obama tallied 365, including 27 from Florida.[115]

In the Sunshine State, Obama won 4,282,074 votes (51.0 percent), while McCain tallied 4,045,624 (48.2 percent).[116]

But in Brevard County, McCain was the victor with 157,536 votes (54.6 percent) while Obama had 127,561 (44.2 percent).[117]

Did the candidates' space policies make a difference?

Any conclusion would be debatable at best.

In 2000, Republican George W. Bush won Brevard County with 52.7 percent over Democrat Al Gore at 44.5 percent.[118]

In 2004, Bush won Brevard County with 57.6 percent over Democrat John Kerry at 41.6 percent.[119]

In the 2008 election, 43.5 percent of Brevard County voters were registered Republican, 37.3 percent Democrat, with 19.2 percent nonpartisan or other.[120] McCain's 2008 54.6 percent–44.2 percent margin of victory fell in between the last two election results, and was consistent with local partisan voter registration, so it would be dubious to argue that "Obama lost Brevard County" because of his space proposals. He won Florida despite Brevard County.

Even though Brevard County didn't make a difference in the election results, the events of the 2008 campaign would be cited time and again during Obama's presidential first term, often falsely, as the forces of OldSpace and NewSpace clashed for the future of the American space industry. The early skirmishes in that battle trace back to an August 2 Obama campaign rally in Titusville.

The Constellation gaffe was still hounding Obama. The McCain campaign clearly saw it as an opening. On July 29, the fiftieth anniversary of President Eisenhower signing the National Aeronautics and Space Act, the McCain campaign issued a statement that read:[121]

> While my opponent seems content to retreat from American exploration of Space for a decade, I am not. As President, I will act to ensure our astronauts will continue to explore space, and not just by hitching a ride with someone else. I intend to make sure that the NASA constellation program has the resources it needs so that we can begin a new era of human space exploration. A country that sent a man to the moon should expect no less.

The Obama campaign also released an anniversary statement, but it did not mention Constellation:[122]

> In recent years, Washington has failed to give NASA a robust, balanced and adequately funded mission. Though the good people of NASA who work day in and day out on new frontiers are doing amazing things, Americans are no longer inspired as they once were. That's a failure of leadership.
>
> I believe we need to revitalize NASA's mission to maintain America's leadership, and recommit our nation to the space program, and as President I intend to do just that. We must revive the American ingenuity that led millions of children [to] look to NASA astronauts and scientists as role models and enter the fields of math, engineering and science. Our leadership in the world depends on it.

The next day, the Obama campaign announced that the candidate would stage a town hall event on August 2 at the Brevard Community College campus in Titusville. A *Florida Today* editorial column called on Obama to "seize the day to chart the cosmos by supporting funding to send Americans back to the moon—this time to stay." The editorial assumed this could only be accomplished by "the sustained budget it will take to build the post-shuttle fleet of Ares rockets and manned Orion spacecraft," with no consideration given to other and perhaps smarter options that were not already entrenched in the local economy.[123]

As the rally began, Bill Nelson introduced Obama. "I must admit, I've been working on Barack," Nelson said. "Telling him that it's this I-4 Corridor[124] of Florida that will make the difference . . . He's here today for a lot of reasons, not the least of which is to tell you about what he wants to do with our space program."

Obama's remarks were largely about the declining economy but, as Nelson promised, the candidate did offer specifics about space policy.[125] His significant talking points:

- "One of the areas where we're in danger of losing our competitive edge is in science and technology, and nothing symbolizes that more than our space program."
- "Today we have an administration that sets ambitious goals for NASA, without giving NASA the support it needs to reach them."
- "That's why I'm going to close the gap. Ensure that our space program doesn't suffer when the Shuttle goes out of service. We may extend an additional Shuttle launch. We're gonna work with Bill Nelson to add

at least one more flight beyond 2010. By continuing to support NASA funding. By speeding the development of the Shuttle's successor. By making sure that all those who work in the space industry in Florida do not lose their jobs when the Shuttle is retired, because we can't afford to lose their expertise."

- "I'm gonna re-establish the National Aeronautics and Space Council, so that we can develop a plan to explore the solar system, a plan that involves both human and robotic missions, enlists both international partners and the private sector."

During the question-and-answer session, Obama acknowledged the Constellation gaffe. "I know it's still being reported that we were talking about delaying some aspects of the Constellation to pay for our early education program. I told my staff, 'We're gonna find an entirely different offset,' because we've got to make sure that the money that's going into NASA for basic research and development continues to go there. That has been a top priority for us."

In future chapters, we'll discuss which promises Obama kept, which he broke, and which were thwarted by Congress. But it should be noted that he never explicitly promised that day to continue Constellation. He did promise to speed "the development of the Shuttle's successor," but he didn't say that successor would be Constellation. He did imply it, though, by saying his administration would no longer delay Constellation by transferring its funding to education programs.

Obama said he would ensure that Florida space workers would not "lose their jobs when the Shuttle is retired." That promise was impossible to keep, but his administration did propose funding a workforce retraining program. That funding was eventually blocked by Senate Republicans in November 2010, so it died.[126] More about that in Chapter 9.

But in the summer of 2008, all that was in the future. Obama first had to win the election.

After Clinton conceded, her space adviser Lori Garver was recruited to join the Obama campaign. Two weeks after the Titusville town hall, the campaign released to the press a revamped space policy. Now representing the Obama campaign, Garver answered *Sentinel* reporter Robert Block's questions. She said that the candidate and his staff had listened to voices in the space and education communities, and that the campaign now "recognized the importance of space."[127]

The new policy paper once again rhetorically skirted the issue of continuing Constellation. Although the paper acknowledged that a minimum five-year gap existed between the shuttle's retirement and "the first elements of the Constellation program," it didn't exactly state that Constellation would continue to be the successor.

> Obama will expedite the development of the Shuttle's successor systems for carrying Americans to space so we can minimize the gap. This will be difficult; underfunding by the Bush administration has left NASA with limited flexibility to accelerate the development of the new systems . . .
>
> Obama will stimulate efforts within the private sector to develop and demonstrate spaceflight capabilities. NASA's Commercial Orbital Transportation Services is a good model of government/industry collaboration.

Sometime around August 9, the McCain campaign updated its online space policy. Unlike Obama, McCain committed to "funding the NASA Constellation program to ensure it has the resources it needs to begin a new era of human space exploration." As did Obama, McCain pledged to "review and explore all options" to minimize the gap and to, "Ensure the national space workforce is maintained and fully utilized." The policy noted that McCain had "sponsored legislation to support the up and coming commercial space industry," but did not specifically pledge to pursue it as Obama did.[128]

McCain spoke August 18 at a Veterans of Foreign Wars convention in Orlando, then rode his campaign bus to Brevard Community College in Cocoa for a private meeting with eighteen aerospace industry leaders. No public event was held in Brevard County, perhaps because Tropical Storm Fay was expected to arrive the next day.[129] He returned to Orlando on September 15 to hold a town hall rally, but he did not visit the Space Coast.[130]

McCain finally held a public rally in the Space Coast on October 17, on the Brevard Community College campus in Melbourne. "If I am elected President, I won't cut NASA like Senator Obama did," he said, even though Obama had done no such thing; Obama now called for a $2 billion increase in NASA's budget. McCain also said he would support a $2 billion increase in NASA's budget, even though he vowed at the rally to freeze all federal spending except for the military, health and veterans care, and Social Security.[131]

Senator Joe Biden (D-DE), Obama's vice presidential running mate, appeared with Bill Nelson on October 28 at Wickham Park in Melbourne. Biden told the crowd, "And as it did in the Kennedy administration, we will create a new generation of engineers, mathematicians and scientists and a few astronauts like Bill Nelson as well in the process."[132]

* * * * *

On November 6, two days after the general election, the GAO posted a web page listing thirteen "Urgent Issues" they viewed as "critical and time sensitive and require prioritized federal action" by whomever won the White House. "Retirement of the Space Shuttle" made the list.[133]

> According to NASA, reversing current plans and keeping the shuttle flying past 2010 would cost $2.5 billion to $4 billion per year. In addition, extending the shuttle will likely be costly and logistically difficult, particularly since it would require restarting production lines and possibly recertifying suppliers as well as the shuttles. On the other hand, the new administration may well decide to extend the shuttle and defer development of new transportation vehicles in light of budgetary constraints, as the new vehicles are expected to cost more than $230 billion[134] to develop and deploy.

The GAO suggested two action items for the next administration: (1) quickly fill NASA key leadership positions to decide on Shuttle's fate, and (2) "retain the workforce, facilities, equipment, and suppliers necessary to continue operating the Space Shuttle."

In refutation of those advocating the extension of the Space Shuttle program, NASA KSC Launch Integration Manager Wayne Hale posted a blog article on August 28 making it clear that was not practical, much less desirable. He described how much of the program's supply chain already had been shut down. Many of the vendors producing specialty parts were no longer available. To certify new vendors would take years and be cost-prohibitive. For those who had forgotten the lessons of the past, he wrote, "Our shuttle history tells us that when we try to cut corners, trouble results." Hale's conclusion: "We started shutting down the shuttle four years ago. That horse has left the barn."[135]

Although the barn was *sine equus*, it didn't stop Brevard County politicians, labor unions, government contractors, and other interests from uniting in the next two years to pressure the Obama administration into extending the Space Shuttle program, in the name of protecting the local workforce.

President Bush had announced on January 14, 2004, the space shuttle's planned retirement by 2010. Although Constellation was planned as its successor, it was never going to replace all the jobs lost by the shuttle's retirement. A March 2008 NASA assessment concluded that "5,700 to 6,400 jobs will be lost at KSC before 2012, after which a few hundred jobs will be added yearly as the new moon-landing program gets started," according to *Florida Today*.[136]

What no one had anticipated in 2004 was the Great Recession, which began in December 2007 and would not end until June 2009. As the Obama administration was about to take office, the economic Category 5 hurricane was making landfall on Brevard County. Double-digit unemployment in the county would coincide with the end of the Space Shuttle program. The gap had arrived.

The commercial space boom so far had fizzled, just worthless promises and portents. No space tourists were flying to suborbit with Virgin Galactic or anyone else; Space Florida had bet on the wrong future. Angel investors—wealthy people willing to provide startup money in exchange for a stake in the company—were generally skittish about NewSpace. Venture capitalists—professionally managed pools of capital looking for a return on investment within a few years—saw the risk as too high and the political risks too unpredictable. Space capitalists—deep-pocketed entrepreneurs passionate about space—seemed the most promising; among them were Elon Musk (SpaceX), Richard Branson (Virgin Galactic), and Jeff Bezos (Blue Origin).[137] But none of them offered any hope of replacing the thousands of jobs about to be lost on the Space Coast.

Those about to join the incoming Obama administration didn't create these problems, but they were about to inherit them. Deserved or not, they were about to feel the wrath of those in Brevard County looking for someone to blame.

8

A JFK Moment

Lori Garver is fond of quoting Ralph Cordiner, the General Electric CEO at the beginning of the Apollo era, who feared the consequences if America's nascent space industry became "increasingly dependent on the political whims and necessities of the Federal government." Cordiner warned, "This would leave the nation almost no choice except to settle for nationalized industry in space."[1]

Garver is also fond of the phrase *self-licking ice cream cone*, an American military term describing programs designed to justify their existence in perpetuity.[2] It was used in 1992 by Pete Worden, at the time a USAF colonel assigned to the Strategic Defense Initiative Organization, to describe how he viewed NASA's annual budget process. In his opinion, the process was designed largely to perpetuate funding for programs in the states and districts of its congressional oversight committees. Worden wrote, "Since NASA effectively works for the most porkish part of Congress, it is not surprising that their programs are designed to maximize and perpetuate jobs programs in key Congressional districts. The Space Shuttle–Space Station is an outrageous example. Almost two-thirds of NASA's budget is tied up in this self-licking program."[3]

As the Obama administration prepared to take office on January 20, 2009, Garver was determined to reform this practice, which traced back to the time of James Webb and Project Apollo.

Garver was asked to head the incoming administration's "transition team." As a new president prepares to take office, transition teams are appointed to coordinate with the outgoing president's administration so the business of government continues uninterrupted.

That's the ideal, but in reality some transitions are more genial than others.

Joining Garver on the transition team were Alan Ladwig, who had run the Space Flight Participant Program during the Reagan Administration; Edward Heffernan, an intergovernmental affairs specialist who became NASA chief of staff during the last months of the Clinton administration; George Whitesides, a chemist who had been executive director of the National Space Society from 2004 through 2008; and Roderick Young, press secretary for NASA Administrator Dan Goldin during the Bill Clinton administration.[4]

Florida Senator Bill Nelson, meanwhile, was lobbying the transition team to keep Michael Griffin as NASA administrator. Griffin was nominated by President George W. Bush and took office on April 14, 2005.[5] He decided to retool Project Constellation with his own vision, declaring it "Apollo on Steroids."[6] Some critics have blamed Griffin for Constellation's design delays and cost overruns. A typical example is Space Frontier Foundation and space entrepreneur Rick Tumlinson, who wrote in an October 27, 2010, *Space News* opinion column that Griffin was "a good man at heart, but his ideas and obsession with doing things his way come hell or high water killed the dream, turning a major new American space initiative into a dead-end jobs program."[7] The *Orlando Sentinel* reported on November 7, just three days after the election, that Nelson had already called Lori Garver to express his preference for Griffin to remain NASA administrator until a "surefire choice" could be found to replace him. According to Nelson's press secretary, Nelson wanted to "stay the course" with Constellation, and extend the Space Shuttle program during the projected five-year gap.[8]

Griffin addressed KSC workers on November 13. He told the assembled that he would continue as NASA administrator if Obama asked him, but "I doubt that will happen." Even if asked, he would remain only if Obama continued Constellation.[9]

Garver wrote in her autobiography *Escaping Gravity* about the difficulties she and the transition team had with Griffin and certain NASA executives. In her opinion, Griffin wanted to remain as NASA administrator. Garver believed that NASA and contract managers were hiding from the team details about the Constellation program. "The attitude to keep us in the dark was pervasive throughout the program's management," she wrote.[10]

The *Orlando Sentinel* reported on December 11 that Griffin and Garver had an "animated conversation" during a book event at NASA headquarters. According to the article, Griffin accused Garver of calling him a liar because she wouldn't trust what he was telling the team about Constellation. The *Sentinel* reporters obtained a copy of an email suggesting that Griffin had directed Constellation contractors to first clear with him any materials they intended to share with the team.[11] Griffin issued that day a memo to NASA employees denouncing the article as "simply wrong . . . We are fully cooperating with transition team members," Griffin wrote. "I am appalled by any accusations of intimidation, and encourage a free and open exchange of information with the contractor community."[12]

Surrogates continued to campaign on Griffin's behalf. Four-time shuttle astronaut Scott Horowitz, who had worked under Griffin from October 2005 to September 2007 as an associate administrator, circulated a petition "in support of keeping Mike Griffin as the NASA Administrator." He praised Griffin as "one of the most technically and managerially competent administrators in NASA's history."[13] Griffin's wife Rebecca sent an email to the aerospace community asking recipients to sign the petition.[14]

In a *Time* article posted the same day as the *Sentinel* report, Horowitz told journalist Jeffrey Kluger, "Lori Garver is not equipped to make technical judgments on the architecture of a space exploration system." Kluger incorrectly described Garver as "the HR rep" who had been a onetime NASA "public affairs officer," a position she never held. He wrote, "NASA is right to be uneasy about just what Obama has planned for the agency since his position on space travel shifted—a lot—during the campaign."[15]

Implicit in their arguments was the assertion that Garver was unqualified. Unlike Griffin, she was not an engineer. That argument was somewhat specious, because Griffin openly admired James Webb, NASA's administrator during the Kennedy and Johnson administrations. Webb was no engineer either. In a 2007 NASA Oral History Project interview, Griffin said, "I certainly am an admirer of Jim Webb's. He did an awful lot of things right." He praised Webb's managerial style, which he claimed to emulate.[16]

Webb published in 1969 a book about large project management called *Space Age Management*. He addressed his personal concern that he was "not the right man for the job," lacking knowledge of "the intricacies of advanced scientific knowledge and advanced technology." Webb wrote that President Kennedy told him "that the job was not one for a scientist or an engineer, but for someone experienced in the broadest aspects of national and international policy." Webb added to his executive staff Hugh Dryden and Robert Seamans, men "of proved worth and with experience in many areas basic to NASA's needs."[17]

Griffin acknowledged that Webb had lacked a technical background, and had "augmented himself" with technical executives. Chris Shank, a NASA spokesperson, nevertheless told the *Orlando Sentinel* that Griffin believed Garver's transition team "lacks the engineering expertise to properly assess some of the information its members have been given."[18]

Webb, however, in a May 1969 Syracuse University project management research interview said that engineers should leave the politics to the politicians. "The engineer . . . mustn't assume that he knows more than the political scientist or the economist. The great curse of much of this specialization is a guy who is a good engineer who thinks he can tell you how to get support in Congress for NASA."[19]

Obama chose not to keep Griffin, so the NASA administrator position became vacant on January 20, 2009, when the Bush administration ended and the Obama administration began.

The new administration began looking for candidates; one rumored name was retired US Air Force General J. Scott Gration, who was part of the "senior officers group" that had advised Obama during the campaign. According to a media report, Gration was one of the authors of Obama's revised space policy released in August 2008 after the Titusville event. The two had grown close in recent years; Gration had accompanied Obama on a trip to Africa in 2006. During the campaign, Gration stumped for Obama in Iowa to burnish the senator's national security credentials.[20] In her book, Garver wrote that Gration's candidacy had come directly from the president-elect.[21] But Bill Nelson made clear his opposition, preferring someone with direct NASA experience. As chair of the 111th Senate's space subcommittee, Nelson was in an influential position to promote or impede a candidate.[22]

Reports surfaced that Nelson championed retired US Marine Corps Major General Charles Bolden, a four-time space shuttle astronaut who was the pilot on Nelson's STS-61C space shuttle mission in January 1986. A January 6, 2009, *Orlando Sentinel* blog article quoted a Nelson spokesperson as saying the senator thought Bolden was a "top notch" individual.[23]

In a March 2009 collective interview with journalists from the *Orlando Sentinel* and other regional newspapers, Obama said that he intended to task his new administrator with "shaping a mission for NASA that is appropriate for the 21st Century." He also acknowledged his awareness of the economic impact of the space program on the Space Coast.[24] Obama noted that his fiscal year 2010 proposed budget would include funding for one more space shuttle flight, directed by Congress in the 2008 NASA Authorization Act (HR 6063). The act included $150 million to fly the Alpha Magnetic Spectrometer to the ISS, and directed NASA to add this flight as one additional mission unless the risk was "unacceptable."[25] That mission eventually became STS-134; it used a spare external tank damaged during Hurricane Katrina in 2005 and restored to flight configuration at the Michoud Assembly Facility in New Orleans, Louisiana.[26]

The Obama interview named individuals considered candidates for administrator, including Nelson's preference for Charlie Bolden. Another candidate was former NASA Comptroller Steve Isakowitz. But Nelson insisted on Bolden; he told the *Tampa Bay Times*, "He's one of my best friends and he's one of the best leaders."[27] Garver wrote in her autobiography that Obama selected Isakowitz for administrator "a few weeks after inauguration," but dropped him after Nelson objected.[28] Bolden's Columbia, South Carolina, hometown newspaper quoted Obama administration officials as saying that "the White House could no longer count on" Nelson's support for Obama initiatives unless Bolden was named administrator.[29]

The White House and NASA finally announced on May 23, 2009, that Bolden would be nominated as the next NASA administrator, with Lori Garver his deputy.[30] Nelson told NBC News, "I trusted Charlie with my life—and would do so again."[31] The *Orlando Sentinel* reported that the White House was concerned about Bolden's ties to NASA contractors, opposition to any possible budget cuts, and the lack of a close relationship with President Obama.[32]

Although unknown to the public at the time, the relationship between Bolden and Garver during the next four years would be strained. Garver wrote in *Escaping Gravity* that "Charlie's friendly and humble manner made him a beloved public figure," yet not finding "a way to develop a more trusting relationship with Charlie" was one of her greatest regrets.[33] In a 2017 interview with public administration professor and author W. Henry Lambright, Bolden said he and Garver "didn't function as a team. Our leadership was dysfunctional."[34]

Garver believed she was legally and ethically responsible for implementing the president's policies. In various roles over the past year, she had been representing Obama's direction for the American space program, and had helped him formulate his space policy after she left the Clinton campaign. No one knew what Obama intended more than Lori Garver.

In his 2017 Lambright interview, Bolden said, "I had learned in the Marines that a leader has to take care of your people, and they'll take care of you." He expressed the concern that White House officials wanted to quickly implement the president's policies without considering the impact to the NASA workforce. "The implication is that people are not that important." He acknowledged, "The President's program became my program," but he also felt "loyal to my employees. They were loyal to me."[35]

* * * * *

The intricacies of the Bolden-Garver relationship during the four years of her service are best told by them. Our interest is in how their dysfunction affected Florida space politics and policy. Their subplot intertwines with ours, because the events of the 2009–2011 time frame fundamentally altered the direction not only of NASA but also the American commercial space industry, which would then alter the direction of the Pentagon's space capabilities. OldSpace and NewSpace were about to battle for supremacy—with Florida the battlefield.

On May 7, more than two weeks before the Bolden/Garver nominations, the White House announced NASA's proposed FY 2010 budget as well as "the launch of an independent review of planned U.S. human space flight activities with the goal of ensuring that the nation is on a vigorous and sustainable path to achieving its boldest aspirations in space."[36]

The review panel informally came to be known as the Augustine Committee after its chair. Norm Augustine was a retired Lockheed Martin CEO, a former under secretary of the Army, and a former president of the American Institute of Aeronautics and Astronautics.[37] Among the committee's nine other members were a national space security corporation CEO, a former Boeing space shuttle orbiter director, the chair of the National Academies Space Studies Board, an engineer, an astrophysicist, a NewSpace entrepreneur, a retired Air Force general, and two retired astronauts.[38]

The Augustine Committee held public hearings throughout the summer of 2009. They met in Washington, DC; Huntsville, Alabama; Houston, Texas; and on July 30 in Cocoa Beach, Florida. The public hearings were broadcast live on NASA Television.[39]

At the Cocoa Beach hearing, Augustine opened the event by explaining that the committee's role was not to make specific recommendations, but to offer President Obama options. "It will be up, I assume, to the president and the Congress to decide which options to choose."[40] That same day, Bolden by television addressed the KSC workforce. Bolden said he would support whatever decision the president made. "I'm in favor of whatever option the president settles on. Whatever is decided, we're going to make it work well."[41]

Senator Nelson addressed the panel via videotape from Washington, DC. He urged the panel to ignore federal budget limitations and propose options they considered best. Regarding the gap, Nelson cited one estimate that the Space Shuttle program could continue for $1.7 billion a year. "Whatever it is, I wish you would consider extending the Shuttle to a point in time that would lessen the gap so that we could have Americans riding American vehicles to get to our Station."[42]

The committee's 157-page report was released on October 22, 2009. Titled *Seeking a Human Spaceflight Program Worthy of a Great Nation,* its executive summary began:[43]

> The U.S. human spaceflight program appears to be on an unsustainable trajectory. It is perpetuating the perilous practice of pursuing goals that do not match allocated resources. Space operations are among the most demanding and unforgiving pursuits ever undertaken by humans. It really is rocket science. Space operations become all the more difficult when means do not match aspirations. Such is the case today.

The executive summary noted that there were "more options available today" than when President Kennedy proposed the crewed lunar program in 1961. "Space exploration has become a global enterprise," which meant "the burden and benefit" could be shared. "A burgeoning commercial space industry" had emerged, with the potential to reduce government costs. The American space industry had the benefit of nearly fifty years of experience.[44]

The committee concluded that Constellation's technical problems could be solved with time and money, but noted the contradiction of its funding scheme. The Ares I was being designed to send crew aboard an Orion spacecraft to the ISS, but it was unlikely to be operational until 2017 or later. Since Ares I was to be funded by ending the ISS in 2015, the Orion would have nowhere to go. "The length of the gap in U.S. ability to launch astronauts into space will be at least seven years."[45]

The space station's actual termination date was dubious. The original spending estimates that accompanied the Vision for Space Exploration in early 2004 assumed the ISS would be sacrificed by 2015 to pay for Constellation.[46] Congress, however, in the 2008 NASA Authorization Act required the administrator to "take all necessary steps to ensure that the International Space Station remains a viable and productive facility capable of potential United States utilization through at least 2020 and shall take no steps that would preclude its continued operation and utilization by the United States after 2015."[47] But if the ISS were extended, how would Constellation be funded? It was yet another riddle left for the Obama administration to solve.

Section 902 of the 2008 authorization act, "Commercial Crew Initiative," directed NASA to "enable a commercial means of providing crew transfer and crew rescue services for the International Space Station." The Augustine Committee recalled the precedent established by the US Post Office in the 1920s, awarding "a series of guaranteed contracts for carrying airmail, stimulating the growth of the airline industry. The Committee concludes that an exploration architecture employing a similar policy of guaranteed contracts has the potential to stimulate a vigorous and competitive commercial space industry."[48]

The committee looked at various scenarios, including the unrealistic money-is-no-object option. They concluded that it was feasible to continue the Ares I and Orion system, with additional funding to resolve technical issues, but those technologies "will not be available until near the end of the decade."[49] Once it began operation, it would be "a very expensive system for crew transport to low-Earth orbit . . . the Orion is a very capable vehicle for exploration, but it has far more capability than needed for a taxi to low-Earth orbit." A commercial crew system had "the potential to be safe, sooner and significantly less expensive."[50]

In its examination of NASA's human spaceflight program, the committee reached three conclusions:[51]

- Human exploration beyond low Earth orbit was not viable within the existing NASA budget.
- An additional $3 billion per year above the existing budget would make "meaningful" human exploration possible.
- The additional funding would allow for a "Moon First" approach, using the moon as a proving ground that could be used for a later Mars program; or a "Flexible Path" approach, visiting various destinations in the inner solar system such as an asteroid to evolve the technologies for an eventual Mars mission.

Augustine represented his team before the Senate Science and Space Subcommittee, chaired by Bill Nelson, on September 16, 2009. In his opening remarks to Augustine, Nelson left no doubt he was placing the onus on President Obama to solve the problems created over many decades by not just presidents but also Congress and the space industry:[52]

> And what you've laid out is a blueprint, a menu for the president to make choices. And, it is my fervent hope that he's going to say we're going to put the juice to it, we're going to have a vision that we're going beyond low Earth orbit, and in the process we're going to nourish that workforce so that we have them ready when we do the next huge leap for mankind.

* * * * *

When James Webb in the early 1960s formulated his "noble vision" for Project Apollo as a national technological stimulus, he didn't intend to create a permanent workforce dependent on government largesse.

In the autumn of 1963, Webb wrote an article for *Business Horizons,* an academic journal published by the Kelley School of Business at Indiana University. The topic was the "long-range impact of the space program on the nation's economy and on the American business community." NASA was only five years old, evolving from its NACA roots as an incubator of aerospace technology into—what? Webb described a vision of technological advancement that not only enhanced "international leadership" but could also result in "practical uses of space technology for the benefit of mankind" that would stimulate "the minds and aspirations of youth." He acknowledged, "The benefits may come slowly in the form of tiny increments of technical skill." The future was unpredictable. But nowhere in the essay did he suggest that government workfare was NASA's raison d'être.[53]

In his 1969 book *Space Age Management,* Webb wrote that NASA had built up from 1962 to 1967 "a work force of 420,000" but in the next two years "reduced this number by 140,000, or one-third. We know today that we will have to make further sharp reductions, but do not know whether we can then plan to stabilize at the lower levels or will have to build up again."[54] According to biographer Piers Bizony, Webb had no concern for "preserving jobs just for the sake of it. Although one of his principal ambitions with NASA was to spur the economy and provide a source of employment, those jobs still had to be justified by the tasks at hand."[55]

Bizony noted that "the fledgling agency" Webb inherited had employed about 6,000 people on civil service salaries. That number increased to an average of about 36,000 in the mid-1960s, in addition to almost 400,000 contractor jobs nationwide. "For every NASA staffer, there would be ten personnel in the private sector . . . NASA was a federal agency, not an industrial manufacturer." Bizony estimated that 90 percent of NASA's budget appropriations during Apollo went to the private sector.[56] To achieve Kennedy's objective, NASA had no choice but to reverse its predecessor agency NACA's practice of keeping work in-house.

We discussed in Chapter 2 the conflicting visions between Webb and his boss, President John F. Kennedy. The president intended NASA as a propaganda tool to promote and enhance American "prestige" on the global stage. Webb envisioned Apollo's scientific advances as a national economic stimulus. But no evidence could be located by this author to suggest that either Kennedy or Webb ever intended to create a standing army of government contract laborers who would be employed in perpetuity. The massive layoffs at Apollo's end confirm that.

Webb, however, had no problem with politicians taking credit for bringing the jobs to their districts and states. In his 1969 oral history interview with Lyndon Johnson's presidential library, Webb said that "we followed a very simple policy in NASA. We did what we thought was right for the program, and we let the politicians take the credit when and where they wanted to."[57]

The implicit corollary is that when jobs were lost, the politicians would want to avoid the blame.

During the decades of the Space Shuttle program, members of Congress sought appointment to committees that would assure NASA contracts went to employers in their districts and states, creating a permanent dependency on federal spending. When President George W. Bush announced on January 14, 2004, that the Space Shuttle program would end upon completion of the ISS, the countdown began on those contractor jobs—which were, by definition, temporary.

But as the calendar turned to 2010, as the job losses became real, many in Brevard County leadership staged protests and public forums to rally the locals. Politicians, editorial page writers, union leaders, even ministers and astronauts claimed that NASA had an implicit moral commitment to permanently employ the contract workforces at Kennedy Space Center and elsewhere. Failure to sustain these workforces was equivalent in their view to surrendering the nation's space supremacy to foreign and perhaps sinister forces. If Space Shuttle program jobs were not perpetuated, either through an extension of the Space Shuttle program or preservation of its successor Constellation, then certainly those responsible were doing the unwitting bidding of rivals in Russia or China.

Others were less melodramatic, interested simply in helping their constituents, their congregants, their union members, their subscribers. What John F. Kennedy, what James Webb intended decades ago was irrelevant. People they cared about were losing their jobs. What could be done to help those caught in the crossfire?

* * * * *

The history of labor protests at the Cape is nearly as old as the military base on the peninsula. Disputes were not only over wages and working conditions, but over issues as basic as what union would represent which workers.

As early as 1955, the International Association of Machinists had petitioned the National Labor Relations Board to let IAM replace the Transport Workers Union as the bargaining agent for machinists on the Cape. *The Cocoa Tribune* edition published on February 5, 1957, reported that Pan American World Airways workers had staged "a short walkout" at Patrick Air Force Base to protest their not being allowed to vote for the union of their choice. The TWU accused the IAM of using "muscle methods" to raid their membership for members. The IAM eventually withdrew its claim.[58]

In May 1957, the Teamsters brought to a stop all construction work at the Cape. Two thousand workers refused to cross the picket line. The strike was called because a construction company refused to recognize the Teamsters as a bargaining agent. A court injunction forced most picketers back to work by the end of June.[59]

Members of the International Brotherhood of Electrical Workers Local 756 in May 1958 led about 500 electricians, pipefitters, iron workers, and carpenters to walk off Titan missile program construction sites, protesting nonunion civil service electricians being used on the job. The wildcat strike, not authorized by the union itself, ended three days later after the IBEW members voted unanimously to return to work.[60]

About 1,500 members of TWU Local 525 staged a wildcat strike in July 1958 to protest what they claimed were Pan Am safety and medical failures at the Cape. The protest was triggered by the death of a worker who fell off an Atlas service tower. Construction employees brought the Titan program to a stop again because they wouldn't cross the TWU picket lines. Most of the strikers were mechanics, truck drivers, janitors, and pump operators. The strike ended four days later, after Pan Am agreed to the workers' demands; the union leaders were facing a contempt of court charge after ignoring a judge's order to end the strike.[61]

IBEW Local 756 in Port Orange, near Daytona Beach, published in 2021 a commemorative history book that looks back at those early days. The book recalls that, with the arrival of the Air Force Missile Test Center in 1951, "Cape Canaveral and the rocket program expanded tremendously, providing Local 756 members with an abundance of employment." Local 756 electricians worked in 1951 on construction of the base's Central Control Building, as well as the earliest launch pads that were built to test cruise missiles.[62]

The local's district encompassed far more than military and civilian operations at the Cape. Electricians could find work on other projects in the area that had nothing to do with space or missiles. Many of the workers were classified as "journeymen" because they were used to traveling the nation to find work. With so many people arriving in Brevard County to work at the Cape, new housing construction boomed, "providing more employment opportunities for the local's members." When layoffs occurred, IBEW members were able to find work on residential and commercial projects in Cocoa Beach, Daytona Beach, and other localities. When electricians were in demand at the Cape, the local sometimes experienced a shortage of "wiremen" for these civilian construction sites. Local 756 officers imported union journeymen from across the country.

Labor relations generally improved at the Cape during the 1960s, although the number of strikes spiked as construction began at the new Kennedy Space Center site in north Merritt Island. In October 1964, the commanding officer of the Eastern Range commended TWU Local 525 for honoring its no-strike, no-lockout pledge, praising TWU and Pan Am for five years of uninterrupted operations.[63] But strikes still occurred. Pipefitters called a two-day work stoppage in March 1965 because of a dispute with a contractor paying nonunion workers less than union wages. About 3,000 construction workers refused to cross KSC picket lines until the assistant secretary of labor agreed to meet with union representatives. KSC Director Kurt Debus reminded the local that their international union had signed a no-strike pledge.[64] The IAM called a strike in September 1965 against The Boeing Company; 270 IAM members walked out, joined by 1,600 from other union locals even though IAM hadn't asked for their support.[65] A *Miami Herald* reporter observed, "Construction workers at the Cape have a reputation for honoring almost any picket line, whether sanctioned by their unions or not."[66]

An October 1965 NASA report looked at "work stoppages and related events" at KSC and the Eastern Range from the proving ground's inception in 1950 through July 1965. It chronicled every known labor action. Although the report found that some records were incomplete or contradictory, it showed a general decline of lost "man days" after the Teamsters strike of May–June 1957. The numbers picked up again in 1964–1965 as KSC construction began.[67]

A second report, for the period July 1965 to July 1967, noted "the absence of labor disputes at Kennedy Space Center or Cape Kennedy since September 1966" due to two factors. One was the drop in demand for labor, since construction was nearly finished. The other was the creation of a Labor Relations Office; "close co-ordination" between that office, labor representatives, and contractors had "prevented major disputes from occurring" as well as "a change in labor representatives."[68]

By the late 1960s, the dominant union locals at the Cape were no longer "construction" but "industrial." Construction laborers typically worked a short-term contract to build a site. Once the site became operational, "industrial" workers typically performed routine daily tasks such as installation, maintenance, and repair.

One industrial laborer strike, against Bendix Corporation, was by employees represented by what was now called the International Association of Machinists and Aerospace Workers (IAMAW). Over 600 machinists walked out on March 15, 1968, after their contract expired on March 1 and no new agreement was reached. The strike was settled on March 20 with the help of a federal mediator.[69]

Strikes became less common over the decades, but workers still walked out from time to time. On June 14, 2007, about 600 IAMAW Local 2061 members went on strike against United Space Alliance, the Boeing–Lockheed Martin partnership managing the Space Shuttle program's service contract. Among the tasks performed by striking workers were operating Vehicle Assembly Building cranes and the crawler that transported the shuttle stack to the launch pad. The members rejected USA's contract offer by a 93 percent margin. They set up a picket line outside the KSC gates and at other local USA operations. USA replaced the strikers with supervisors, nonunion employees, and temporary help. By July 10, about a hundred IAMAW members had crossed the picket line and returned to work. USA and IAMAW settled on November 4.[70]

The IAMAW, the IBEW, the TWU, and other unions to this day still represent Cape workers. The unions are allied through the American Federation of Labor and Congress of Industrial Organizations, commonly known as the AFL-CIO. This author reached out to representatives of these union locals and the Florida AFL-CIO to better understand not only these events of the Apollo era, but also the labor protests that were staged in Brevard County as the Space Shuttle program approached its denouement.

Unionists may have protested in 2010, but it doesn't appear that workers protested as Apollo came to an end.

Steve Williams joined IBEW Local 756 in 1968 as an eighteen-year-old apprentice. He learned his trade working on various projects at CCAFS, KSC, and throughout the Local's jurisdiction. His training was through an apprenticeship program funded by the IBEW and the National Electrical Contractors Association. Williams became a licensed journeyman in 1972, and found himself assigned in its early days to the space shuttle. As a journeyman, he traveled the nation to work on various construction projects, such as the Trans-Alaska Pipeline System. Over the decades, Williams recruited electrical workers to join the IBEW, challenged illegal employment practices, and organized picket lines at construction sites. He was elected by fellow members as the local's business manager, a position he held until he retired in 2012.

This author interviewed Williams in September 2023.[71] When asked if members of union locals had protested the loss of their jobs as Apollo came to an end, Williams didn't recall any. "If there would have been any protests, it would have been service contract workers," Williams said. The industrial laborer contracts are an example of service workers, who have long-term contracts to perform ongoing functions. "There was nothing to protest. If the program has gone away, it's gone away. They protested about things like their contract negotiations and not negotiating in good faith. Most workers understand that they have no control over the program."

During the 2008 presidential general election, the AFL-CIO supported candidate Barack Obama. The national organization voted unanimously to support Obama, which gave him access to the AFL's political machinery and financial resources.[72] But for the union locals, struggling with the early months of the Great Recession, their main concern was finding work.

"In '08, we were struggling just to stay alive," Williams recalled. "We couldn't keep our doors open. The least of my concerns was an Obama speech in Titusville, to be honest with you." As business manager for his union local, Williams was trying to find jobs anywhere for his members—on the Cape, in Brevard County, anywhere in the nation. But none could be found.

* * * * *

While Space Coast union locals paid little more than cursory attention to the 2008 presidential campaign, other elections were on the ballot that November.

Robin Fisher bench-pressed 505 pounds when he was a nose guard for the Florida Gators college football team in the early 1980s. He would later cite his football experience, and his college degree in public relations, as skill sets he applied to his political career.[73]

Fisher built a successful insurance business in Titusville, becoming a community leader. He had served terms on the Palm Bay and Titusville City Councils, and two terms as chair of the Central Florida YMCA's board of directors. He sponsored a sportsmanship program at Titusville High School to teach honesty and integrity through sports. Proud of his Space Coast home, Fisher liked to brag to other insurance agents, "I'm from where they send people to the moon."[74]

Brevard County is divided into five supervisorial districts. Each district elects a representative to a board of commissioners. Unlike some US counties, these positions are partisan, with party candidates chosen in a primary before the general election later in the year.

Fisher decided in early 2008 to run for the Democratic nomination for District 1 commissioner. In a July 2023 interview with this author, Fisher said that the county Democratic party did not support his candidacy, preferring another candidate.[75] One reason was his refusal to sign a pledge promising to support Democratic candidates in the November general election. Fisher won the August 26 primary anyway, with 52.5 percent of the vote among three Democratic candidates.[76]

In November 2008, as it did at the time of this writing, District 1 leaned Republican. According to county records, on election day 43.6 percent of registered voters were Republican, 38.7 percent were Democratic, and 17.7 percent were third-party or nonpartisan.[77] Fisher beat the odds to win by 308 votes out of 52,403 votes cast.[78] He was the first African American elected to the county's board of commissioners.[79]

Fisher was part of a Brevard County delegation that traveled to Washington, DC, in early December 2008 to meet with Lori Garver's NASA transition team. Upon his return, Fisher told *Florida Today,* "They welcomed us with open arms and, basically, wanted to be briefed on some of the concerns we have here in Brevard County. And the loss of jobs is one that is near and dear to my heart."[80]

After the release of the Augustine Committee report, Fisher started a movement called Save Space to advocate for Brevard County's post–Space Shuttle program space interests. "I was worried about jobs," Fisher recalled in his 2023 interview. "People that lived in our community were going to be affected by it when the wheels stopped rolling. I was truly hoping to get some type of delay, to get the Shuttle to fly longer, and a clear plan for what was going to be funded next and what was going to happen to the whole space program."

Fisher began with a letter-writing campaign, hoping to send a half-million letters to the White House by the end of October. He also started a website,

SaveSpace.US, that was hosted by the Brevard County government. On the Letters page, the website said, "Between now and the end of October, send a letter to the President. Tell President Obama why America's Space program is important to you." The site offered six sample letters, including one written for students or family members of KSC employees.[81] In the last week of October, Fisher delivered 2,000 letters to the county's Economic Development Commission, far short of the 500,000 target, but told *Florida Today* that he hoped people across the nation were writing to Obama.[82]

On February 1, 2010, NASA released its proposed fiscal year 2011 budget. It was the most audacious and radical overhaul of the agency since President Kennedy transformed NASA into a demonstration of American soft power in the 1960s. The budget proposal would return NASA to its roots as a crucible of aerospace research and development. The OMB issued a press release that proclaimed, "The President's Budget cancels Constellation and replaces it with a bold new approach that invests in the building blocks of a more capable approach to space exploration." The proposal would have increased NASA's budget by $700 million from FY 2010, extending the ISS beyond 2016, invested $500 million to begin the commercial crew program, and provided $1.2 billion "for transformative research in exploration technology that will involve NASA, private industry, and academia, sparking spin-off technologies and potentially entire new industries." The budget overview provided by NASA said the budget proposed a "Top line increase of $6.0 billion over 5-years compared to the FY10 Budget, for a total of $100 billion over five years."[83]

"LOST ON SPACE," declared the *Florida Today* front-page headline the next day. "Under Obama budget, Constellation's demise likely to trigger more than the predicted 7,000 layoffs at KSC."[84] But the subheading was misleading; Constellation was never going to employ more than a few thousand contract workers for several years. A subsequent *Florida Today* editorial published on February 3 opined that "Obama broke his campaign promise to minimize post-shuttle reductions."[85]

Robin Fisher felt betrayed. "President Obama clearly didn't keep his word," he told the *St. Petersburg Times*. "Now we've got to hope we can get some congressional support to keep people working."[86]

Bill Nelson was already on the job. The day after the budget release, during a February 2 Senate Budget Committee hearing, Nelson told OMB Director Peter Orszag: "You have cut off the testing and development of an alternative rocket. There is no fail-safe position."[87]

Rep. Suzanne Kosmas (D-New Smyrna Beach), whose district included KSC, was on the House Space and Aeronautics Subcommittee that met the next day on February 3 to hear a GAO report on "Key Issues and Challenges Facing

NASA." In her opening remarks, Kosmas said: "I am extremely concerned about some of the lack of direction that we might have in the policy as put forth in the president's budget. I don't see a vision, I don't see an inspiration, and I see a major loss of workforce and workforce skills."[88]

NASA already had a workforce transition strategy in place. It had been mandated by the Consolidated Appropriations Act that Congress passed in December 2007. The law directed the NASA administrator to "prepare a strategy for minimizing job losses when [NASA] transitions from the Space Shuttle to a successor human-rated space transport vehicle." The act also required the administrator to "update and transmit to Congress this strategy" every six months until the successor vehicle was "fully operational."[89]

The third edition, published in July 2009, showed that no full-time NASA civil servants were going to lose their jobs, but the Space Shuttle program's contractor workforce was forecast to shrink in upcoming years. In FY 2010–2011 3,900 net contractor jobs would be lost as the Space Shuttle program ended, then 500 would be added for Constellation in FY 2011–2012 and another 500 in FY 2012–2013.[90]

On March 1, a month after the budget proposal was released, NASA opened the Workforce Transition Office at KSC. The local human resources office had partnered with the county's Brevard Workforce to help employees find new jobs. Two job fairs were already planned for June.[91] But for all the efforts to find new jobs for the workforce, the problem remained that no jobs were to be had because of the national recession.

Kosmas and Rep. Bill Posey (R-Rockledge) were two of twenty-seven congressional representatives who sent a letter to Administrator Bolden on February 12 asking him to disband "tiger teams" rumored to be shutting down Constellation-related contracts. The letter reminded Bolden that the 2010 Consolidated Appropriations Act passed in December 2009 forbade the administrator from terminating any Constellation-related programs and from creating any new programs.[92] The letter also asked for "your personal assurance that there will be no instructions to contractors or to Center Directors to slow down or to terminate contracts related to the Constellation programs."[93]

Posey's district didn't include KSC, but it did include CCAFS as well as central and south Brevard County, where many NASA employees and contractors lived. Posey and Kosmas attended a statewide space industry summit hosted by Governor Crist on February 18, 2010, at Orlando International Airport. Newspapers quoted Posey as accusing Obama of having broken a campaign promise to close the gap between "Shuttle and Constellation," although what Obama actually said in Titusville on August 2, 2008, was that he would close the gap by "speeding the development of the Shuttle's successor." He didn't say that Constellation would be the successor. Posey declared, "Well, it appears he hasn't closed the gap. He's made it eternal."[94]

Crist, who was running in the Florida Senate Republican primary against Marco Rubio, issued a statement before the event claiming, "By cutting this program, President Obama is putting an end to significant investment in moon exploration and costing Florida's Space Coast thousands of jobs."[95]

Space industry leaders at the summit tried to convince the politicians in attendance to embrace the NewSpace future. A *Florida Today* opinion column noted, "There were entrepreneurs who said President Obama's plan to shift manned launches to commercial rockets was correct, and Florida had to pull itself out of denial fast and get with the new program or be left in the dust."[96]

Local union representatives, meanwhile, organized a protest event. The Save Our Space Exploration rally was held February 27 near Brevard Community College in Titusville.

Fernando Rendon, a business agent for the IBEW Local 606 in Orlando, was one of several Brevard County residents recruited by *Florida Today* to write occasional guest opinion columns. On February 25, the newspaper published his column inviting the public to attend the rally. Rendon wrote: "Our elected officials must ensure there is continued federal funding for NASA and that federal protections apply to all workers on jobs where federal dollars are spent."[97]

By "federal protections," Rendon was referring to two federal laws. The Davis-Bacon Act, a 1931 federal law requiring payment of prevailing wages on federally funded or assisted construction projects, applied to contracts such as the electrical work performed by IBEW members, and plumbing work performed by the United Association of Plumbers & Pipefitters members.[98] The McNamara-O'Hara Service Contract Act (known as SCA) of 1965 applied similar requirements to contractors and subcontractors performing services on prime contracts such as United Space Alliance. USA had collective bargaining agreements with the IAMAW as well as the IBEW and other union locals.[99]

The initial Space Act Agreement awarded SpaceX in 2006 to demonstrate commercial cargo delivery capabilities to the ISS did not require the company to comply with Davis-Bacon or the SCA. It did not require SpaceX to use union employees at all.[100] Federal labor laws would not apply to strictly private contracts SpaceX might have with commercial customers or other nations.

As for the commercial crew program, Davis-Bacon did not apply because it was not a construction project. According to an October 2023 NASA Office of Communications email to this author, the SCA (now known as Service Contract Labor Standards) was applicable only "when the principal purpose of the contract is to furnish routine services to the Government through the use of service employees." The commercial crew contracts "are highly technical efforts not performed through the use of service employees," but by "professional employees that possess high level education degrees" or by "exempt professionals such as engineers, executives or administrators."[101]

The implications for union locals were sobering—even if NewSpace created jobs, those would not be union jobs, meaning lower wages, fewer members, and possibly unsafe working conditions.

In a September 2023 interview, Rendon said the Save Our Space Exploration rally was organized by the Space Coast Central Labor Council—an alliance of various union locals, which was one of ten AFL-CIO central labor councils in Florida. To his recollection, the labor council came up with the idea for the rally about two months before the event. The council invited Brevard Workforce to participate, but the agency declined. The labor team mobilized its membership to promote the event; Rendon recalled, "We hit every business establishment on the Space Coast with flyers and posters to advertise this. These businesses, from barber shops to mom-and-pop stores, they were all on board. They knew their jobs, their businesses depended on it. The community really was on board."[102]

A hyperbolic *Florida Today* front-page report the day after the rally opened with, "The coming collapse of the space industry." The front page showed protestors waving signs demanding that both the Shuttle and Constellation programs continue. Many signs read, "JOBS is Job One." The article estimated that about 1,500 protestors attended.[103]

District 1 Commissioner Robin Fisher spoke at the rally. "My family is worth fighting for," he said. "My community is worth fighting for. And these jobs are worth fighting for." Fisher pointed at the gymnasium where Obama spoke in August 2008. Fisher wanted to "help remind the president what he said in that building, that he's going to help save jobs." He called on Florida's elected representatives to "hold up all votes on everything until Florida is taken care of. If that stops Washington, DC, that's OK."

National labor representatives spoke at the event.[104] IAMAW General Vice President Bob Martinez told the attendees, "To me, the Constellation program is all about our national integrity and security. This is about our community, our country, and our jobs." AFL-CIO President Richard Trumka told the crowd, "At a time when our Great Recession has left us 11 million jobs short, does it make any sense at all to kill off another 7,000 jobs and endanger tens of thousands of more jobs? It doesn't make any sense! Enough is enough! It's time we keep it and make it right here in America! Our jobs should come first of all!"

The article quoted a United Space Alliance worker as saying he didn't want "to see our astronauts get suited up and get on Russian or Chinese spaceships," even though American astronauts and Russian cosmonauts had been flying on each other's spacecraft since 1994.[105] No one at NASA, much less the White House, had plans for a crewed space partnership with China. Another worker, a machinist, told the reporter, "We voted for the president, and this is what he's done to us, and we're pretty angry about it."

The implicit message was that perpetuating the local NASA contract workforce equated to American national security and prestige.

The IAMAW started a website, Vision4Space.com, that called for continuing the Space Shuttle program as well as the Constellation program. "President John F. Kennedy gave America a vision for space in 1962. Now we need a new vision for space today. The current plans to shut down the Space Shuttle program and cancel the Constellation program cannot stand, for the sake of our national prestige, our national security and our nation's future."[106]

* * * * *

Amid all the hysteria and accusations, the White House issued a press release on March 7 announcing that Obama would visit Florida on April 15 "to host a White House Conference on the Administration's new vision for America's future in space."[107]

> The President, along with top officials and other space leaders, will discuss the new course the Administration is charting for NASA and the future of U.S. leadership in human space flight. Specifically, the conference will focus on the goals and strategies in this new vision, the next steps, and the new technologies, new jobs, and new industries it will create. Conference topics will include the implications of the new strategy for Florida, the nation, and our ultimate activities in space.

Two days later, on March 9, a "Space Forum" was held at the Brevard Community College campus in Cocoa. The event was hosted by the college, *Florida Today*, and the Canaveral Council of Technical Societies. It was streamed live on the *Florida Today* website, and aired tape-delayed on the BCC community cable channel. The moderator was *Florida Today* "watchdog" columnist Matt Reed. None of the four panelists represented NASA or the Obama administration, which Reed acknowledged.[108]

Former Rep. Dave Weldon (R-Palm Bay) accused President Obama of "for all intents and purposes, killing the manned spaceflight program." Weldon resurrected the Obama campaign's early Constellation gaffe, waving the November 2007 education policy paper in the air, and claimed it was issued in August 2008. But Obama had renounced that plan during his August 2, 2008, Titusville rally. The education policy paper on his campaign website as of July 31 still referred to Constellation, but the Constellation reference had been deleted by August 7.[109] Weldon acknowledged that the George W. Bush administration was responsible for creating "the gap," but claimed Obama had said during the campaign that "he wanted to take that five-year gap and make it a ten-year gap. And now we know, he would just like to do away with the whole thing." Weldon concluded, "If you don't get rid of Barack Obama in 2012, this is going to be an ongoing problem, because he is not a space supporter. He has clearly laid out his agenda, and I think it's very bad news for us here in central Florida." Much of the audience cheered, whistled, and applauded.

During a discussion about the feasibility of commercial space, IBEW Local 2088 business manager Dan Raymond expressed the concern that this author heard from several union labor representatives interviewed for this book—NewSpace companies were likely to pay less than what government programs paid.

Space shuttle–era astronaut Winston Scott, another panelist who more recently had served as director of the Florida Space Authority, said he believed the solution was to increase NASA's funding. "I believe that the current direction of the administration where we get rid of our indigenous capability and all the fallout will cause us to relinquish our leadership in space."

Because the panel did not include a White House or NASA representative, the moderator quoted from a February 3, 2010, *Huffington Post* guest column written by Apollo 11 astronaut Edwin "Buzz" Aldrin, who endorsed the Obama administration's proposed budget. Aldrin called the proposal a "JFK moment."[110]

* * * * *

What is a "JFK moment"?

In the context of space politics, a classic example would be Kennedy's speech to Congress on May 25, 1961.

> I believe that this nation should commit itself to achieving the goal, before this decade is out, of landing a man on the moon and returning him safely to the earth.

This single sentence has been cited often over the years by those for whom Apollo is the paradigm. It has a goal, a destination, and a timeline.

Another example would be Kennedy's Rice University speech on September 12, 1962.[111] Many passages are quoted from that day, but perhaps this is the one oft-quoted:

> We choose to go to the moon in this decade and do the other things, not because they are easy, but because they are hard, because that goal will serve to organize and measure the best of our energies and skills, because that challenge is one that we are willing to accept, one we are unwilling to postpone, and one which we intend to win, and the others, too.

A common criticism of the Obama administration's proposed fiscal year 2011 NASA budget was that it didn't fit the Kennedy paradigm. What is the goal? What is the destination? What is the timeline?

By 2010, over 150 million Americans had been born since the Apollo 11 landing.[112] They had grown up on Apollo mythology. Many space advocates from these generations yearned for another president to deliver an inspirational speech calling for a bold exploratory adventure to the stars. This, they believed, would unite the nation, taxpayer dollars would flow, and American prestige would be satisfied. They knew no other paradigm, because that was what they were taught in school, in popular culture, and by NASA public relations.

Astrophysicist Neil deGrasse Tyson coined this yearning "Apollo necrophilia." Addressing the National Space Society on May 29, 2008, Tyson reminded the audience that Kennedy was motivated not by a passion for space exploration, but by the Cold War. "We have this memory of the period where a visionary leader challenges us to go to space, rather than is scared to death of the Commies and wants to beat them out to the Moon."[113]

Presidents since Kennedy have proposed similar visions and failed. President George H.W. Bush delivered such a speech July 20, 1989, on the steps of the National Air and Space Museum. President George W. Bush went to NASA Headquarters on January 14, 2004, to announce his Vision for Space Exploration. Neither was taken all that seriously by Congress.

And now the same trap was being set for Barack Obama.

James Webb had warned against setting such a trap. During his May 1969 Syracuse University interview, Webb said, "I think it's utterly foolish to ask a political leader to commit himself for some ten to twenty years in advance to some goal that's going to involve a lot of money and that can easily be attacked but is hard to defend."[114]

The interviewer referred to this as "the mandate idea, on the moon by such and such a date." Webb replied that he had never used the word *mandate* in his congressional testimony. "The public image that we had to get on the moon in this decade was not the real concept of the management of NASA. This was sort of a political forward thrust . . . What we had in mind was to build all elements of a total space competence." Project Apollo was the justification for developing "a full and complete set of space tools." Apollo was possible only because it built on technologies developed by earlier programs; Webb cited Project Gemini as an example of evolving "space competence."[115] Webb's view was aligned more with the role of NASA's predecessor, the NACA, than the political propaganda organ Kennedy created.

On February 24, 2010, Bill Nelson chaired a Senate Science and Space Subcommittee hearing titled, "Challenges and Opportunities in the NASA FY 2011 Budget Proposal," with Administrator Bolden the first witness. In his opening remarks, Nelson said he had told Obama, "There's only one person that can lead the space program, and that is the president, and that the president . . . would have to make that declaration of support." Nelson acknowledged that the budget proposal "has a lot of very forward thinking and cutting edge stuff," but it had been "misinterpreted" with some parts "completely overlooked." He contended that, because Obama "did not make a declaration himself" explaining why Constellation was being terminated at a time of "angst" due to pending layoffs, "It gave the perception that the president was killing the manned space program."[116]

Both *Florida Today* (February 26) and the *Orlando Sentinel* (March 2) ran editorials expressing a similar opinion. *Florida Today* said that Bolden's testimony was missing "the biggest detail of all—the ultimate goal and vision that goes with it." The editorial called for Congress to "withhold funding until policy details are clear and they have a chance to refine them." (Withholding funding would have halted all NASA operations, probably resulting in furloughs if not layoffs.) The *Sentinel* editorial opined, "The White House's plan would undermine America's prospects of maintaining its historic leadership in manned space exploration and continuing to reap the benefits it has yielded," such as "international prestige." No specific measurements for these benefits were cited by the editorial, just assumed. Both columns complained about the lack of protection for the KSC contractor workforce; *Florida Today* claimed that "killing Constellation would further decimate a KSC workforce that will suffer about 7,000 job losses when the shuttle stops flying."[117]

The Wall Street Journal reported on March 4 that Bolden "has asked senior managers to draw up an alternate plan for the space agency" that "threatened to undercut White House efforts to get its proposed NASA budget through Congress." The *Journal* cited a March 2 memo written by Johnson Space Center director Michael Coats, a 1980s-era astronaut and space shuttle mission commander. The email told senior managers at the other NASA centers that Bolden had "agreed to let us set up a 'Plan B' team."[118] Bolden released a statement that said, "I did not ask anyone for an alternative to the president's plan and budget."[119] In an August 5, 2015, NASA Oral History Project interview, Coats acknowledged that Bolden had not authorized a "Plan B." Coats had taken the initiative himself during a senior management meeting with Bolden. In the interview, Coats said he believed that Bolden and perhaps Nelson had intervened with the White House to keep him from being fired.[120] As JSC director, Coats objected to the commercial crew program being based at KSC. He thought it should be at JSC because "we happen to know a little bit about crews." Coats claimed that Bolden told him, "We have to do it for political reasons. Bill Nelson wanted it there, and the White House said, 'Florida's a purple state. We're going to put it down there.'"[121]

Nelson met with Obama and Vice President Joe Biden at the White House on March 16 to discuss Constellation and KSC job losses. After the meeting, he told reporters they had an "excellent conversation" and, "We'll see the fruits of that conversation when the president visits on April 15."[122]

Two days later, on March 18, Nelson's space subcommittee held another hearing titled, "Assessing Commercial Space Capabilities." In his introductory remarks, Nelson stated that he expected Obama "to lay out his vision and what the goals and what the timelines are for America's manned space program" during the president's KSC visit.[123]

The next day, on March 19, at yet another space forum in Cocoa cohosted by Brevard Community College and *Florida Today*, Nelson announced that he would have the US Senate draft legislation ordering NASA to develop a super-heavy lift rocket for missions beyond low Earth orbit, as well as directing the agency to hire Space Coast workers for commercial cargo or crew missions to the ISS. "It is my hope that we're going to get additional work that is going to cushion the blow after the last space shuttle mission is flown. It's time we get out of low Earth . . . and that's what we intend to do."[124]

While Nelson advocated for a JFK moment, another delegation of six Brevard County representatives planned a trip to Washington, DC, to lobby Congress.[125] District 1 Commissioner Robin Fisher told *Florida Today*, "The goal is to try to get those guys to come up with some kind of a timeline, a destination and a vision for human spaceflight." Joining him was New Life Christian Fellowship Pastor Larry Linkous, who said, "If you live and breathe in this area, you need to be very, very concerned about the proposed budget in its present state." Cocoa Beach Area Chamber of Commerce President Melissa Stains said, "We are not telling them how to do it. We're telling them it needs to be done."[126]

After they returned, Fisher began to organize another rally, called Save Space after his earlier letter-writing campaign. The rally was to be held April 11 at the Cocoa Expo Sports Center in the city of Cocoa. *Florida Today* reported that retired astronauts Jon McBride, Bob Springer, and Winston Scott would speak at the rally, along with Florida Lieutenant Governor Jeff Kottkamp. "This is a pro-space rally," Fisher told the paper. "It is about jobs in the community."[127] The SaveSpace.US website's home page declared, "We need a **BOLD GOAL**, a **DESTINATION**, and a **TIMELINE!"** (**Bold** emphasis in the original.)[128]

The Cocoa Expo event began with the national anthem.[129] All three retired astronauts in attendance wore their iconic NASA-issued blue flight suits. Speaking first, Pastor Linkous said, "Many still remember the call of President John F. Kennedy, inspiring and uniting America, in a grand mission that took our astronauts to the moon. . . . President Obama, we say it's time for another JFK moment for America's human space exploration."

Fisher had received an invitation to attend Obama's KSC event on April 15. He told those at the rally what he intended to ask the president if given the chance: "I'm going to say, 'Mr. President, can you tell me what your vision is? Can you tell me what the goal is? Give me a destination. I need a timeline. We need human spaceflight.'"

Rep. Posey claimed that Obama had broken his promise to "close the gap between Space Shuttle and Constellation and to keep America first." As noted earlier, Obama never promised to continue Constellation.

Rep. Kosmas added, "I have said from the beginning that what was proposed in the President's budget is not acceptable for the Space Coast or for the future of America's space program." She called for continuing the Space Shuttle program as well as to "take what is useful in the Constellation program and maximize it for future space exploration." Kosmas said Obama should provide "a destination, a NASA-led vehicle, and a timeline, so that our workforce will be focused and ready to do what it is designated that they should do for the next phase of space exploration."

Lieutenant Governor Jeff Gottkamp invoked memories of the Apollo program and said, "We cannot allow the Chinese or the Russians or anyone else to seize superiority in space exploration away with us." He then asked Obama to "go back to the moon by 2015, and we need to go to Mars by 2020," ignoring the reality that such technologies didn't exist, nor did the funding or the political support for it—but the line got a round of applause. Gottkamp concluded, "Mr. President, don't take away these jobs! Save them!"

* * * * *

The *Orlando Sentinel*'s online space blog "The Write Stuff" reported on March 24, 2010, that Lori Garver had been in town "scoping out possible venues" for Obama's town hall. According to the report, "Garver really liked the O&C building because it was designed to be energy-efficient and could be used to assemble commercial capsules that would launch from nearby pads."[130] The Operations and Checkout Building was completed in 1964. It was used to process Apollo spacecraft and lunar modules, then the Skylab space station, and then space shuttle payloads. Space Florida contributed $35 million to LockMart in 2006 to help the company refurbish the O&C for the Crew Exploration Vehicle, now known as Orion.[131] An Orion capsule mockup would serve as a symbolic backdrop for Obama's address.

Garver wrote in her book *Escaping Gravity* that by this time the Obama administration "in an effort to court Senator Nelson" had decided the president would visit KSC and "offer concessions" if Nelson would guarantee the administration's top priorities, including commercial crew. The administration offered to continue the Orion crew vehicle, but it would now be used only as an ISS "lifeboat" in case of emergency. The president would also identify a "deep-space destination" for crewed flight, but only if "the next intended place astronauts would go [was] a meaningful destination [that] could be achieved realistically" by 2025.[132]

In his 2017 Lambright interview, Charlie Bolden recalled that he asked Obama's cabinet secretary Christopher Lu to arrange a meeting for him with the president. "I believed strongly that he would need to deliver a message that gave hope to the workforce that we would continue to pursue human space flight and go beyond low-Earth orbit." He also recommended that Obama "meet Congress half way," working with Nelson and Senator Kay Bailey Hutchison (R-TX) to develop a compromise that would be a foundation for NASA's authorization bill.[133]

On April 13, two days before the KSC event, the White House began a media blitz to boost support for the president's NASA budget proposal. The White House Office of Science and Technology Policy released a three-page "Fact Sheet on the President's April 15th Address in Florida." The fact sheet claimed that the Obama plan "leads to more than 2,500 additional jobs in Florida's Kennedy Space Center area by 2012, as compared to the prior path." The word *jobs* appeared four times on the first page. The document touted creation of "a new commercial space transportation industry . . . projected to create over 10,000 jobs nationally over the next five years."[134] The job numbers appear to have come from a Tauri Group analysis released April 9. The Tauri Group was an analytical consulting firm retained by government agencies and contractors to provide independent analysis. They had been hired by the Commercial Spaceflight Federation, an advocacy group largely comprised of NewSpace companies, "for an objective estimate of jobs" that would be created by Obama's budget proposal.[135] The analysis assumed that Congress would fund the budget as proposed, which would turn out to be its fatal flaw.

The OldSpace community responded with heroes of the Apollo era. Apollo astronauts Neil Armstrong, Jim Lovell, and Gene Cernan signed a letter calling

Constellation's cancelation "devastating" and suggesting it "destines our nation to become one of second or even third rate stature."[136] The *Orlando Sentinel's* "The Write Stuff" blog published a second letter, signed by twenty-seven men, including Lovell, Cernan, and other astronauts of NASA's golden age, as well as former NASA Administrator Michael Griffin. The letter stated, "We are very concerned about America ceding its hard earned global leadership in space technology to other nations. We are stunned that, in a time of economic crisis, this move will force as many as 30,000 irreplaceable engineers and managers out of the space industry." The authors did not cite where they got the 30,000 number, much less how these engineers were being forced out of the space industry.[137]

Air Force One touched down on KSC's runway that mid-April day at 1:25 p.m. The president was not alone when he descended its air stair. With Obama was Buzz Aldrin, signaling that at least one Apollo astronaut supported his proposal. Rep. Kosmas and Senator Nelson were also aboard.

Protestors gathered alongside NASA Causeway, the road crossing the Indian River from Titusville to the space center. WESH-TV Channel 2 Orlando reporter Michelle Meredith estimated that there were about 200 protestors on the causeway. She said that none of the protestors were willing to identify themselves as "NASA workers" and concluded that they were "mostly Tea Party protestors who are taking on the space issue." When offered an audio feed to listen to Obama's speech, the protestors yelled, "No!"[138] *Florida Today* noted that among their protest signs were ones that read "Obama Lied, NASA Died," "Where is JFK?" and "Obama Kills KSC." The newspaper reported that the protest had been organized by Space Coast Patriots, whose members were aligned with the Tea Party movement.[139]

Obama never saw the protestors. His motorcade took him to Cape Canaveral Air Force Station's Launch Complex 40, where a SpaceX Falcon 9 was vertical on the pad. This booster was scheduled for the first test flight in June of a SpaceX rocket, and it would be from Cape Canaveral. The booster's payload would be a mockup of the cargo Dragon spacecraft.[140]

The SpaceX tour yielded a trove of historic photos showing President Obama striding across the pad, his suit jacket slung over his left shoulder, with Elon Musk at his side, and a Falcon 9 erect before them. That was all the public ever saw, until 2023 when a journalist filed a Freedom of Information Act request.

In response, the Barack Obama Presidential Library on June 7, 2023, posted on YouTube video clips from the SpaceX tour.[141] The video shows Obama entering the pad's horizontal integration hangar, where he is met inside by Musk. The two then walk out to the launch pad. Parts of their conversation can be clearly heard. After thirteen years, the public now knows what they talked about.

President Barack Obama tours Launch Complex 40 with SpaceX CEO Elon Musk on April 15, 2010. Image source: NASA/Bill Ingalls.

The two men shook hands. Obama said to Musk, "Tell me about SpaceX."

Elon appeared a bit nervous. "Well, our goal is to make a huge difference in space exploration, try to make space a lot more like air transport."

Musk praised the "revolution taking place" at NASA but said that "on the space side" the FAA had yet to adapt. "Well that's good to know," Obama replied. The president turned to his science adviser, John Holdren. "John?! FAA. We're going to have to get them involved."

The entourage exited the hangar and headed for the pad, where the erect Falcon 9 waited. Obama removed his jacket. Musk walked alongside. "How are you?" Obama called out to four pad workers watching a president inspect their work. He shook hands with each of them.

Obama and Musk stood before the Falcon 9. Pointing up at the fuselage, Musk explained the thermal protection system that would allow the booster design one day to land so it could fly again. "The real big breakthrough that's needed with rocketry is a truly reusable rocket," he said. "Y'know, if we're to make a revolution in space, it's gotta be reusable. If you think about any mode of transport—bicycles, planes, every—"

"It's just too expensive to have a disposable rocket," Obama interjected.

"Exactly," Musk replied. "And, thus far, it's been an elusive dream in space to create a truly reusable rocket, but that's what we're gonna try to do with that," he said, gesturing again at the fuselage. Musk explained how, "in a later phase," he intended the Falcon 9 to fly back and land on a drone ship at sea. Obama could be heard saying, "It's going to be exciting!"

The brief tour ended. Obama turned for his motorcade. He and Musk discussed distant plans for crew Dragon.

"Obviously," Obama said, "we're gonna try to see if we can accelerate this kind of, uh, this kind of collaboration."

"It's the right plan," Musk assured him.

"Well, congratulations. This is terrific." They shook hands again.

That gesture conjoined their fates. Elon Musk intended to risk his financial fortune to create a reusable rocket. Barack Obama intended to risk his political capital to make it happen. They sealed their bargain with a handshake.

The Cape's comeback began in this moment, with a handshake, on this pad, that decades ago had been no more than swamp and scrub.

But now the president had to go sell it.

* * * * *

Robin Fisher was part of a small select group invited to a private meeting with Obama and Bill Nelson before the speech. They were gathered in the Mission Briefing Room of KSC's cavernous O&C Building.

"Nelson joked with President Obama," Fisher recalled in his July 2023 interview, "about me being a Florida Gator, and I was a pretty bad dude. The president kicked back and said, 'Yeah, I know.' He had read all my stats, which I was really impressed by. He said, 'Well, I have some bad dudes next to me.' He pointed to his Secret Service guys. They opened their coats up and smiled."

Nelson asked those in the room to tell Obama "who he's getting ready to face in this crowd and what are some of their concerns." Fisher stood and told the president what he thought, as he had promised the Cocoa Expo crowd four days earlier. "I didn't tell him anything he hadn't already heard. He knew the issues. He knew what people were saying. He was well read on stories and facts and what was being said."

In Fisher's words, Obama told him, "I know nobody will believe me now, but soon as we turn this over to the private sector, they will do it better, faster,

President Barack Obama is greeted by Senator Bill Nelson and NASA Administrator Charles Bolden on April 15, 2010. Image source: NASA/Jim Grossmann/DVIDS.

and cheaper. Commercial space will be good for this country. I know you don't see it now, Robin, but trust me, it will be."

Fisher now acknowledges, "He was right."

Also in the room were Rep. Kosmas; space activist Dale Ketcham, who was now the vice president of government and community relations for Space Florida; Al Neuharth, founder of *Florida Today* and later *USA Today;* and Malcolm Kirschenbaum, a business lawyer who was a close friend of both Nelson and Neuharth.[142]

In a November 2023 phone interview with this author, Kirschenbaum said that he met Nelson in 1958. They attended the University of Florida together in 1960 and had been good friends ever since. Malcolm had also been Al Neuharth's lawyer for nearly fifty years; they drove together to the event.[143]

Kirschenbaum recalled that the private meeting was an opportunity for Obama to privately meet some "Democratic locals" and tell them about what he intended to say in his speech. Obama in particular was interested in the privatization of the government's space operations. "He wanted to know a little bit more about the history of what was going on out here."

The speech lasted about twenty-five minutes.[144] The president was introduced by Administrator Bolden, with Senator Nelson also standing on the dais.

Obama acknowledged his awareness of local angst:

> I know there have been a number of questions raised about my administration's plan for space exploration, especially in this part of Florida where so many rely on NASA as a source of income as well as a source of pride and community. And these questions come at a time of transition, as the space shuttle nears its scheduled retirement after almost 30 years of service. And understandably, this adds to the worries of folks concerned not only about their own futures but about the future of the space program to which they've devoted their lives.

The president proposed increasing NASA's budget by $6 billion over the next five fiscal years, despite the freeze on discretionary spending and budget cuts elsewhere. He also called for investing $3 billion to conduct research on an advanced heavy lift rocket based on new technologies to reach deep space; he expected NASA to start building it by 2015. He declared his intent to reverse the Bush administration's plan to end the ISS by 2015, extending it to at least 2020 and most likely beyond.

Obama then gave the critics what they wanted—a goal, a destination, and a timeline.

The goal: "Our goal is the capacity for people to work and learn and operate and live safely beyond the Earth for extended periods of time, ultimately in ways that are more sustainable and even indefinite."

The destination and timeline: "Early in the next decade, a set of crewed flights will test and prove the systems required for exploration beyond low Earth orbit. And by 2025, we expect new spacecraft designed for long journeys to allow us to begin the first-ever crewed missions beyond the moon into deep space. So we'll start by sending astronauts to an asteroid for the first time in history. By the mid-2030s, I believe we can send humans to orbit Mars and return them safely to Earth."

"And a landing on Mars will follow."

9

The Grand Compromise

In her memoir *Escaping Gravity,* Lori Garver recalled that, after his KSC speech, Obama came up to her and said, "Do you think this will help?" She wrote, "I responded honestly that if it doesn't, nothing would. Unfortunately, I was right. Nothing helped."

The next day, *Florida Today* founder Al Neuharth published in the morning edition his personally penned opinion column. It was titled, "A 'Devastating' Plan: Obama Doesn't Get It; Space Is Last Frontier."[1] The column began: "President Obama in effect pulled the plug on our space program in a speech here Thursday, although he masked it with some vague long-term suggestions."

One might think Neuharth hadn't been there. Evoking Apollo mythology, he wrote, "The late President John F. Kennedy must have turned over in his grave."

Neuharth passed away in 2013, so he can't tell us why he wrote what he did. Some of those who commented on his passing described him as "contrarian." John Seigenthaler, who was *USA Today*'s founding editorial director, wrote, "His ideas often struck us as unexpected, unrealistic, even contrarian; opposite to what opinion polls said public opinion reflected."[2] NBC News anchor and journalist Tom Brokaw commented, "To the end of his life, he was a contrarian in how he tweaked the journalistic establishment." Howard Kurtz of CNN's *Reliable Sources* said, "He took great relish in his finger-in-the-eye approach."[3] Neuharth may have intended a provocation. Why, we'll never know.

Florida Today also published a guest column by Elon Musk. The SpaceX founder concluded, "Today, the president articulated an ambitious and exciting new plan that will alter our destiny as a species. I believe this address could be as important as President Kennedy's 1962 speech at Rice University."[4]

In his speech, Obama called for a $40 million jobs retraining program to help KSC workers, but that wasn't enough for IAMAW District 166 directing business representative Johnny Walker. "For what jobs?" he said to *The Palm Beach Post.* "I commend him for what he said he wants to do down the road." But those jobs were not immediate.[5] Later that summer, Walker would release a statement blaming Obama for the end of the Space Shuttle program, and for NASA's partnership with Russia.[6]

Senator Bill Nelson issued a statement saying that Obama had "announced at least two significant changes" he had sought.[7]

> One is having NASA proceed with building a new space capsule and another is the development of a so-called heavy-lift rocket . . .
>
> The president called for making a decision on a heavy lift rocket design in 2015. "I think we can make the decision much sooner," Nelson said. "We're going to keep testing the monster rockets at Kennedy Space Center."

The term *monster rocket* would be employed by Nelson in the months and years to come, as he and other senators took it on themselves to write legislation mandating that NASA build such a rocket.

Working in tandem with other senators, primarily Kay Bailey Hutchison (R-TX), Nelson began to craft a compromise legislation that could pass both houses of Congress.

On March 3, Hutchison introduced the Human Space Flight Capability Assurance and Enhancement Act.[8] The bill ultimately went nowhere—it was referred to the Senate's science committee, which never took it up—but it laid the foundation for the compromise Hutchison and Nelson were about to seek.

A March 22 memo to Nelson from his space policy adviser briefed the senator for his meeting the next day with Hutchison. "Sen. Hutchison requested this meeting and would like to discuss the NASA budget, the legislation she just introduced, and your position on future space shuttle operations." The memo listed the high points of Hutchison's bill; among them were extension of the Space Shuttle program by limiting it to two flights a year, support for "commercially developed crew transportation systems," and "continued development of a government heavy-lift vehicle" that would be "shuttle-derived" and "begin development immediately."[9]

Around this time, Nelson and the rest of the Florida congressional delegation sent a joint letter to President Obama expressing "deep concerns" with the NASA budget proposal. The letter, dated March 4, described as "especially worrisome" the lack of a "specific heavy-lift program" to replace Constellation. It also worried about the "major upheaval" for "the highly skilled workforce" with no "hope of recovery for many years."[10]

The Senate Committee on Commerce, Science and Transportation held a hearing on May 12, 2010, titled "The Future of U.S. Human Space Flight."[11] It would become infamous for the scathing testimony by Apollo astronauts Neil Armstrong and Gene Cernan. In his testimony that day, Charlie Bolden said he'd spent hours with the two men trying to explain NASA's new course to them, but it didn't matter. In the end, Bolden said, they chose to "disagree."

In his written testimony, Armstrong claimed that the budget proposal "was likely contrived by a very small group in secret who persuaded the President that this was a unique opportunity to put his stamp on a new and innovative program. I believe the President was poorly advised."[12] Cernan testified, "Either the Administration and the originators of this budget proposal are showing extreme naiveté or, I can only conclude, they are willing to take accountability for a calculated plan to <u>dismantle</u> America's leadership in the world of Human Space Exploration. In either case, the proposal is a <u>travesty</u> which flows against the grain of over 200 years of our history and, today, against the will of the majority of Americans." (Underline emphasis in the original.)[13]

In a *60 Minutes* interview that aired on March 18, 2012, SpaceX founder Elon Musk said of Armstrong and Cernan, "I wish they would come and visit and see the hardware we're doing here. And I think that would change their mind."[14] Neither ever did. Armstrong passed away in August 2012. Venture capitalist and SpaceX board member Steve Jurvetson invited Cernan to tour SpaceX, but Cernan declined. Jurvetson finally flew to Houston to personally meet with Cernan. In a July 2012 post on Flickr, Jurvetson wrote, "As I told him these stories of heroic entrepreneurship, I could see his mind turning. He found a reconciliation: 'I never read any of this in the news. Why doesn't the press report on this?'"[15]

Cernan passed away in January 2017, apparently having never visited SpaceX or publicly retracted his allegations.

In any case, their criticisms made national headlines. "Astronauts Attack Obama's NASA Plan," was the *New York Times*'s website headline.[16] "Astronaut Legends Criticize Obama Space Plan" was the headline on the CBS News website.[17] *PBS NewsHour* reported that the "reclusive Neil Armstrong" had come to Capitol Hill to tell the senators that "President Obama's plan jeopardizes America's place as a space pioneer."[18] Similar headlines appeared in east-central Florida newspapers. "Space Pioneers Criticize Budget" was the *Florida Today* front-page headline.[19] "Ex-Astronauts Call Plan for NASA a 'Pledge to Mediocrity,'" was the page 4 headline in the *Orlando Sentinel*.[20]

It's unclear who decided to invite Armstrong and Cernan to the hearing, and why their testimony was considered more relevant than, say, Buzz Aldrin's, who had endorsed Obama's proposal. But if the intention was to divert attention from Obama's reformist plans, that intent succeeded.

The May 12 hearing began the process of the Senate writing its own NASA budget proposal. This is no different from any other agency's budget request. The Budget and Accounting Act of 1921 requires the president to submit an annual budget proposal to Congress. But the US Constitution gives Congress the ultimate control over federal purse strings.[21] There's an old adage in federal politics: "The president proposes, but the Congress disposes."

The committee's senators used the hearing to stake their positions. Many members of the committee represented states with NASA centers and/or legacy contractors. Bill Nelson and Republican George LeMieux represented Florida. Kay Bailey Hutchison (R-TX) had Johnson Space Center in Houston. David Vitter (R-LA) had NASA's Michoud Assembly Facility near New Orleans, where the shuttle's external tank was manufactured. Roger Wicker (R-MS) had NASA's Stennis Space Center and its many rocket engine test stands. Mark Warner (D-VA) had NASA's Wallops Flight Facility.

The Commerce Committee drafts the legislation authorizing NASA operations, but the real power lay with the Senate Appropriations Committee, which disburses funding to government agencies. Hutchison sat on the Appropriations Committee as well, giving her far more influence than other members. Many senators on this committee also had NASA interests as well as its legacy contractors. Barbara Mikulski (D-MD) had NASA's Goddard Space Flight Center. Patty Murray (D-WA) had Boeing. Robert Bennett (R-UT) had solid rocket booster maker ATK. Thad Cochran (R-MS) also had Stennis. Mary Landrieu (D-LA) also had Michoud. Richard Shelby (R-AL) had NASA's Marshall Space Flight Center in Huntsville as well as ULA's manufacturing operations in Decatur. In February 2023, Shelby was named Porker of the Month for the sixth time by Citizens Against Government Waste.[22]

Whatever legislation emerged from this sausage grinding would then have to reconcile with the sausage ground by the House. If and when both houses finally passed it, the reconciliation bill would then be sent to the president for signature or veto.

In May 2020, Lori Garver wrote a guest column for the CNBC website documenting her personal account of what happened that summer of 2010. Garver, Charlie Bolden, OMB Director Jack Lew, and White House Director of Legislative Affairs Rob Nabors met on Capitol Hill with Senators Nelson and Hutchison. Nelson and Hutchison said that they also represented the interests of Senators Mikulski and Shelby. The two senators offered a compromise—the committee would authorize commercial crew only if the White House agreed to support a NASA-designed and owned super-heavy lift vehicle built by the Constellation contractors. The Orion capsule, renamed the Multi-Purpose Crew Vehicle, would carry NASA crew atop this new booster to unspecified destinations beyond Earth orbit, but some legacy contractor jobs would be saved. With minor amendments, the White House agreed. Garver had her doubts, concerned that the compromise lacked specificity about funding amounts and timelines.[23]

A July 23, 2010, *Orlando Sentinel* article reported that, in this meeting, Bolden and Garver warned Nelson and Hutchison that "NASA could not finish the proposed rocket before 2020, according to three sources present at the meeting."[24] But the final legislation that the Senate would soon pass required that the new super-heavy lift vehicle and its Orion spacecraft have "operational capability" by December 31, 2016.

In a December 2023 interview with this author, Senator Nelson gave his perspective. Many key members in both houses of Congress opposed the White House NASA budget proposal. He and Senator Hutchison were attempting to craft a NASA authorization bill that could pass Congress. "We did it not only in a bipartisan way, but also in a way we could pass it," he said. The idea was to order a new NASA super-heavy rocket based on space shuttle legacy technology, which would satisfy the opponents of the Obama budget. "We had to take all that practicality into effect," he explained, "because there were a lot of members of Congress, first senators, who had a stake in this, because they wanted the employment to continue in their states." The Space Shuttle program spinoff would be on a "dual track" with the commercial cargo and crew programs. "Commercial companies would become our partners."[25]

Nelson's insight reminds us of the realpolitik of Congress. One can argue all the merits of a NewSpace NASA but, if Senate and House majorities are unwilling to support the reform, it's doomed to fail.

A July 16, 2010, *Orlando Sentinel* article supports Nelson's recollection that members of both houses of Congress insisted on NASA developing its own rocket, distrusting commercial companies and intent on protecting jobs in their home states and districts. "Nelson said he was forced to compromise," according to the article, which opined that NASA's future was to be determined by "political science," not "rocket science." Former NASA astronaut John Grunsfeld told the reporters that the bill had mandated a "politically constructed" rocket, not one designed by expert engineers.[26]

Nelson also dismissed NASA's warning that the super-heavy lift vehicle couldn't be operational by 2016. The article quoted Nelson as saying that the 2016 deadline in the draft bill was "doable . . . If anyone tells you that's it not, then . . . I would question their particular agenda."

Florida newspaper accounts that summer reported that Space Florida and other Space Coast leaders were concerned that "Nelson—Florida's main space supporter—would take away billions of dollars from commercial rocket and technology development that over the next decade would have diversified the aerospace industry in Florida and provided KSC with new jobs and prestige," according to a July 12, 2010, *Orlando Sentinel* article.[27]

The super-heavy lift vehicle came to be known as the Space Launch System (SLS). The term first appears in Senate Bill 3729, the National Aeronautics and Space Administration Authorization Act of 2010. S.3729 was introduced on August 5 by Senate commerce committee chair Jay Rockefeller (D-WV). It was rushed to the Senate floor that day so it could pass before the August recess.[28]

While the Senate crafted S.3729, the House debated their own version, H.R.5781. That bill was approved by the House Committee on Science and Technology on July 28 and was sent to the House floor, but no final vote was taken.[29]

H.R.5781 was never brought to the House floor because of objections by members of the California delegation to provisions that would continue Constellation while underfunding commercial crew, offering those vendors loans instead of competitive prizes. The House adjourned for the month of August without acting on the bill.[30]

When Congress reconvened after the August recess, Nelson approached Steny Hoyer (D-MD), the House majority leader whose district included Goddard Space Flight Center. Nelson and Hoyer were friends from Nelson's time serving in the House. In his December 2023 interview, Nelson said that Hoyer agreed to place S.3729 on the House suspension calendar, meaning that the rules may be suspended so a bill can be brought directly to the floor for a vote. Suspension of the rules requires a two-thirds approval of the members present. On September 29, House members Dutch Ruppersberger (D-MD) from Baltimore and John Culberson (R-TX) from Houston escorted Nelson on the House floor to their respective parties' members so he could secure votes. Senator Hutchison also worked the floor to secure Republican votes. Late that night, the House approved S.3729 by a 304–118 (72 percent Yea) margin.

S.3729 was signed by President Obama on October 11, 2010.

* * * * *

S.3729 "authorized to be appropriated" funding levels for each of the next three federal fiscal years. But what is "authorized to be appropriated" doesn't mean it will actually be that amount. Nor does it mean that NASA can't ask for more (or less) in its annual budget requests.

In subsequent budgets, NASA would request more than authorized, hoping to speed up commercial crew development and operation, which might have soon liberated American astronauts from the Russian Soyuz spacecraft. But Congress said no.

Congress was far more generous with SLS, but that funding also fell short of what was needed to stay on schedule using the old approach with cost-plus contracts.

Section 304 of the 2010 authorization act all but forbade NASA from competing SLS contracts. Titled "Utilization of Existing Workforce and Assets in Development of Space Launch System and Multi-Purpose Crew Vehicle," the act required NASA "to the extent practicable" to use "existing contracts, investments, workforce, industrial base, and capabilities from the Space Shuttle and Orion and Ares 1 projects." NASA was required to use "Space Shuttle–derived components that use existing United States propulsion systems" and "minimize the modification and development of ground infrastructure."[31]

A July 2014 GAO report found flaws with the congressional design.[32]

> The use of heritage hardware . . . was prescribed in the NASA Authorization Act of 2010, but the hardware was not originally designed for SLS. Therefore, the SLS program must ensure each heritage hardware element meets SLS performance requirements and current design standards . . . each heritage hardware element shares the common issue of operating in the SLS environment that is likely to be more stressful than that of its original launch vehicle as well as unique integration issues particular to that element.

The GAO also called out the required continued employment of the legacy workforce "without contract definitization"—meaning the contracts lacked specific requirements. "Contract actions such as these authorize contractors to begin work before reaching a final agreement with the government on contract terms and conditions. The government is thus in a weaker position to control costs."[33]

The report implicitly criticized the awards of the no-bid contracts. "We have found that promoting competition increases the potential for acquiring quality goods and services at a lower price and that noncompetitive contracts carry the risk of overspending because, among other reasons, they have been negotiated without the benefit of competition to help establish pricing."[34]

The GAO found that NASA had not "developed an executable business case" that would match available funding to the agency's updated target launch date of December 31, 2017. The report concluded that NASA was already $400 million short of what would be needed to have a shot at achieving that objective.[35]

* * * * *

Another casualty of the compromise was Obama's $40 million jobs retraining program to help KSC workers.

The same day as Obama's KSC event, NASA released a fact sheet titled, "Florida's Space Workers and the New Approach to Human Spaceflight." The release stated, "The Administration is launching a $40 million, multi-agency initiative to help the Space Coast transform its economy and prepare its workers for the opportunities of tomorrow." The president directed the creation of a task force to deliver to him an economic development plan within 120 days. The team would be comprised of senior officials from NASA, Labor, Commerce, and Defense. The initiative proposed using $40 million from Constellation funds "to transform the regional economy and prepare its workforce for these new opportunities."[36] Obama issued a memorandum on May 3, 2010, appointing the task force members, with NASA Administrator Charlie Bolden and Commerce Secretary Gary Locke named its cochairs.[37]

Bolden and Locke led a June 4 town hall at the Orlando International Airport. Bolden said he wanted to "hear from the local communities things they think they have to offer. Things they are willing to do. It is critical for the local communities to decide that they want to diversify. They will have to change."[38] On June 24, NASA launched a web page for the task force to inform the public on its progress and to solicit comments.[39]

Separate from the task force initiative was a $15 million Department of Labor grant to Brevard Workforce. The grant, from the department's National Emergency Grants program, would fund services such as career guidance and job search skills training.[40] Labor Secretary Hilda Solis came to KSCVC on June 2 to announce the award, accompanied by Rep. Kosmas and Lori Garver.[41] Brevard Workforce started a new website, LaunchNewCareers.com, "to help aerospace workers transition smoothly from Space Shuttle/Constellation jobs to new employment both inside and outside of aerospace."[42] NASA and Brevard Workforce partnered to hold a job fair at KSC on June 24 and 25 that hosted sixty government and private employers recruiting workers. The SpaceX vendor booth was the most popular, according to one media account.[43]

Commerce Secretary Locke toured KSC's SLSL on August 4. He met with twelve NASA contractor employees and assured them that help was coming. "A lot of resources will be coming to this area in the months ahead," he told *Florida Today.*[44]

The task force report was delivered to the president on August 15. It assumed that Congress would approve the president's $40 million funding request, which turned out to be a fatal error.[45]

While the task force deliberated, the two chambers of Congress were working on their respective versions of NASA's 2010 authorization act. As the Senate wrote S.3729, the House debated their own version, H.R.5781. Section 223 of H.R.5781 authorized a Post-Shuttle Workforce Transition Initiative Grant Program "to make grants for the establishment, operation, coordination, and implementation of aerospace workforce and community transition strategies."

NASA Deputy Administrator Lori Garver at a Kennedy Space Center Visitor Complex event on June 2, 2010. The event announced a $15 million grant to help displaced aerospace workers find new employment. Image source: NASA.

The bill would have authorized $60 million for FY 2011, $40 million for FY 2012, and $40 million for FY 2013.[46]

If it had passed.

The Senate bill contained no specific provision for funding the workforce grant program. Section 602(b)(2) authorized the NASA administrator to work with other federal agencies "to assist displaced workers with retraining and other placement efforts" but beyond that offered no specifics or requirements. Funding was to come from space shuttle flight operations; the original White House proposal would have used diverted Constellation funds.

When the two chambers of Congress pass different versions of a bill, the differences typically are resolved through a process called reconciliation. The two chambers appoint representatives to a temporary ad hoc panel to reconcile their differences.

But the Senate bill went straight to the House floor thanks to suspension of the rules. Suspension does not allow for any amendments to a bill; it's a yea-or-nay vote. So when S.3729 passed the House on September 29, the original House language regarding the workforce program was lost.

One last opportunity remained to provide the workforce program funding. That was through the appropriations bill.

The federal budget cycle begins on October 1. If Congress has not passed an appropriations bill by then, the government shuts down as it has no funding to operate. To keep the government running, Congress passes a continuing resolution, which typically continues funding existing programs but not new ones.

The Senate Appropriations Committee on July 22 sent to the Senate floor S.3636, which included $50 million in NASA funding for transfer to the Department of Commerce "to spur regional economic growth in areas impacted by Shuttle retirement and exploration programmatic changes," and another $15 million for transfer to the Department of Labor "for job training activities in areas impacted by job losses associated with Shuttle retirement and exploration programmatic changes."[47]

But the House failed to pass its own appropriations bill by October 1, so the chambers began passing continuing resolutions to keep the government open.

The November 2 congressional elections were a bloodbath for the Democrats. The Republicans picked up a net of sixty-three House seats to win a 242–193 majority. The Democrats held on to the Senate, but lost six seats for a 53–47 majority (two independents caucused with the Democrats).[48]

With no incentive to cooperate in the lame-duck days until the next session began in January 2011, Senate Republicans announced they would filibuster any legislation that didn't extend George W. Bush–era tax cuts. A number of bills brought to the Senate floor in December by the Democratic majority were filibustered by the Republicans.[49]

Any bill not passed by Congress by the end of its legislative session dies. The 111th session of Congress ended on January 3, 2011. No legislation passed that would have funded NASA's workforce transition program. The final continuing resolution, passed by the new Congress on April 14, 2011, deleted $63 million in "Cross Agency Support" from the NASA budget that could have been used to fund the workforce program.[50]

None of Brevard County's elected federal representatives—its two House members or its two senators—introduced legislation in the new Congress to fund the program.

* * * * *

Does the government owe you a job?

That question was the crux of the debate during the events of 2010. Never before in NASA's history had a major program been proposed with a primary objective of protecting legacy contractors and their workforce.

On September 14, 2011, the members of Congress responsible for crafting the legislative compromise gathered on the ground floor of the Dirksen Senate Office Building to unveil the SLS design.[51] Senator Bill Nelson led the event, with Kay Bailey Hutchison at his side. Charlie Bolden was there. Also present were representatives from both chambers' space-related authorization and appropriations committees, Democratic and Republican.

Nelson stood with an easel that had an artist's concept of the SLS. He pointed out the monster rocket's significant features, in particular its height and thrust. But what would it be used for? That was rather vague; Nelson said it was "to get out beyond low Earth orbit and start to explore the heavens."

After Nelson came Bolden, who spoke from prepared remarks for five minutes. Bolden said that SLS "will take American astronauts further into space than any nation has gone before, and create jobs right here at home." He predicted that future flights would take crews to an asteroid and one day to Mars.

Bolden turned over the podium to Senator Hutchison. She spoke of modifying Constellation contracts "so that our experienced people will be kept to help modify and design the vehicle that will take us beyond Earth orbit . . . I want to see the contracts modified right away."

Next to the podium was Rep. Eddie Bernice Johnson (D-TX), ranking member of the House's science committee. "What we have been concerned about," she said, "is preserving our expertise and making sure that the people who had trained—the scientists that inspired our young people for the next level—will remain in place."

After her came Senator John Boozman (R-AR), the ranking member of the Senate's space subcommittee. "We have a decision now that will allow us to provide certainty, and that's so important, especially as we seek to retain the best and the brightest as we move forward, those employees that will help us get to the next stage that we go to."

Nelson then returned to the podium.

> It's important to underscore the administrator's comment about the workforce. This is the most skilled workforce that is a national asset to this country. And with the phase-out of the Space Shuttle, with the building of the commercial rockets scaling up as the space taxi, and then now with the big rocket, the monster rocket being set, and the contracts as Kay said, within a week the contracts being modified, you will see that workforce then start to scale up so that there's not just the precipitous drop, that you phase from one right into the other. That was

> most important to all of us, it's important to the president, and the administrator made that comment in his opening line, and I just wanted to underscore it.

And so it was that the nationalized space industry so cautioned by Ralph Cordiner in the 1960s was enshrined into law. A workforce of thousands was wedded to a government that pledged to have and to hold them from this day forward until parted by death—or at least by retirement.

That workforce was aging, and aging rapidly. A 2009 study by a group of Jet Propulsion Laboratory analysts found that as of June 2009, "The average NASA full-time, permanent scientist or engineer is 47.9 years of age" and that "37 percent of all scientists and engineers at NASA are eligible to retire today."[52] As for KSC's shuttle contractor workforce, a January 10, 2010, *Florida Today* article found that, "Many of the people who will lose their jobs likely would have left them anyway in coming years. About a third of the shuttle workforce is eligible for retirement, according to most estimates."[53]

Many no doubt were motivated by personal pride in their work and a sense of patriotism, but these aging workers also had a financial reason to stay—they'd been offered generous bonuses and severance packages to remain until the end. United Space Alliance offered workers, with a minimum four years of service, "completion bonuses" and "enhanced severance payments" that paid the worker one week of salary for each year of service, up to twenty-six years.[54] NASA and USA implemented a $100 million "critical skills" bonus in 2008, covering about 6,500 USA employees, or about 80 percent of the shuttle workforce.[55] NASA also agreed to cover the balance of USA's pension fund; Obama's proposed FY 2012 NASA budget included $547.9 million to cover about half of the pension fund's obligations.[56]

Eligibility didn't necessarily equate to financial security. The county was experiencing record homeowner foreclosures, with 10,000 homes lost in 2009. According to the January 10, 2010, *Florida Today* article, the average KSC contract employee at the time earned an annual wage of $65,000 and was age forty-nine. With a national recession and local unemployment rates forecast to reach 15 percent, it was unlikely that shuttle contract workers could find work elsewhere with pay as generous as their current employment. The *Florida Today* article quoted an FIT economist who said, "Unless there is increased government spending . . . I don't know how we are going to avoid the downside."

This was the consequence of Ralph Cordiner's prophesied nationalized space industry or, to quote Pete Worden's more snarky phrase, the self-licking ice cream cone. The workforce had become dependent on government largesse.

The politicians they elected relied on their votes to remain in office. Their relationship was a symbiosis.

If the aging workforce would not retire, where could a young engineer, scientist, electrician, or skilled laborer find a job in the aerospace industry?

Those opportunities were in the NewSpace community.

In a May 2012 National Public Radio interview, Elon Musk said that the average age of a SpaceX employee was about thirty.[57] Unlike USA and many other NASA contractors, SpaceX was not a union shop; Musk's distaste for labor organizing is well-known, not just at SpaceX but also at his Tesla electric car company.[58] SpaceX had a pipeline of inexpensive labor through its internship program. According to the SpaceX website, "Interns and graduate engineers are integral to the success of the company and tackle some of the hardest challenges on the planet . . . This is not an ordinary 'internship'—you will be given as much responsibility as our full-time engineers and will be an integral part of the team."[59]

In a 2018 interview with the technology business website *Fast Company*, then-45th Space Wing commander Wayne Monteith gave his impressions of the innovative SpaceX operations at Cape Canaveral. He recalled observing a SpaceX launch team with a "relatively young" engineer. He was told, "Yeah, he's an intern. He's an intern, but he's the most knowledgeable guy about this system, so he's in charge."[60]

Because SpaceX was not a union shop, it lacked not only the organized labor protections that lead to higher compensation, but also a protected work environment. In May 2017, SpaceX reached a $4 million settlement with employees who alleged they'd been denied mandatory breaks.[61] A November 2023 Reuters investigative report found that "Musk's rocket company has disregarded worker-safety regulations and standard practices at its inherently dangerous rocket and satellite facilities nationwide, with workers paying a heavy price." Reuters "documented at least 600 injuries of SpaceX workers since 2014," including a death at the company's McGregor, Texas, testing facility.[62] In January 2024, the National Labor Relations Board filed a complaint against SpaceX, alleging that eight employees were publicly reprimanded and then fired for publishing an open letter complaining about sexual harassment and Musk's social media behavior.[63] As of this writing, SpaceX has successfully blocked the NRLB ruling on appeal.[64] Reuters reported in April 2024 that Occupational Safety and Health Administration records showed in 2023 workplace injuries at the SpaceX Starbase facility near Brownsville, Texas, increased to 5.9 injuries per 100 workers from 4.8 in 2022; the space industry average was 0.8.[65]

A nonscientific survey retrieved in January 2024 from the workplace website Glassdoor.com concluded that "United Launch Alliance salaries averaged $25,681 higher than SpaceX." But respondents rated their employee satisfaction equally, 3.8 out of 5.0 for both SpaceX and ULA. SpaceX scored much higher on "Career Opportunities" (4.0 vs. 3.6) while ULA scored much higher on "Work-Life Balance" (3.6 vs. 2.6).[66]

Throughout the 2010s, SpaceX continued to add jobs, at Cape Canaveral and nationwide, as the company grew to become the world's dominant commercial launch company. The NewSpace icon achieved one historic first after another. As Barack Obama told Elon Musk that day they toured Launch Complex 40, "It's going to be exciting!"

But SpaceX and the NewSpace movement still had to compete for, well, the right to compete.

In April 2014, Musk announced that SpaceX would file a lawsuit against the US Air Force protesting a noncompetitive contract award to ULA. The contract gave ULA thirty-six missions through 2030, worth $70 billion.[67] SpaceX dropped the lawsuit in January 2015 after the USAF agreed to expand competitive opportunities for SpaceX and other companies.[68]

Perhaps the most symbolic NewSpace triumph was the lease of KSC's historic Pad 39A to SpaceX, which intended to use the launch site for commercial crew, the Falcon 9, and a planned successor, the Falcon Heavy. Reuse of 39A by a commercial company was part of KSC's twenty-first-century Launch Complex modernization, first funded in NASA's FY 2011 budget. Pad 39B would be dedicated to SLS and its legacy contractors, while 39A would be leased to a "private-sector, commercial space partner." SpaceX was awarded the lease despite a protest filed by Blue Origin, which was backed by ULA.[69]

Delayed by underfunding, the commercial crew program progressed, just not fast enough to relieve Russia of transporting American astronauts to the ISS anytime soon. By 2014, NASA had three finalists—the SpaceX crew Dragon, the Boeing CST-100 (named Starliner in 2015), and the Sierra Nevada Dream Chaser. Dragon and Starliner were familiar conical-shaped capsules, while Dream Chaser was a lifting-body spaceplane somewhat similar in shape to the shuttle orbiter, although much smaller. In the end, NASA awarded fixed-price contracts to SpaceX and Boeing; SpaceX was awarded $2.6 billion, while Boeing was awarded $4.2 billion.[70] Both would launch from the Space Coast.

Congress considered sole-sourcing commercial crew to just one vendor, reportedly Boeing. Rep. Frank Wolf (R-VA), who in 2012 chaired the House appropriations subcommittee controlling NASA's budget, introduced legislation trying to force the agency to "downselect to a single competitor or, at most, the execution of a leader-follower paradigm in which NASA makes one large award to a main commercial partner and a second small award to a back-up partner." Wolf was supported by Apollo astronauts Armstrong, Lovell, and Cernan, who wrote a letter to Wolf declaring, "An early downselect would seem to be prudent in order to maximize the possibility of developing a crew-carrying spacecraft in time to be operationally useful." There were reports that, behind the scenes, Boeing was lobbying NASA and Congress to select only Starliner. Wolf finally allowed NASA to select two full partners and one partial partner, then dropped the legislation. Bill Nelson called the proposal "silliness" and "anti-competitive."[71]

In November 2013, NASA's Office of the Inspector General released an audit report that looked at NASA's management of the commercial crew program. The report concluded:[72]

> The Program received only 38 percent of its originally requested funding for FY's 2011 through 2013, bringing the current aggregate budget shortfall to $1.1 billion when comparing funding requested to funding received. As a result, NASA has delayed the first crewed mission to the ISS from FY 2015 to at least FY 2017 . . . The combination of a future flat-funding profile and lower-than-expected levels of funding over the past 3 years may delay the first crewed launch beyond 2017 and closer to 2020, the current expected end of the operational life of the ISS.

The OIG's forecast was accurate. The first SpaceX uncrewed demonstration mission launched in March 2019. The first crewed demonstration mission launched in May 2020. Until that flight, NASA had no choice but to rely on Russia for crew taxis to the ISS.

The audit report also believed that "a lack of funding will require them to 'down select' to a single partner . . . Moving forward increases the risk that NASA could be left without a viable commercial option to transport crew to ISS should issues arise that either significantly delay or render inoperable the selected company's systems."[73]

The folly of a down-select to a single partner was proven in 2024 when the Starliner experienced thruster anomalies during its Crew Flight Test. Its two NASA astronaut test pilots were reassigned to come home on a SpaceX crew Dragon. By the end of 2024, Boeing had lost nearly $2 billion on the fixed-price contract.[74] The taxpayers didn't have to pay Boeing's cost overruns, but neither did they have the operational redundancy envisioned when the contracts were awarded in 2014. Without SpaceX, NASA might have been forced to rely on Russia's Soyuz to return the flight test crew, during a time of strained relations due to Russia's invasion of Ukraine.

By the time the Obama administration ended on January 20, 2017, true competition had been accepted in federal government space circles. The Space Launch System's funding was secure, as were the commercial cargo and crew programs. SpaceX demonstrated reusability—launching, landing, and then launching a rocket again—to bring down costs. As SpaceX matured, US military officials became more comfortable with awarding contracts to the company. Legacy contractors were forced to evolve to remain competitive. The old dogs learned new tricks.

The grand compromise, however imperfect, worked.

10

In with the New

Depend upon it, Sir, when a man knows he is to be hanged in a fortnight, it concentrates his mind wonderfully.

—English Essayist Samuel Johnson (1777)[1]

In his 1965 senior thesis for Yale University, political science student Bill Nelson wrote, "Over the course of interviews for this study, the writer gained the distinct impression that Brevard politics rests on a base of modified economic conservatism."[2]

> It seems a paradox that such political feeling runs rampant in a county where federal spending is at an all-time high and where there is much room for criticism as to the frugality of this extraordinary spending . . . He opposes excessive federal spending in theory, yet the life-blood of his community and county depend primarily on this spending.

In the early twenty-first century, Brevard County remained largely conservative. Despite their conservativism, thousands attended rallies protesting the Obama administration injecting capitalistic competition into the federal government's socialistic space programs. Many believed that the government owed them a job, a well-paying job, until retirement. Some conflated their personal pride and patriotism with these government programs, the rockets' red glare symbolic not only of American exceptionalism but also military supremacy and prestige. President John F. Kennedy had told them so in the 1960s, and few since then had dared to suggest otherwise.

Their protests made local headlines, but beyond Brevard County they seemed to wield little political influence. Jobs and homes were being lost nationwide due to the Great Recession; by one account, about 3.8 million homes were foreclosed between 2007 and 2010.[3] In late 2009, more than 15 million people were unemployed.[4] Brevard County's trauma was not unique; thousands of communities suffered across the nation.

During this time, Bill Nelson was the Florida politician most successful at influencing federal space policy. He partnered with Senator Kay Bailey Hutchison to craft the "grand compromise" that prioritized preservation of legacy contracts, and therefore some of those legacy jobs, with the Space Launch System. Crewed commercial space, although authorized, had little political support and therefore was underfunded. The 2010 NASA authorization act required Orion to be designed as a backup for ISS service should commercial crew fail, but few took that seriously. Either option meant continued reliance on Russia for crew rotations until later in the decade.

Space Florida was not a significant player in the national events of 2010. Its own house was in disorder. Frank DiBello, named interim president and CEO in May 2009, did not receive a permanent appointment until four months later. The agency was well aware of the pending job losses as the Space Shuttle program neared its denouement, but its problems were more fundamental. Too often, Space Florida had been misled by government programs and their legacy contractors into building infrastructure that stood vacant after those programs were canceled.

In a January 2010 interview, DiBello told the *Orlando Sentinel* that Space Florida had fired its Washington, DC-based lobbying firm and shelved for now its plans for a public-private spaceport at Launch Complex 36. The State of Florida had reduced the agency's budget, from $7 million in 2007 to $3.8 million in 2010. DiBello intended to decouple Space Florida from NASA, diversifying its business by seeking customers and revenue elsewhere. He also urged NASA to support emerging commercial space enterprises. "You cannot hold back the tide on commercial space," he told the *Sentinel*. "It's going to happen. We want it to happen, and we need it to happen." In a February 28, 2010, *Florida Today* guest column, DiBello wrote, "Florida is taking control of its own destiny."[5]

Nelson's thesis cited a seminal work by Johns Hopkins University political scientist V. O. Key, *Southern Politics*, published in 1949. Key described the Florida politics of his era as "an incredibly complex mélange of amorphous factions . . . In its politics it is almost literally every candidate for himself." Key raised the question "whether a government propped up by so tenuous a political underpinning as that of Florida's can act, on a broad front at least, either for good or for ill."[6]

Key wrote largely in the context of partisan elections, from state to local but, as Nelson wrote in his 1965 thesis, this "every man for himself" political culture still existed in the mid-1960s. This phenomenon continued to typify Florida's space politics for decades. The formation of the Spaceport Florida Authority in the late 1980s was opposed by some politicians, perhaps for partisan reasons, perhaps because it wasn't in their district, perhaps because it was closely identified with an unpopular governor. In the 1990s, the SFA nearly folded due to a lack of funding; the agency was extended a loan from the state's transportation department, a loan that apparently went unpaid. The authority won Air Force grants to help pay for infrastructure upgrades at the Cape, but likely customers were heading overseas.

Entering the first decade of the twenty-first century, the authority's bookkeeper was arrested for embezzlement, further eroding the agency's credibility with the state legislature. The agency's name was tweaked as a marketing gimmick, but that didn't seem to make a difference. The loss of *Columbia* and its crew on February 1, 2003, fundamentally altered the course of not just NASA, but also the renamed Florida Space Authority. President George W. Bush's policy shift toward human deep space flight planned a premature demise for the ISS, which meant the FSA's recently opened Space Life Sciences Lab lost potential customers who saw no long-term commitment by the federal government to microgravity research.

There were a few notable successes. The Apollo-era LC-37 was resurrected thanks to the authority financing a new Horizontal Integration Facility for Boeing's Delta IV rocket program. An eccentric billionaire named Elon Musk chose the Space Coast for his first operational Falcon rocket launches, with financial and bureaucratic aid from the authority.

But the political intrigue persisted. The authority was disbanded in 2006, replaced by Space Florida with a new board of directors, all chosen by the governor. Jeb Bush appointed a space neophyte as the agency's first president, who rankled many in the state government and the aerospace industry. Frank DiBello became Space Florida's president in 2009, and remained in the position until he retired in 2023, finally providing the agency with a degree of stability and continuity.

"Florida is taking control of its own destiny," DiBello had declared. A significant first step was to resurrect the vision for a commercial space research park outside KSC's gates. As discussed in Chapter 6, the SLSL was to have been the anchor for Space Commerce Park, but those plans fell apart as NASA turned its focus away from microgravity research to Project Constellation. The idea survived the transition as the FSA became Space Florida in the fall of 2006. Opening the park for business would signal Space Florida's independence and competence.

Under DiBello's predecessor, Steve Kohler, the agency revived the park idea. Space Florida's website first mentioned the park in December 2007. It described the SLSL as "the initial phase of a new 400-acre Research and Technology Park at Kennedy Space Center. The park will provide an ideal location for businesses and research groups with a need for close proximity to the launch and landing facilities and technical capabilities at the Cape Canaveral Spaceport." The name "Exploration Park" was being used on the agency's website by January 2009.[7]

The park's environmental assessment was published in December 2008. The concept foresaw development of eight buildings, starting with a two-story "cornerstone facility" owned by Space Florida that would feature offices and classroom spaces.[8]

In April 2009, Space Florida signed an agreement with The Pizzuti Companies to construct the cornerstone facility. According to a *Florida Today* article, Pizzuti was responsible for funding construction and finding tenants. Space Florida was to find funding for utilities infrastructure.[9]

But with the Great Recession raging, finding funding was all but impossible. The Space Coast Transportation Planning Organization requested the state legislature earmark funding for the estimated $7.3 million in infrastructure.[10] Brevard County Commissioner Robin Fisher proposed using $2.1 million in unspent Obama administration road construction stimulus funds, but faced opposition from local Republicans who wanted to send back the money to the federal government as a political protest. The panel on February 11, 2010, voted 11–7 to send $800,000 to Space Florida for the project; all the no votes came from south county members.[11] The state legislature chipped in $7.5 million for the park in the FY 2011 budget.[12]

On June 25, 2010, Space Florida officials broke ground on Exploration Park's infrastructure improvements, but Pizzuti executives had problems finding potential tenants. One told *Florida Today* in March 2011 that customers were concerned about the future of the American space industry as the Space Shuttle program came to an end.[13] As of this writing, the cornerstone facility remains unbuilt.

Another bill enacted during the 2010 state legislative session terminated Space Florida's existing board of directors. HB 451, introduced by north Brevard County state representative Steve Crisafulli, created a new board with thirteen voting members and two nonvoting members. Nine voting members reflecting "the statewide presence of Florida's aerospace industry" were to be appointed by the governor. The remaining four were ex officio members, including the governor, the state secretary of transportation, the president of Workforce Florida, and the president of Enterprise Florida. The two nonvoting members represented the leaders of the state House and Senate.[14]

The legislature passed several other bills affecting the Florida aerospace industry. The Space Transition and Revitalization Act (HB 1389) authorized the use of OTTED's Quick Action Closing Fund to mitigate the impacts caused by the Space Shuttle program's end.[15] HB 969 allowed Space Florida to reallocate to other facilities funds encumbered for LC-36 improvements "in order to attract new space vehicle testing and launch business to the state."[16] SB 1752 created a Manufacturing and Spaceport Investment Program within the OTTED that allowed up to $50,000 a year in state sales and use tax refunds for eligible equipment purchased by companies involved in spaceport activities.[17] The *Tallahassee Democrat* published a guest column by Frank DiBello praising the legislature's accomplishments. "Regardless of the direction of shifting federal policies," he wrote, "we can steer our own trajectory into the next decade and toward our goal of tripling the aerospace related activity in our state."[18]

The federal Department of Commerce on September 22, 2010, awarded Space Florida $400,000 "for development of a strategic economic development strategy for Florida's Space Coast Region." The award from the department's Economic Development Administration was in response to a recommendation by President Obama's retraining task force that had suggested public-private partnerships to spur creation of new businesses on the Space Coast. Because Congress had not passed the FY 2011 budget that was to fund Obama's $40 million workforce program, the grant money came from unused federal stimulus funds.[19]

Republican venture capitalist Rick Scott was elected governor in November 2010, and was sworn into office on January 4, 2011. Scott proposed that Space Florida and other state public-private partnerships be consolidated into one economic development agency he would control. The proposal created fear in the aerospace industry that Space Florida would lose all autonomy and control over its funding. The final compromise left Space Florida as a separate entity with its own board of directors, but the agency now had to answer as well to the Enterprise Florida board of directors. In exchange, the agency would have access to a $120 million business recruitment "seed fund" Scott had requested. The OTTED was eliminated and was replaced by the new Department of Economic Opportunity.[20]

Florida's FY 2012 budget provided $16 million from the State Transportation Trust Fund to Space Florida for spaceport launch and infrastructure projects.[21] SB 634, signed by Governor Scott on February 16, 2012, modified the state's legal definition of "launch support facilities" to mean facilities that are "located at launch sites or launch ranges that are required to support launch activities, including launch vehicle assembly, launch vehicle operations and control, communications, and flight safety functions, as well as payload operations, control, and processing."[22] The change not only aligned state code with the federal definition, but also broadened the options for which Florida Department of Transportation funding could be used.[23]

FDOT created the Spaceport Improvement Program (SIP) as a mechanism for processing Space Florida capital improvement funding requests. According to the program's 2016 Project Handbook, "FDOT provides support and funding to Space Florida for high-priority spaceport projects through the Spaceport Improvement Program. This funding stimulates public and private investment into emerging and growing aerospace enterprises while advancing a safer and secure spaceport transportation system." In 2019, as the program evolved, FDOT created a separate Spaceport Office to process financing requests. The updated 2023–2024 Project Handbook stated that, during FY 2019–2023, FDOT provided $318 million in spaceport investments.[24]

* * * * *

During the second decade of the twenty-first century, Space Florida, FDOT, NASA, and the USAF evolved a unique partnership. With Space Shuttle and other programs coming to an end, both federal government agencies had to decide whether to keep antiquated and obsolete facilities or demolish them. Although some might protest the demolition of historic sites, the reality is that taxpayer dollars rarely are appropriated to preserve for posterity an obsolete rusting hangar built decades ago for a long-canceled program. Space Florida and FDOT evolved into a new option where a rusty hangar might be good enough for a private company looking for operating space.

An example is the complex once known as the Orbiter Processing Facility (OPF), three hangars near the Vehicle Assembly Building that were originally constructed to maintain the space shuttle orbiters. As the Space Shuttle program came to an end in July 2011, NASA granted Space Florida the rights to operate, maintain, and improve the third hangar, known as OPF-3. Space Florida then signed a fifteen-year agreement with Boeing, which would use the hangar to assemble and maintain its CST-100 Starliner. The State of Florida offered Boeing $50 million in economic development incentives and was responsible for refurbishing the hangar. Space Florida spent $45 million to demolish the Space Shuttle-era infrastructure, completing that phase in April 2014. The agency renamed OPF-3 to the Commercial Crew and Cargo Processing Facility, or C3PF, perhaps an oblique nod to a certain *Star Wars* droid with a similar name. By transferring OPF-3 to Space Florida and not directly to Boeing, NASA avoided the appearance of playing favorites with potential commercial vendors.[25]

Space Florida also played a role in upgrading OPF-1 and OPF-2 for the USAF's X-37B. In January 2014, the agency provided $9 million from FDOT to refurbish the hangars. Boeing added $4.5 million. The hangars were transferred in October 2014 from NASA to the USAF, which had contracted with Boeing to build and operate the military spaceplanes.[26]

On June 29, 2013, NASA announced that Space Florida had been selected to manage the Shuttle Landing Facility, the runway used to land the orbiters. Space

Florida changed the name to the Launch and Landing Facility (LLF), expanding its potential uses to commercial horizontal launch and landing missions—a vision foreseen by Space Florida's predecessor in its 2002 master plan.[27]

In the next few years, Space Florida found several tenants for the runway. Negotiations were typically confidential, given a code name until an agreement was reached. One example is "Project Kraken." According to an April 22, 2021, Space Florida board meeting agenda, Project Kraken was the "establishment and operations of a spacecraft manufacturing and support facility" at the LLF. The company planned to invest $300 million and said it would create 2,100 jobs by 2025, with estimated average annual wages of $84,000 per employee. Space Florida would "pursue conduit financing" and seek funding through FDOT's SIP.[28]

The company eventually was revealed to be Terran Orbital, a small satellite manufacturer. Joined by Florida governor Ron DeSantis, Space Florida and Terran Orbital on September 27, 2021, announced plans to build at the LLF the world's largest space vehicle manufacturing facility.[29] But on October 31, 2022, Terran pulled out of the deal after a $100 million investment in the company by Lockheed Martin. As part of the deal, Terran agreed to expand its existing facility in Irvine, California, rather than build a new plant at the LLF.[30]

Project Oz was another proposal for "development of new facilities for spacecraft manufacturing and refurbishment" at the LLF. According to the January 26, 2023, meeting agenda, Space Florida would have sought $14 million in funding from FDOT and $14 million from the mystery company.[31] But as of this writing, the project has not proceeded nor has the company been revealed.

Another project, code-named Project Comet, was revealed to be Amazon's plans for a $120 million facility to process its Project Kuiper satellites. Similar in concept to the SpaceX Starlink network, Kuiper is a project to create a constellation of communication satellites for its customers. Amazon announced on July 21, 2023, that the company would build a satellite processing facility at the LLF; the satellites could launch from the Cape on Blue Origin New Glenn or ULA Vulcan rockets.[32]

Multiple companies explored Space Florida agreements to use the LLF for horizontal launch and/or landing. XCOR, Swiss Space Systems, Virgin Orbit, and Stratolaunch were all companies that at one time or another negotiated, and perhaps signed, agreements with Space Florida. Of the four, only Stratolaunch still exists as of this writing, but the company has restricted its operations to the Mojave Air & Space Port in California.[33]

Sierra Nevada Corporation, denied one of the first two commercial crew contracts, was awarded a NASA commercial cargo contract in January 2016. SNC had sought Space Florida subsidies for years, but, without a NASA contract, the company was unlikely to proceed with completing their Dream Chaser spaceplane. Space Florida and SNC finally signed an agreement in May 2021 for Dream Chaser to land at the LLF, as well as to use the nearby Reusable Launch Vehicle Hangar.[34] Around that time, SNC spun off its space technologies into a subsidiary, Sierra Space Corporation. Dream Chaser became part of Sierra Space.

Several Cape Canaveral sites also benefited from state funding. Space Florida in August 2012 contributed $5 million to renovate Launch Complexes 25 and 29, old Navy submarine missile test sites within the Naval Ordnance Test Unit (NOTU) at the Cape. The sites were refurbished for Strategic Weapons System (SWS) Ashore, a facility that provided the Navy with a single, land-based simulation operation for testing submarine missile systems.[35]

Space Florida found a tenant for LC-20, originally built in the 1950s for the Titan ICBM program. The old Spaceport Florida Authority had targeted LC-20 in the late 1990s for small commercial rockets but had never found any customers. On February 22, 2019, Space Florida announced that Firefly Aerospace would establish launch operations at LC-20, and a manufacturing facility in Exploration Park. Space Florida agreed to match Firefly's investments up to $18.9 million, using funds from FDOT's SIP.[36] As of this writing, Firefly is conducting launches from Vandenberg Space Force Base but not from LC-20.[37] The company has yet to break ground on any facilities in Exploration Park.

Relativity Space moved into another old Titan pad, LC-16, but chose to partner directly with the USAF instead of Space Florida. Relativity cofounder and CEO Tim Ellis told *Spaceflight Now* on January 18, 2019, that the company preferred a direct agreement with the USAF, wanting exclusive rights to the site for twenty years. On March 23, 2023, Relativity conducted its first test flight of its Terran 1 rocket from LC-16; although the booster worked, the second stage failed to achieve orbit.[38]

Space Florida sought for years to find a tenant for Launch Complex 36. SpaceX in its early years considered LC-36 for its Falcon boosters before settling on LC-40. As discussed earlier, Space Florida had reached a five-year agreement in 2008 with the USAF to lease LC-36, but shelved its plans in 2009 after state auditors faulted the lack of a master plan for the complex, losing political support in the state legislature.

Blue Origin, having lost its bid for KSC's Pad 39A to SpaceX, finally settled at LC-36. Local newspapers had reported Space Florida was negotiating with a "high-tech aerospace company" that Senator Bill Nelson confirmed was Blue Origin. *Florida Today* reported on May 17, 2015, that "Project Panther" was seeking $8 million in incentives over ten years from the North Brevard Economic Development Zone for a 330-employee facility at Exploration Park and a launch site at LC-36. Space Florida's fiscal year 2016 annual report stated, "FDOT infrastructure funding assistance to Space Florida in support of the Blue Origin project is $26.4 million. Space Florida is leveraging these funds through a public-private partnership with Blue Origin, which will match the infrastructure funding on a dollar-for-dollar basis and fund the remaining project costs."[39]

The Blue Origin manufacturing facility was built across Space Commerce Way from the original Exploration Park site. The original site was designated Phase 1, while Blue Origin's complex became Phase 2.[40]

Although Space Florida never built its planned Phase 1 cornerstone facility, the agency found another tenant for a nearby site. "Project Sabal" turned out to be OneWeb Satellites, a partnership of OneWeb and Airbus. Space Florida issued the contract to construct the spacecraft assembly building, then lease it to OneWeb. FDOT contributed $17.5 million in matching funds.[41]

Where once the Space Life Sciences Lab stood alone awaiting tenants, new neighbors began to arrive. Thanks to the Obama administration extending the service of the International Space Station, the Lab was about to fulfill its potential.

* * * * *

The 2005 NASA Authorization Act designated the US orbital segment (USOS) of the ISS as a national laboratory. Section 507 of the act directed the NASA administrator to seek increased use of USOS by "other federal entities and the private sector through partnerships" and cost-sharing agreements. The administrator was authorized to "enter into a contract with a nongovernmental entity" to operate the laboratory.[42]

As NASA neared completion of the ISS, a November 2009 GAO report to Congress noted that the agency intended to use only about half of the USOS for its own research, leaving the other half available for US National Laboratory researchers. The GAO recommended that NASA, "Establish a body that oversees U.S. ISS research decision making, including the selection of all U.S. research to be on board and ensuring that all U.S. ISS research is meritorious and valid." The report also recommended that NASA reach out to potential commercial customers and that ISS users have access to internal NASA expertise.[43]

These recommendations found their way into the 2010 NASA authorization act. Section 504 required the NASA administrator to "provide initial financial assistance and enter into a cooperative agreement with" a nonprofit organization "to manage the activities of the ISS national laboratory . . . This organization shall develop the capabilities to implement research and development projects utilizing the ISS national laboratory and to otherwise manage the activities of the ISS national laboratory."[44]

On December 2, 2010, NASA announced that the agency was "seeking an independent, nonprofit research management organization to develop and manage the U.S. portion of the station . . . The selected organization will capitalize on the unique venue of the orbiting laboratory as a national resource; and develop and manage a diversified research and development portfolio based on U.S. needs for basic and applied research in a variety of fields."[45]

In response to the request, Space Florida formed a 501(c)(3) nonprofit called the Center for Advancement of Science in Space, or CASIS for short. On July 13, 2011, NASA announced that CASIS had been selected from four bidders to negotiate a contract to manage the lab. CASIS would operate from Space Florida's SLSL just outside the KSC gate. The final agreement, announced on September 9, gave CASIS a ten-year contract to manage the labs on the US segment of the station. CASIS was to receive $15 million annually from NASA.[46]

As with many startups, the early years were bumpy for CASIS. The first director, Jeanne Becker, submitted her written resignation on February 29, 2012, to Frank DiBello, who was serving as the interim CASIS board chair. Becker cited CASIS's business relationship with ProOrbis, a for-profit consulting firm hired by Space Florida to write the proposal that led to NASA awarding the contract to CASIS. Becker believed that ProOrbis was creating for itself an open-ended for-profit role in CASIS that could jeopardize the fledgling agency's nonprofit status. "The Space Florida interim board persists in pursuing engagement of ProOrbis on behalf of CASIS, with CASIS management forced to bear the responsibility of mitigating ensuing organizational risks occurring as a result of the interim board's actions." She also cited "unrealistic expectations" by congressional staffers, NASA, and ProOrbis.[47] The Ohio congressional delegation tried to exploit the chaos by demanding that NASA strip CASIS of the contract and give it instead to a Cleveland group, but nothing came of their scheme.[48]

The first CASIS chief economist (director of economic valuation) was charged in April 2019 by federal prosecutors with wire fraud for having submitted reimbursement requests for prostitutes and female escorts on CASIS business trips from 2011 through 2015. Once CASIS learned of his behavior, the economist was dismissed, and CASIS turned over the matter to NASA's OIG. He pleaded guilty in February 2020 to one charge of tax fraud, and agreed to pay restitution to CASIS.[49]

Despite the distractions, CASIS established relationships with clients that continue to this day. Perhaps the most prominent example is Nanoracks, a Houston-area startup formed in 2008 to provide small payload logistics services on the ISS. On April 12, 2012, Nanoracks and CASIS announced that CASIS would be a customer for a research platform Nanoracks intended to deploy outside the Japanese Kibo module. CASIS would then solicit the research community to use the platform. CASIS paid Nanoracks $1.5 million, which enabled the company to start construction of the platform.[50] Space Florida and Nanoracks partnered to host the Space Florida International Space Station Research Competition. Space Florida paid for eight payloads to fly with Nanoracks to the ISS. The two entities held a workshop in Cocoa in October 2012 to help researchers apply for the available slots.[51]

Nanoracks partnered with Boeing to develop Bishop, the first permanent ISS commercial module. Bishop addressed a growing bottleneck aboard ISS—deploying experiments outside. The only way at the time to deploy external experiments was through Kibo; the other two airlocks were for crew egress. Attached to Node 3 in 2020, Bishop offered five times the capacity of Kibo. Bishop could be detached by the station's robot arm for remote deployment. Payload deployments were coordinated through CASIS; according to the agency's fiscal year 2024 annual report, the ISS at that time had twenty-three National Lab commercial facilities supported by fourteen commercial service providers, including seven Nanoracks facilities.[52] Voyager Space Holdings acquired Nanoracks in May 2021.[53]

Another early CASIS partner was Made in Space, Inc., a company attempting 3D manufacturing in microgravity. The technology would give NASA and ISS partners the ability to 3D-print custom tools on demand aboard the ISS rather than having to launch them into orbit aboard a cargo craft. Their first 3D printer, built in collaboration with NASA's Marshall Space Flight Center, launched to the ISS in September 2014 aboard a SpaceX cargo Dragon. In December 2014, Made in Space uploaded to the device the design for a ratchet that had not existed before on the station. It was printed and certified for use within a week.[54]

In June 2020, Redwire Space, a Jacksonville-based company, acquired Made in Space. According to its website, Redwire seeks to deliver space infrastructure solutions to its customers. In January 2023, Redwire installed on the ISS a 3D bioprinter. In September that year, Redwire used the device to print a human knee meniscus. In May 2024, Redwire announced that the company had 3D bioprinted live human heart tissue samples on the ISS.[55]

CASIS sponsors an annual science competition called Genes in Space. The program challenges students in grades seven through twelve to design DNA experiments for the ISS. Genes in Space was founded in 2015 by miniPCR bio and Boeing. Among the experiments in the early 2020s were investigations into how space travel affects the DNA of astronauts.[56]

In December 2025, NASA announced that it had extended its cooperative agreement with CASIS through 2030. The press release noted that more than 940 payloads had been launched to ISS since NASA first signed the agreement with CASIS in 2011. "The ISS National Lab has created a robust pipeline of commercial research that has pushed advancements in pharmaceutical development, advanced communications, consumer goods, and more."[57]

* * * * *

We discussed in Chapter 3 the blowback from the governor's 1989 Spaceport Florida feasibility study recommendation that the state consider a commercial launch site at the abandoned farm town called Shiloh. Even though Florida acquiesced to political pressure from local environmentalists, Shiloh nonetheless remained on the books as NASA property subject to future development. Although the US Fish and Wildlife Service managed Shiloh as part of the Merritt Island National Wildlife Refuge, their 1963 agreement specified that the land remained subject to NASA use in the future.

NASA revisited the idea in 2008, exploring a possible 150-acre commercial launch site with two pads at KSC—either south of Pad 39A and north of LC-41—or somewhere in MINWR south of the Mosquito Lagoon in the vicinity of Shiloh. Although NASA would provide the land for free, either the commercial tenant or the State of Florida would have to fund the complex. NASA held four hearings to gauge public reaction.[58]

The public reaction, as you might surmise, was the same as in 1989. A standing-room-only crowd of about 300 people packed Titusville City Hall for two public hearings on February 25 to oppose the Mosquito Lagoon idea. Due to the overwhelming opposition, NASA shifted its sights to CCAFS. The agency and the USAF never reached an agreement for a joint commercial launch complex.[59]

The 2002 *Cape Canaveral Spaceport Master Plan* spoke of a horizontal launch and recovery facility to address the anticipated future demand for such missions. A planning map showed possible runways north of LC-39 and east of the existing space shuttle runway, encroaching into MINWR, south of Shiloh and State Route 402, the "beach road" into Canaveral National Seashore.[60] NASA's 2011 *Agency Master Plan* showed the same area designated for "Horizontal Launch & Landing."[61] The 2012 *Kennedy Space Center Master Plan* showed "an east-west corridor just south of Beach Road" to support horizontal launch and landing.[62]

On September 20, 2012, Florida Lieutenant Governor Jennifer Carroll sent a letter to NASA Administrator Charles Bolden and USDOT Secretary Ray LaHood requesting 150 acres of what she called "excess launch property" near Shiloh "to develop and operate this site as a commercial launch complex independent of the neighboring federal range and spaceports."[63] The response from the local environmental community was immediate and predictable. "Some of these bad ideas have a way of being reborn," Charles Lee of Audubon Florida told the *Orlando Sentinel.* "It is in an area where the losses to the national wildlife refuge could be severe."[64]

Seth Statler, the NASA associate administrator for legislative and intergovernmental affairs, responded to Carroll with a letter dated November 30, 2012. The letter corrected her assumption that the land was "excess launch property"—"The property identified in your request has not been reported as excess." Statler suggested that his agency work with Frank DiBello at Space Florida to pursue other options.[65]

SpaceX and Blue Origin both expressed interest in the Shiloh site. SpaceX was looking for a site independent of government operations at the Cape; the company was also looking at a site near Brownsville, Texas. Blue Origin was looking for its first orbital launch site.[66]

To complicate matters, the *Tampa Bay Times* ran a front-page article on July 5, 2013, reporting that the proposed site "contains the ruins of an eighteenth-century English plantation, complete with slave villages, a sugar factory and a rum distillery." A local historian told the paper that the ruins had been fully documented five years before by archaeologists. "They should have known," she said. A Space Florida representative said that the site would be noted in the project's Environmental Impact Statement (EIS).[67]

Space Florida initiated the EIS process that month.[68] The FAA published its Notice of Intent in the *Federal Register* on December 26, 2013. Public scoping meetings were held in New Smyrna Beach on February 11, 2014, with 445 attendees, and at the Eastern Florida State College (formerly Brevard Community College) campus in Titusville on February 12, 2014, with 328. The vast majority of comments submitted by the public opposed the Proposed Action, citing adverse impacts to wildlife habitats, closures of State Route 3, and impacts on nearby culture resources, among other reasons.[69]

SpaceX chose to locate its new commercial launch site at Boca Chica, a beach near Brownsville in Cameron County, Texas. Governor Rick Perry announced on August 4, 2014, that the state was providing $2.3 million from the Texas Enterprise Fund to SpaceX for what is now called Starbase, with another $13 million from the Spaceport Trust Fund to the Cameron County Spaceport Development Corporation. The Texas legislature passed a bill amending the state's Open Beaches Act, which guaranteed the public's free access to state beaches, that created an exception allowing county commissioners to close a public beach in proximity to a launch site. As of this writing, joint plaintiffs are proceeding with a lawsuit challenging the beach closures for SpaceX launches.[70]

SpaceX has used Starbase to develop Starship, its next-generation launch system that Elon Musk hopes one day can be used to transport colonists to Mars. The first Starship test flight left debris scattered over a wide area, although no debris was found in a nearby wildlife refuge. Environmental groups have filed lawsuits against the FAA charging the agency failed to properly analyze the environmental impacts caused by Starbase and Starship test flights.[71] The *New York Times* reported in July 2024 that a June 6, 2024, Starship flight test "unleashed an enormous burst of mud, stones and fiery debris across the public lands encircling Mr. Musk's $3 billion space compound. Chunks of sheet metal and insulation were strewn across the sand flats on one side of a state park."[72] In September 2024, SpaceX agreed to a consent agreement fine of nearly $150,000 for repeated discharges of deluge water into nearby wetlands.[73]

With SpaceX testing at Boca Chica, and Blue Origin settling in at LC-36, Space Florida lacked a likely tenant for Shiloh. On April 8, 2020, the FAA published a notice in the *Federal Register* stating that the EIS had been rescinded, although "Space Florida is conducting further analysis of the proposed site."[74]

* * * * *

Elon Musk survived a near-bankruptcy experience in 2008, having burned through most of the capital invested in his startup twins, SpaceX and Tesla. Musk filed for divorce that year from his first wife Justine. The first three Falcon 1 launch attempts at Kwajalein failed. The fourth launch, on September 28, 2008, was successful. As the Great Recession collapsed the US economy, it became all but impossible to raise more investment capital. Musk borrowed from SpaceX to keep Tesla afloat. NASA finally tossed SpaceX a lifeline on December 23, 2008, awarding a $1.6 billion contract for twelve cargo Dragon deliveries to the ISS. (Orbital Sciences received $1.9 billion for eight deliveries.)[75]

The first Falcon 9 launch from Cape Canaveral—the first Falcon 9 launch from anywhere—was from LC-40 on June 4, 2010. The *Florida Today* front-page headline the next day declared, "Falcon 9 Nails Test—Rocket's Roar into Orbit Bodes Well for Obama's Plan to Privatize Space." For the hometown paper, the launch was as much a vindication for Barack Obama as it was for Elon Musk. The article quoted Musk as saying, "I think this bodes very well for the Obama plan. It really helps vindicate the approach that he's taking—that even a small company like SpaceX can make a real difference."[76]

Florida Today editor John Kelly wrote that "we, here in Brevard County, are big winners" because the American launch industry now had "a new, cheaper" rocket, and it would be launching from the Cape. Kelly predicted:[77]

> A decade from now, the launch business—for science, for the military, for NASA, and for private spacecraft makers—is going to look a lot different. This will be one of the days people will point to as a watershed moment in the American space program.

Musk filed lawsuits in his company's early years, trying to break the ULA monopoly. After Boeing and Lockheed Martin formed ULA in May 2005, Musk filed a lawsuit in October 2005 charging the companies with "violations of antitrust, unfair competition and racketeering laws." A federal district court dismissed the lawsuit in February 2006, finding that SpaceX had not suffered any loss because the company lacked any launch services with which to bid for government contracts. As discussed earlier, SpaceX filed a lawsuit again after the USAF awarded ULA a noncompetitive contract; this time, having launched multiple Falcon 9 missions, Musk had a much stronger argument, so the USAF agreed to expand bid opportunities for SpaceX and other companies.[78]

A decade later, by the mid-2020s, SpaceX was the company with the near-monopoly in US launch services. According to one count by astrophysicist Jonathan McDowell, in 2024 there were 145 orbital launch attempts in the United States; 132 of those were the Falcon 9, and another two were the Falcon Heavy. Four were Starship test flights. China had 68 launch attempts, and Russia 17.[79] The *Orlando Sentinel* counted 93 launches in 2024 from the Cape—26 from KSC, and 67 from CCAFS; 88 of the 93 launches were by SpaceX, with the other 5 by ULA.[80] Of the 132 Falcon 9 launches in 2024 from either the Cape or Vandenberg, 89 were launches supporting the SpaceX Starlink network.[81]

SpaceX is the only company as of this writing to figure out routine reusability. The first Falcon 9 landing was at Cape Canaveral on December 21, 2015. SpaceX maintains two autonomous spaceport drone ships at Port Canaveral. If a booster can't make it back to the Cape, it will land at sea on a drone ship. In January 2025, SpaceX achieved its 400th orbital landing.[82] As of this writing, one Falcon 9 has launched and landed thirty-two times.[83]

Two anticipated competitors plan to progress into reusability. Blue Origin's New Glenn rocket, which launches from the Cape's LC-36, has successfully demonstrated the ability to land downrange on a seafaring platform. The NG-2 mission, which launched on November 13, 2025, landed on the drone ship *Jacklyn* after deploying two Mars-bound satellites for NASA. ULA's new Vulcan rocket may one day have a partially reusable system based on ULA's Sensible Modular Autonomous Return Technology (SMART). The first stage engine section would separate from the expendable fuselage and parachute to an ocean landing for recovery.[84] Both New Glenn and Vulcan use Blue Origin's BE-4 engines.[85]

ULA launched its final Delta IV mission on April 9, 2024, from LC-37. SpaceX is looking at the complex as a possible site for Florida-based Starship launches. The site would complement a Starship launch site already under construction at KSC's Pad 39A.[86]

SpaceX has come to so dominate the American commercial launch business that, by the end of 2023, some in the industry had begun to speak of SpaceX as an inadvertent monopoly.[87] The plucky startup dismissed, if not disdained, by so many on the Space Coast has vindicated the few who risked their political careers to fight for competition and innovation in Florida's future.

* * * * *

The other half of the "grand compromise" was the Space Launch System. When Congress mandated in 2010 that NASA build the SLS, it did not tell NASA what it should be used for, other than "as a follow-on to the Space Shuttle that can access cis-lunar space and the regions of space beyond low-Earth orbit in order to enable the United States to participate in global efforts to access and develop this increasingly strategic region."[88]

To find a use, the Obama administration proposed what was first called the Asteroid Retrieval Mission (ARM). The proposal had its roots in a 2012 Keck Institute study advocating the capture and return of a near-Earth asteroid.[89] NASA adapted the idea in its fiscal year 2014 proposed budget, which asked for funding to robotically capture a near-Earth asteroid and redirect it to cislunar space where astronauts could visit and explore it, using SLS and Orion. NASA argued that such a mission was a step toward fulfilling Obama's 2010 proposal to send crew to Mars in the 2030s.[90] Senator Nelson supported ARM, as did other politicians representing Florida, Texas, and Alabama. Some politicians lobbied for a crewed lunar mission instead, which NASA Administrator Bolden said the agency couldn't afford.[91] ARM lacked any widespread political support within Congress. After Donald Trump became president on January 20, 2017, his administration proposed canceling the program, and those working on it were notified in April that ARM had been defunded.[92]

Some of the ARM technology transferred into Gateway, a cislunar space station where Orion could dock.[93] Gateway's origins trace back to another Obama administration program called NextSTEP (Next Space Technologies for Exploration Partnerships). NASA sought to create public-private partnerships that would create "deep space exploration capabilities to support more extensive missions in the proving ground around and beyond Cislunar space" for "significant commercial applications beyond NASA."[94] In August 2016, NASA selected six companies to develop concepts and ground prototypes for deep space habitats—the next step, so to speak, after development of SLS and

Orion.[95] One of those companies, Northrop Grumman, was selected in June 2020 to contribute the crew module (Habitation and Logistics Outpost, or HALO) for Gateway.[96]

President Trump issued Space Policy Directive-1 on December 11, 2017. The directive modified President Obama's 2010 National Space Policy by redirecting NASA back to the moon "with commercial and international partners" as a step toward sending humans "to Mars and other destinations." It deleted one paragraph from the 2010 document, which set 2025 as a date for crewed missions beyond the moon, including an asteroid, and for sending crew to Mars orbit by the 2030s. That paragraph was replaced by one that simply directed NASA to "lead the return of humans to the Moon" with no timeline to be met.[97]

On September 1, 2017, Trump nominated Rep. Jim Bridenstine (R-OK) to be NASA administrator. For the first time, a politician would run the agency, the implication being that only a politician could speak the language of politicians who determine NASA's programs and finances. Florida's two senators, Bill Nelson (a Democrat) and Marco Rubio (a Republican), criticized the selection. Rubio told *Politico* that he worried Bridenstine's selection would inject partisanship into the administrator's office. "I just think it could be devastating for the space program. Obviously, being from Florida, I'm very sensitive to anything that slows up NASA and its mission." Nelson said in a written statement, "The head of NASA ought to be a space professional, not a politician."[98]

But by the time he left office on January 20, 2021, at the end of the first Trump term, Bridenstine arguably had proven to be one of NASA's most effective administrators. Perhaps his signature achievement was the creation of Project Artemis, which pulled together NASA's disparate deep space endeavors into one coherent and articulate policy objective. In Greek mythology, Artemis was the twin sister of Apollo. Project Artemis was announced on May 13, 2019, as part of a White House proposed $1.6 billion addition to NASA's fiscal year 2020 budget, trying to accelerate plans for landing crew on the south pole of the moon by 2024. The name reflected a mission objective to land the first woman on the moon.[99] Artemis continued to rely on SLS and Orion to deliver crew to lunar orbit, but, once there, commercially developed spacecraft would house those crews in orbit and transport them to and from the surface.

Ten days later, on May 23, Bridenstine spoke to an audience at Florida Tech. The event was broadcast on NASA TV. Bridenstine detailed the agency's Artemis plans, and announced that Maxar would build Gateway's Power and Propulsion Element (PPE) under a commercial fixed-price contract.[100]

On April 16, 2021, NASA announced that SpaceX had been selected under a firm fixed-price milestone-based contract to build a Human Landing System

(HLS) version of Starship as part of NextSTEP.[101] A second HLS contract was awarded on May 19, 2023, to Blue Origin, also funded through NextSTEP.[102]

While commercial companies design and build the Project Artemis infrastructures and technologies, SLS has become a totem for another time. In September 2010, Bill Nelson worked to convince skeptical House members to support the NASA authorization act passed by the Senate. House science committee chair Bart Gordon (D-TN) doubted the Senate's estimate that SLS could be built and operational in five years for $11.5 billion. Nelson told *Florida Today,* "If we can't do a rocket for $11.5 billion, we ought to close shop."[103] An April 2017 GAO report found that NASA had slipped the first test flight of SLS and Orion to no earlier than November 2018, but that date was "likely unachievable." The cost to achieve that first uncrewed test flight, including design and development, was estimated to cost $23.8 billion—$9.7 billion for SLS, $11.3 billion for Orion, and $2.8 billion for ground systems.[104] SLS finally launched its test flight, Artemis I, on November 16, 2022. A September 2023 GAO report concluded that NASA had "spent $11.8 billion to develop the initial SLS capability."[105] A May 2023 GAO assessment of major NASA projects estimated that the agency had spent $13.8 billion on Orion, including a test flight on a ULA Delta IV in December 2014 and the Artemis I test flight in November 2022.[106]

As of this writing, Artemis II will launch no earlier than April 2026, according to the NASA website. Artemis II will be the first crewed test flight for SLS and Orion. The ten-day mission will fly a crew of four (three Americans and one Canadian) on a trajectory that will take them 4,600 miles beyond the moon, the farthest humans will have traveled in the solar system.[107] A May 2024 NASA OIG report estimated that the agency will have spent more than $55 billion on SLS, Orion, and associated ground systems by the time Artemis II launches.[108]

Artemis III, the first crewed landing attempt, is targeting no earlier than mid-2027.[109] An April 2024 report by *Ars Technica* suggested that the mission objective of a crewed landing may be jettisoned if the SpaceX Starship HLS is not ready. Among the options being considered are a crewed Orion-Starship rendezvous in low Earth orbit, or a crew docking with Gateway in lunar orbit.[110]

* * * * *

Lori Garver continued to serve as NASA's deputy administrator until September 2013, when she accepted a position as general manager of the Air Line Pilots Association. Garver was arguably the most influential deputy administrator in the agency's history, at least since the Apollo era.

Garver was the right person in the right place at the right time. She came aboard the 2008 Obama presidential campaign at a time when it had no space policy expert. The candidate, like most politicians, was enamored by NASA's mythology but had no expertise in the subject matter. Obama listened to Garver's advice and risked his political capital to implement her reforms, which were at odds with many in Congress and the NASA bureaucracy.

After she left office, Garver became one of the Space Launch System's most vocal critics. Unchained from the responsibilities of office, Garver spoke openly about the program's shortcomings. One example is her January 2, 2014, appearance on National Public Radio's *The Diane Rehm Show.* Garver said that NASA shouldn't be "just trying to relive Apollo and doing it in a way like those we beat did it in a socialist way." Garver called for cutting NASA programs building on "previous technology" citing SLS as an example.[111]

Bill Nelson was reelected in November 2012, defeating Republican candidate Connie Mack IV by about 1.1 million votes, a 55.2 percent–42.2 percent margin.[112] In Brevard County, Nelson beat Mack by about 15,000 votes, 50.9 percent to 45.6 percent.[113] Six years later, in November 2018, Florida politics were in a different place, and Nelson lost to Republican governor Rick Scott by about 10,000 votes statewide, 50.1 percent to 49.9 percent.[114] In Brevard County, Nelson lost to Scott by about 39,000 votes, 56.7 percent to 42.9 percent.[115] Brevard County voters could have saved Nelson. They did not.

On December 10, 2018, on the Senate floor, Nelson gave a farewell address about "the future of our space program." He reflected on his career, including his flight aboard *Columbia* in January 1986. Nelson said, "I appreciate the steady hand and transformative contribution of NASA leaders like Charlie Bolden and Bill Gerstenmaier and Bob Cabana." (Lori Garver didn't make the list.) Nelson "celebrated the long overdue emergence of female superstars like [Lockheed Martin CEO] Marillyn Hewson and [SpaceX COO] Gwynne Shotwell among the space industry leadership." He applauded Jim Bridenstine for keeping NASA out of partisan politics.[116]

Nelson turned to the events of 2010.

> Senator Kay Bailey Hutchison, she and I recognized this back then, that when we set NASA's human space flight program on its current dual path—to build private sector capabilities in low Earth orbit, and a government-led program for deep space and ultimately Mars—we recognized that out of some of the misdirection, and lack of direction, that the space program had had, you needed that direction. And once Kay Bailey Hutchison and I passed that NASA authorization of 2010, that dual path approach is now bearing fruit, including our recapturing of a majority of global commercial launch market, a market we had almost completely lost to overseas competitors.

Nelson recalled the bureaucratic resistance to change within the federal government. "A few years ago, business at the Cape was much different than it is today. Commercial launch companies were looking elsewhere to take their business, despite all of the available infrastructure and the amazing workforce on the Space Coast. Too much bureaucracy stood in the way of progress." Nelson said he convened "top leaders" from NASA, the USAF, and the FAA to look at an aerial photo of "all the abandoned launch pads at the Cape, and got their commitment to work together with the private sector to bring these pads back to life . . . They are roaring back to life, with launches and landings on those very same pads."

"Quite simply," Nelson said, "jobs and ingenuity are soaring because rockets are soaring, and as goes Florida's Space Coast and the Houston area, so goes the US space industry as a whole."

Nelson was in retirement when President Joe Biden asked him in March 2021 to succeed Bridenstine as NASA administrator.[117] When Bridenstine was nominated in December 2017, Senator Nelson had said, "The head of NASA ought to be a space professional, not a politician." That criticism was forgotten; indeed, Nelson invited Bridenstine as well as Charlie Bolden to his oath-of-office ceremony. (Bridenstine participated via remote video.)

In an April 2021 *Scientific American* guest column, Lori Garver wrote, "Senator Nelson's imprint on the space program has thus been to perpetuate a system that rewards legislators whose states and districts have existing space facilities and jobs to protect."[118] As administrator, Nelson protected SLS as part of Project Artemis, but he also continued NASA's trend toward competition and fixed-price contracts with milestone payments. An example is Nelson's testimony on May 3, 2022, before the Senate appropriations subcommittee responsible for NASA's budget. Nelson was asked about the agency's plans to acquire a commercial lunar lander. He answered that President Biden and the Congress had expressed the desire for NASA to run a competition for the lander's fixed-price contract.[119]

> I believe that that is the plan that can bring us all the value of competition. You get it done with that competitive spirit, you get it done cheaper, and that allows us to move away from what has been a plague on us in the past, which is a cost-plus contract, and move to an existing contractual price.

On September 2, 2022, *PBS NewsHour* aired a segment about the pending SLS Artemis I launch, anchored by correspondent and independent journalist Miles O'Brien.[120] The segment interviewed Lori Garver, who recalled that SLS was sold to Congress in 2010 as a shuttle derivative that would launch sooner for less money. "I don't believe these people thought it would be true, but they knew they could sell that to Congress."

O'Brien interviewed Nelson while they stood in a Vehicle Assembly Building high bay across the transfer aisle from the stacked SLS Artemis I awaiting rollout to Pad 39B. The journalist asked the former senator about those early promises that SLS would launch by 2016. Nelson replied, "It was a seducing argument, because there were certain technologies that we were comfortable with."

* * * * *

History tends to repeat itself, so it seems logical that Cape Canaveral will follow the evolution of military airfields and navy yards to one day become a commercial spaceport.

Many airports began as military airfields. Orlando International Airport, for example, began as McCoy Air Force Base. (Its airport code, to this day, is MCO.) The City of Orlando received the property as surplus after McCoy closed in 1974.[121] Former US Navy yards in Philadelphia and Brooklyn are operated today by various forms of public-private partnerships. A closed naval shipyard in San Francisco is being transferred parcel by parcel to the city for possible commercial use once toxic cleanup is completed.[122]

One such model may be the Cape's future.

It may not happen soon. The USAF has studied the idea, but so far has resisted letting go of Cape Canaveral.[123] As of this writing, Florida has no Bill Nelson in the Senate with the stature and seniority to mandate change through sheer force of political will.

NASA has been more progressive about the commercial spaceport idea. KSC's *Future Development Concept*, published in 2012, spoke of "commercial operations zones" within an independent authority operating a multiuser spaceport.[124]

Space Florida president Frank DiBello in June 2016 presented his "Vision 2025," which included an independent authority running spaceport operations at the Cape. Such an authority would operate beyond the whims of federal and state legislatures and their annual budget cycles. The authority could seek outside funding and respond to market trends.[125]

An update of Space Florida's *Cape Canaveral Spaceport Master Plan* released in January 2017 called for an independent spaceport authority by 2025, "with NASA and the USAF becoming mission focused users." Among the benefits the plan foresaw was, "Provide access to private capital markets to finance needed common infrastructure."[126]

This trend was reported in May 2019 by the GAO. "While the federal government has not directly funded the construction of infrastructure at launch sites in recent years, state and local governments have done so . . . state and local governments are investing in infrastructure to obtain the economic benefits of attracting space-related businesses to their areas."[127]

Around this time, elements of the USAF (later spun off into the US Space Force) were studying how to make their space launch locations "more efficient and less bureaucratic." US Space Command General John "Jay" Raymond in 2018 created a "Range of the Future" task force to better understand how the space economy would impact the branch's operating environment. The task force issued a white paper in 2020 that recommended the development of a National Spaceport Strategy. The report recommended the creation of a National Spaceport Development Corporation to assume daily operations of the spaceports at the Eastern and Western Ranges. Congressional legislation would be necessary to create the NSDC and transfer that authority property from the DOD and NASA.[128]

As of this writing, the FAA's Office of Spaceports is hosting a working group "to develop a National Spaceport Strategy to leverage the full network of domestic spaceports to the benefit of the space transportation industry and the nation as a whole." Members of the group include representatives from the FAA; the Departments of Defense, State, and Commerce; the Space Force; and NASA.[129]

As for the Space Force, at General Raymond's direction the branch is proceeding with its Range of the Future 2028 initiative. Its vision statement is to "Preserve and advance national security interests through globally competitive Ranges with capacity to support launch and test operations on demand." Among the initiative's priorities are to "Establish a new Range operations business model emphasizing public-private partnerships" and "Transform the Ranges into National Spaceports."[130]

To further blur the line between spaceport and seaport, Space Florida in April 2024 released a study looking at Port Canaveral's support of the vertical launch and landing business at Cape Canaveral. Its key finding: "Current facilities at Port Canaveral and surrounding areas are insufficient to meet the projected demand for maritime operations related to space launches, necessitating over 9,000 linear feet of dedicated wharf space." As a first step, the study recommended dredging and building a new wharf complex just outside the CCSFS main gate, where a SpaceX launch control center and a space museum are located. The long-term recommendation is to extend the middle turning basin north along the eastern shore of the Banana River.[131]

* * * * *

The few residents of Cape Canaveral in the late nineteenth century tied up at a dock not far from where twenty-first-century spaceships now return to port.

Land speculators bought up Cape properties in the first half of the twentieth century, hoping to get rich by building a port that might welcome passengers and cargo from around the world. The port was finally built, but the property owners never got rich, as their land was acquired by a state port authority using eminent domain.

War brought a military presence to Brevard County, first to Cocoa Beach, then north to Cape Canaveral. When the war ended, the local economy was left economically dependent on federal spending to thrive.

Salvation came from the arrival of missile men testing their weapons on the tip of the Cape. Two ports—one sea, one space—grew in parallel, not quite twins but siblings of a sort.

The Cold War rivalry between two superpowers nurtured a nascent space race. As has happened throughout human history, nations raced to establish a technological superiority dependent on advances in transportation—in this circumstance, rocketry. American advances—and setbacks—launched from the Cape.

As the 1960s began, the Soviet Union launched the first human into space. The young American president, embarrassed by this triumph and a failed coup attempt in Cuba, proposed his nation send a man to the moon by the end of the decade and return him safely to the Earth.

Thousands of engineers and their families answered the call, moving on faith to remote Brevard County, where there wasn't enough housing to shelter them all. The housing industry caught up but, after the slain young president's legacy was fulfilled and prestige was reaped, the raison d'être for the whole enterprise was lost in the past. Missile treaties were signed, astronauts and cosmonauts shook hands in space, and the Cold War no longer seemed so warlike. Thousands of jobs were lost at the Cape, home mortgages could no longer be paid, and the local economy seemed on the verge on collapse.

For the next half century, the Space Coast economy remained addicted to federal funding. Memories of Apollo's aftermath haunted the people of Brevard County. Every time there was a historic turn of events—the loss of *Challenger*, the end of the Cold War, the loss of *Columbia*—locals panicked that another apocalypse was nigh.

But how to kick the habit?

An idealistic handful of space evangelists, based in Brevard, conceived the idea of a "space chamber of commerce" that would promote commercial space enterprises. Met with antipathy at home, they approached Florida's governor in

Tallahassee with the radical idea to create a "commission on space" that would recommend how the state could compete with other states for aerospace business and investment. The commission recommended creation of a state spaceport authority, along the lines of the Port Authority of New York and New Jersey.

And so the nation's first state space authority was born.

Spaceport Florida struggled for much of its existence. The exigencies of state politics drove the authority into the personal control of whomever occupied the governor's office. Over the years, funding went up, funding went down, and politicians scrapped to redirect tax dollars to their constituencies. Other states formed their own space authorities, but much of the commercial satellite industry went overseas for launch services, where they could find cheaper prices.

Because no robust commercial launch industry emerged, federal revenues became the primary source of funding for the Cape's few tenants. Fearing that the companies might go out of business if they had to compete, in 2005 the federal government allowed two of the surviving vendors to form a legal monopoly. The Pentagon had mission assurance, but prices went higher, driving even more of the commercial satellite industry to launch from Russia, Europe, or China.

After the loss of *Columbia,* NASA began to wind down the Space Shuttle program. The agency was directed by the White House in 2004 to complete the International Space Station, honoring American commitments to their spacefaring partners but, to pay for a new project, NASA was told to plan for deorbiting the ISS five years after its completion. That project, called Constellation, had grandiose visions but only enough funding to perpetuate the status quo.

Florida's space authority was folded by the state legislature in 2006 and replaced by a new agency called Space Florida. It inherited its predecessor's dependency on federal funding. Facilities were built in anticipation of new NASA tenants, only to find those programs canceled. With the Space Shuttle program's retirement planned for 2010, Space Florida and Brevard County elected officials watched as an economic time bomb ticked down to zero. Many yearned for another "JFK moment," believing the myth that a young president who articulated a bold vision would save their jobs and yet again prime the local economy.

Such a young president took office in 2009. As did his early 1960s predecessor, this young president had no real knowledge of space policy. During his 2008 campaign, he'd turned to a space evangelist who was a kindred spirit with those Brevard County space evangelists of the late 1980s who had envisioned a state space chamber of commerce as a way to break free of federal dependency. She helped align the young president's space policy with progressive visions of this NewSpace movement; after he took office, he appointed her to be deputy administrator of NASA.

The young president came to Brevard on April 15, 2010, to give them the "JFK moment" many had demanded. He gave them a bold vision, but it wasn't the vision many of them wanted to hear. His critics wanted the OldSpace past to continue—legacy contractors awarded cost-plus contracts that guaranteed a profit, with generous wages for the contractors' workforce but no penalty for delays or cost overruns. It's not easy to kick an addiction.

Florida's senior senator, the Democratic scion of a south Brevard cattle rancher, found himself in the unique position of determining the fate of the American space program. He'd flown on the space shuttle in his early days, when as a congressional representative of the Space Coast he'd been invited by NASA to fly on a mission as an observer. By 2010, as chair of the Senate's space subcommittee, he not only had power and influence, but friends in both houses on both sides of the aisle. Working with his friend, a Republican senator from Texas, they crafted legislation that became a grand compromise of sorts—a new rocket based on shuttle technology built by OldSpace contractors. In exchange, some of the reforms proposed by the young president could proceed, only with less immediate priority.

Space Florida's new president declared his agency independent of NASA. The agency sought its own commercial tenants for aging KSC and Cape Canaveral facilities. The formula worked; as NASA planned to convert KSC to a multiuser spaceport, Space Florida helped provide financial support for companies looking to move in.

By the end of the decade, the Space Coast's economy was booming. A NASA study for fiscal year 2021 concluded that KSC had generated $5.25 billion in economic output within the State of Florida, with 27,004 jobs linked to its space activities. NASA contractors provided 49.7 percent of the jobs, but commercial launch companies provided 22.3 percent of the jobs. Within Brevard County, commercial launch providers had created 6,790 jobs and generated $2.1 billion of economic output. The launch companies directly employed 2,744 people—1,135 by SpaceX which, thanks largely to its Cape operations, had become the dominant player in the global launch industry, government or commercial.[132]

Jobs had returned, the Space Coast economy flourished, and the addiction to federal spending had been all but broken.

The comeback was complete.

ACKNOWLEDGMENTS

"No man is an island," John Donne wrote in 1624, and most certainly that is true of all history writers.

This book would not have been possible without many life experiences and the assistance of many friends, colleagues, contemporaries, librarians, and researchers.

Many of those who participated in the events chronicled in this book still walk the earth with us. Stephen Morgan, Chris Shove, Ben Everidge, David Teek, and Edward Ellegood contributed their recollections about the early days of the Spaceport Florida Authority, the Astronauts Memorial Foundation, the Space Research Foundation, and other relevant entities. Pam Dana provided insights to her time while serving as director of the Office of Tourism, Trade, and Economic Development (OTTED).

Others involved did not respond to inquiries, so I relied on contemporary media accounts, in particular *Florida Today* and the *Orlando Sentinel.* Tim Walters at *Florida Today* located videos from their archives of their space news coverage in 2008–2010. Reporter John Kennedy with the Gannett chain recalled the circumstances that led to his investigating the use of *Challenger* license tag revenues.

Newspapers.com is an invaluable resource; the first news reports are not always accurate, but Newspapers.com was my first stop for most subplots that weave through this tale.

Alan Ladwig provided a copy of the task force report that led to the Space Flight Participant Program (SFPP) and Teacher in Space. He also helped me understand the politics behind the SFPP and the Reagan administration flying members of Congress on the space shuttle. Alan later served on President Obama's NASA transition team, and provided his insights to events during later 2008 through early 2009.

I'm endlessly grateful to those who contributed their recollections of Space Coast events during the critical years of 2008–2010. Dale Ketcham with Space Florida offered critical insights into the local politics of the time. Robin Fisher shared his time to discuss his Save Space efforts and his meeting with President

Obama. Malcolm Kirschenbaum, a lifelong friend of Senator Bill Nelson, recalled the events of April 15, 2010, and also put me in touch with the senator.

When I interviewed him, he was NASA Administrator Bill Nelson, but he prefers to be addressed as "Senator." He was more than gracious with his time, granting two lengthy telephone interviews to help me understand his perspectives about the 2010 NASA authorization act. He also provided his Yale University senior thesis, which is cited at the beginning of Chapter 10. Hopefully Senator Nelson writes his own book about these events.

I've known Lori Garver going back to her days as executive director of the National Space Society. Lori provided her own invaluable insights. She was at the epicenter of these events that forever changed the course of American space policy. Without Lori Garver, Barack Obama's space policies would have been very different. Her memoir, *Escaping Gravity*, is the authoritative reference for the topic. *Return to Launch* is "the rest of the story."

Steve Robinson, who preceded Lori as candidate Obama's space policy advisor, was kind enough to not only provide his time but also a window into the early days of the Obama 2008 presidential campaign.

A big thank-you to those in organized labor who helped me understand not only labor's history at the Cape but also their perspectives during 2008–2010. In particular, Steve Williams and Matt Nelson at IBEW Local 756, Fernando Rendon at IBEW Local 606, and Rich Templin with the Florida AFL-CIO provided their time and insights. This book is better for their contributions.

Several libraries and archives were more than helpful. Holly Baker at the Library of Florida History contributed several documents recounting the early days of Cape Canaveral and Brevard County. The State Archives of Florida scanned and emailed me several critical documents. Other libraries contributing to this project include the University of Florida in Gainesville and the University of Central Florida in Orlando. The University of Florida has Senator Bill Nelson's personal papers; thank you to Steve Hersh, Michele Wilbanks, and in particular Matt Kruse for tolerating my incessant requests.

Federal agencies also helped with either document requests or for clarifications of policy. The Federal Aviation Administration helped search for documents from the early days of the Office of Commercial Space Transportation. Joshua Finch at NASA explained the intricacies of Commercial Crew contracting. Aimee Crane at NASA helped with media usage guidelines and image searches.

From 2011 through 2021, I was a communicator at the Kennedy Space Center Visitor Complex. Communicators lead tours, deliver lectures, and on occasion escort retired astronaut speakers. Those ten years were the foundation for this book. Therrin Protze, KSCVC's chief operating officer, helped with understanding the relationship between the attraction and state agencies. Ken Romer, one of my former communicator colleagues, has been working on his own book, on astronaut genealogy. Ken helped with Bill Nelson's family history and offered various insights for the Nelson chapter.

Astronaut Winston Scott, who later became the final executive director of the Florida Space Authority, not only provided recollections of his time with the FSA but also helped with my researching FSA documents. Astronauts are a special breed; it was always a privilege to work with him at KSCVC.

John M. Logsdon, founder of the Space Policy Institute at George Washington University, is arguably the nation's foremost space policy analyst. He's been an eyewitness to US space history going back to Project Apollo. His trilogy of presidential space policy books are cited throughout this work. John has been a mentor and a friend; he's the Obi-Wan Kenobi of this project.

Science fiction author M. L. Clark also provided advice and support, experienced in the academic press world, but also a valued colleague and talented writer.

I've saved the best for last.

Sian Hunter is the senior acquisitions editor at the University of Florida Press. She has been alongside this project from the day I emailed the proposal to UFP. Sian dispels every notion one might have about editors. This book doesn't happen without Sian. Thank you from the bottom of my heart.

Carlynn Crosby is an assistant editor at UFP. Carlynn helped take this project across the finish line. This book doesn't happen without Carlynn, either.

My wife, Carol, has suffered through this project, long before I typed the first character. She had a ringside seat for the book's evolution, revisions, and completion. Carol endured all the times I talked out an obstacle or writer's block until I had the solution. Way back in 2009, she agreed to move here to the Space Coast to chase our dream to be part of the US space program. We've had an incredible run. I love you.

ABBREVIATIONS FOR SOURCES

Some newspapers merged or changed their names over the years—for example *Today* became *Florida Today* in 1985. Deprecated names such as *Today* remain in the endnotes, while current names typically are abbreviated. Older newspapers that no longer exist, such as *Florida Star*, are not abbreviated. Newspapers cited only once are not abbreviated.

Some older newspapers published under various names—for example, the *Orlando Evening Star* and *Orlando Sunday Sentinel-Star*—those names are also preserved in the endnotes, although they may have had the same publisher.

AJ	*Albuquerque Journal*
AP	Associated Press
AT	*Ars Technica*
DP	*Daily Press* (Newport News, VA)
DT	*Daily Times* (Salisbury, MD)
EPT	*El Paso Times*
FMNP	*Fort Myers News-Press*
FT	*Florida Today*
IA	Internet Archive
LAT	*Los Angeles Times*
LR	*Lompoc Record*
MH	*Miami Herald*
NYT	*New York Times*
OS	*Orlando Sentinel*
PBP	*Palm Beach Post*
PNJ	*Pensacola News Journal*
SB	*Sacramento Bee*
SBI	*Santa Barbara Independent*

SFE	*San Francisco Examiner*
SFNM	*Santa Fe New Mexican*
SLOT	*San Luis Obispo Tribune*
SFSS	*South Florida Sun Sentinel*
SMT	*Santa Maria Times*
SN	*Space News*
TBT	*Tampa Bay Times*
TD	*Tallahassee Democrat*
UPI	United Press International
WM	*Wayback Machine*
WAPO	*Washington Post*
WSJ	*Wall Street Journal*

NOTES

Chapter 1. Space Canaveral

1 The circumstances for the Bumper 8 launch were drawn from: Stanley O. Starr, "The Launch of Bumper 8 from the Cape: The End of an Era and the Beginning of Another," *Volume 32 of the AAS History Series*, ed. Christophe Rothmund, Proceedings of the 35th History Symposium of the International Academy of Astronautics (IAA), Toulouse, France, 2001, (pub. 2010), 75–97. "Bumper Q&A," https://www.nasa.gov/missions/highlights/webcasts/history/bumper-qa.html, archived November 5, 2008, at the *WM*. John Uri, "70 Years Ago: First Launch from Cape Canaveral," NASA, July 24, 2020, https://www.nasa.gov/history/70-years-ago-first-launch-from-cape-canaveral/.

2 OS first used the phrase in a December 6, 1964, real estate ad on 11D, https://www.newspapers.com/image/224853470/. Florida newspapers began to use the term with more regularity by the end of the 1960s.

3 Jerrell H. Shofner, *History of Brevard County, Volume 1* (Brevard County Historical Commission, 1995), 12–14.

4 Generally from Paul A. Schmalzer et al., *Geology, Hydrology and Soils of Kennedy Space Center: A Review* (NASA, August 1990), https://ntrs.nasa.gov/api/citations/19910001129/downloads/19910001129.pdf.

5 Shofner, *Volume 1*, 114.

6 Generally from Richard S. Levy et al., *An Archaeological Survey of Cape Canaveral Air Force Station, Brevard County, Florida* (Resources Analysts, Inc., 1984). Rick Neale, "Space Force Station Yields Prehistory for UCF Students," *FT*, April 10, 2022, 1A, 16A, https://www.newspapers.com/image/831898030/.

7 Johanna Fantova, "Cape Cañaveral to Cape Kennedy," *The Princeton University Library Chronicle* 26, no. 3 (Autumn 1964): 57, https://www.jstor.org/stable/26402928.

8 Peter John Ferdinando, "Atlantic Ais in the Sixteenth and Seventeenth Centuries: Maritime Adaptation, indigenous Wrecking, and Buccaneer Raids on Florida's Central East Coast" (PhD Diss., Florida International University, 2019), 9–10, https://digitalcommons.fiu.edu/dissertations/AAI3721473/.

9 "History & Culture," Canaveral National Seashore, https://www.nps.gov/cana/learn/historyculture/index.htm. Vera Zimmerman, "The First Settlers, 10,000 BC to 1820," Brevard GenWeb, https://usgenwebsites.org/flgenweb/FLBrevard/History/10K-1820.html. The use of "Merritt Island" without the possessive first appears in various *Florida Star* newspapers archived on Newspapers.com that were published circa 1900.

10 "A County Called Mosquito," Florida Memory, December 12, 2018, https://www.floridamemory.com/items/show/342049. "History Summary," Brevard County, https://www.brevardfl.gov/HistoricalCommission/HistorySummary.
11 Lesa Lorusso, "Cape Canaveral Lighthouse, Cape Canaveral Air Force Station, FL," Florida Historical Society, https://myfloridahistory.org/preservation/cape-canaveral-lighthouse-cape-canaveral-air-force-station-fl.
12 Levy et al., *Archaeological Survey*, 48–49.
13 Jerrell H. Shofner, *History of Brevard County, Volume 2* (Brevard County Historical Commission, 1996), 20–21.
14 "History at a Glance," City of Cocoa Beach, https://www.cityofcocoabeach.com/162/History-at-a-Glance.
15 Levy et al., *Archaeological Survey*, 53.
16 George E. Buker, *Sun, Sand and Water: A History of the Jacksonville District U.S. Army Corps of Engineers, 1821–1975* (U.S. Army Corps of Engineers, 1981), 155–156, https://books.google.com/books/download/Sun_Sand_and_Water.pdf?id=weOqDIop0WYC.
17 Canaveral Harbor ad, *MH*, October 14, 1925, 6-C–7-C, https://www.newspapers.com/image/616508663 and https://www.newspapers.com/image/616508675.
18 "Canaveral Group Buys 8,000 Acres," *The Sunday Sentinel-Star*, February 27, 1938, 1, 3, https://www.newspapers.com/image/313707221/.
19 "Canaveral Property Sale Recommended," *MH*, February 18, 1938, 13, https://www.newspapers.com/image/617915157/.
20 "Canaveral Would Be Aid in Case of War," *The Sunday Sentinel-Star*, August 27, 1939, 5, https://www.newspapers.com/image/313698839/.
21 Buker, 192–193.
22 "Canaveral Deal Pushed," *Orlando Morning Sentinel*, August 22, 1939, 12, https://www.newspapers.com/image/313697368/.
23 "Canaveral's $1,661,000 Harbor Wins Army Engineer Approval," *Orlando Morning Sentinel*, March 19, 1941, 1, https://www.newspapers.com/image/313706275/.
24 "Decision on Canaveral Nearing," *Orlando Reporter-Star*, May 8, 1941, 3, https://www.newspapers.com/image/343365611/.
25 Frank A. Kennedy, "Big Sum Planned for Waterways," *PNJ*, March 19, 1944, 4, https://www.newspapers.com/image/352803129.
26 "Projects by Inlets Signed by President," *PBP*, March 3, 1945, 3, https://www.newspapers.com/image/134142403/.
27 "$830,500 Cash Allocated for Canaveral Harbor," *The Orlando Star*, July 15, 1946, 1, https://www.newspapers.com/image/343361507/.
28 "Canaveral Pushed," *The Orlando Star*, July 16, 1946, 1, https://www.newspapers.com/image/343361655/.
29 "Harbor Bond Election Set for Nov. 25," *MH*, October 23, 1947, 1-B, https://www.newspapers.com/image/617739837/.
30 "Vote Assures Canaveral Harbor," *MH*, November 27, 1947, 1-B, https://www.newspapers.com/image/617750703/.
31 "City of Romance: Canaveral Harbor, Town That Never Was," *OS*, February 18, 1962, 16-F, 21-F, https://www.newspapers.com/image/223165503/, https://www.newspapers

.com/image/223165511/. The freezing of English funds is discussed in "Conway Kittredge Closes $115,000 Brevard Project," *OS*, May 31, 1959, 4-C, https://www.newspapers.com/image/223401383/.

32 "Canaveral Court Test Is Concluded," *MH*, April 24, 1949, 1-B, https://www.newspapers.com/image/618189304/.

33 "Jury Decides Land Costs," *MH*, July 30, 1949, 1-B, https://www.newspapers.com/image/617975481/.

34 Melissa Williford Euziere, *From Mosquito Clouds to War Clouds: The Rise of Naval Air Station Banana River* (master's thesis, Florida State University, 2003), 15, https://diginole.lib.fsu.edu/islandora/object/fsu:168623.

35 Ewart Hendry, "Brevard County Bits," *The Sunday Sentinel-Star*, October 8, 1939, 7, https://www.newspapers.com/image/313546749.

36 "Banana River Base Approved as Senate Votes Naval Funds," *Orlando Morning Sentinel*, April 20, 1939, 1, https://www.newspapers.com/image/313677428.

37 "Realty Notes," *The Miami Daily News*, May 7, 1939, 3-D, https://www.newspapers.com/image/298587298/.

38 "Banana River Base Survey Under Way," *Orlando Morning Sentinel*, May 17, 1939, 3, https://www.newspapers.com/image/313684541/.

39 "Navy Seeks Land in Jacksonville," *The Miami Daily News*, October 1, 1939, 6-A, https://www.newspapers.com/image/298479077/.

40 "Banana River Land Goes to Government," *Orlando Morning Sentinel*, November 1, 1939, 3, https://www.newspapers.com/image/313548510/.

41 "Banana River Naval Air Station Placed in Commission Tuesday," *The Cocoa Tribune*, October 3, 1940, 1, https://www.newspapers.com/image/779795725/.

42 "Bombing," *The Sunday Sentinel*, October 19, 1941, 1, https://www.newspapers.com/image/313699419/.

43 Euziere, 30–31.

44 "City Briefs," *The Sunday Sentinel-Star*, June 22, 1941, 21, https://www.newspapers.com/image/313542599/.

45 Euziere; John Wilds, "Sub Attacks Off Florida Coast Halted as Navy Flying Patrol Goes into Action," *The Miami Daily News*, August 13, 1942, 7-A, https://www.newspapers.com/image/297383495/.

46 David J. Castello, "When a German U-Boat Attacked Palm Beach County," WestPalmBeach.com, https://www.westpalmbeach.com/when-a-german-u-boat-attacked-palm-beach-county/.

47 Euziere, 40.

48 An example is "Air Station Seeks Mechanical Help," *MH*, June 21, 1942, 9-A, https://www.newspapers.com/image/617681126/.

49 "Navy Keeps 5 Florida Air Bases," *MH*, February 1, 1946, 8-B, https://www.newspapers.com/image/617011080/.

50 "Base Personnel Now at 2,200," *MH*, August 24, 1946, 2-B, https://www.newspapers.com/image/617531678/.

51 John T. Farquhar, *Arctic Linchpin: The Polar Concept in American Air Atomic Strategy, 1946–1948*, 2–3, Department of Defense Technical Information Center, https://apps.dtic.mil/sti/pdfs/AD1015392.pdf. Frank Kennedy, "Banana River Air Base

May Be Closed," *PNJ*, June 15, 1947, 4, https://www.newspapers.com/image/352877015/.

52 "Curtailment Seen in BRNAS Aerial Training Program," *Orlando Evening Star*, May 30, 1947, 1, https://www.newspapers.com/image/340980078/.

53 "Drive Opens to Retain Navy Base," *MH*, June 2, 1947, 1-B, https://www.newspapers.com/image/617403047.

54 "Base to Close," *MH*, June 15, 1947, 12-C, https://www.newspapers.com/image/617733955/.

55 Edith Voss, "Such Scenes Will Soon Be Only Memories," *Orlando Evening Star*, July 14, 1947, 3, https://www.newspapers.com/image/340981366/.

56 "Navy Delays Close at Banana River," *MH*, August 1, 1947, 1-B, https://www.newspapers.com/image/617736062/.

57 "Evolution of the 45th Space Wing," Patrick Air Force Base, http:/www.patrick.af.mil/library/factsheets/factsheet.asp?id=4514, archived June 13, 2011, at the *WM*.

58 "Banana River Base Opening Under Study," *MH*, July 23, 1948, 1-A, https://www.newspapers.com/image/617961838/.

59 "Air Force Gets Base," *Orlando Evening Star*, September 3, 1948, 2, https://www.newspapers.com/image/340913415/.

60 John T. Carlton, "Banana River Air Base to be Proving Ground," *The Miami News*, January 25, 1949, 2-A, https://www.newspapers.com/image/297705118/.

61 "7,500 Men for Banana River," *Orlando Evening Star*, January 26, 1949, 1, https://www.newspapers.com/image/340984766/.

62 Chris Mathisen, "Lack of Testing Range Delays Guided Missile Plan, Officers Testify," *The Evening Star*, A1, A6, https://www.newspapers.com/image/868171359/.

63 "Banana River Guided Missile Experimentation Project Set," *The Palm Beach Post-Times*, March 6, 1949, 17, https://www.newspapers.com/image/129973430/.

64 "Cocoa AF Project Hailed," *Orlando Evening Star*, May 4, 1949, 1, https://www.newspapers.com/image/340995757/.

65 "Cocoa Area Readies for Building Boom," *Orlando Evening Star*, May 5, 1949, 1, https://www.newspapers.com/image/340995839/.

66 "U.S. Prepares to Take Cape Canaveral Area," *MH*, April 19, 1950, 4-C, https://www.newspapers.com/image/624657324/.

67 "Huge Land Condemnation Suit to Be Heard Here This Month," *Orlando Sunday Sentinel-Star*, September 2, 1951, 8, https://www.newspapers.com/image/222162495/.

68 "Jury Gives Canaveral Landowners Twice as Much as Original Award," *OS*, July 19, 1955, 13, https://www.newspapers.com/image/223409863.

69 "Canaveral Harbor, Town That Never Was," *OS Florida Magazine*, February 18, 1962, 16F, 21F, https://www.newspapers.com/image/223165503/, https://www.newspapers.com/image/223165511/.

70 Cape Canaveral was renamed Cape Kennedy by President Lyndon B. Johnson on November 29, 1963, to honor his predecessor slain one week before. The Florida Legislature restored the original name in 1973. For consistency, this book will use the term *Cape Canaveral* unless historical circumstances require otherwise.

71 1960 US Census, Florida population, US. Department of Commerce, 11-11–11-12,

https://www2.census.gov/library/publications/decennial/1960/population-volume-1/vol-01-11-c.pdf.

72 1970 US Census, Florida population, US Department of Commerce, 11–20, https://www2.census.gov/prod2/decennial/documents/1970a_fl1-01.pdf.

73 Cape Canaveral Space Force Station has undergone many name changes over the years. In 1955, it was named the Cape Canaveral Missile Test Annex. From 1964 to 2020, it was an "Air Station" or "Air Force Station." On December 9, 2020, the station became part of the new Space Force branch of the US Air Force. This book will use *Air Force Station* for the base until the 2020 name change, then use *Space Force Station.*

74 *Analysis of the Brevard County, Florida, Housing Market as of February 1, 1966* (Department of Housing and Urban Development, Federal Housing Administration, August 1966), i, https://www.huduser.gov/portal/publications/pdf/scanned/scan-chma-BrevardCountyFlorida-1966.pdf.

75 Charles D. Benson and William Barnaby Faherty, *Moon Launch! A History of the Saturn-Apollo Launch Operations* (University Press of Florida, First Paperback Printing, 2001), 502.

76 *Environmental Impact Statement for the John F. Kennedy Space Center* (NASA, Final, October 1979), 3-54, 3-56, https://ntrs.nasa.gov/api/citations/19800013404/downloads/19800013404.pdf.

77 An excellent documentation of this very complex subject is Arnold S. Levine, *Managing NASA in the Apollo Era* (NASA Scientific and Technical Information Branch, SP-4102, 1982), https://www.nasa.gov/wp-content/uploads/2023/03/sp-4102.pdf.

78 An *FT* spreadsheet of Kennedy Space Center employment history starting in 1964 was provided to this author by an *FT* reporter in an email, September 3, 2013.

79 *EIS for KSC*, 3-56.

80 Tom Myers, "Brevard Jobless Rate Likely to Remain High," *Florida Today*, May 23, 1975, 14C, https://www.newspapers.com/image/125392154/. *Florida Today* began on March 21, 1966, as just *Today*, owned by the Gannett Newspaper Group. Gannett around that time bought other local publications—the *Cocoa Tribune*, *Titusville Star-Advocate*, and *Eau Gallie Courier.* The other publications eventually became part of *Today.* On August 26, 1985, *Today* changed its format and became *Florida Today.* Gannett began publishing *USA Today* in 1982.

81 "Strategic Arms Limitations Talks/Treaty (SALT) I and II," US Department of State, https://history.state.gov/milestones/1969-1976/salt.

82 Mark C. Cleary, *The 6555th: Missile and Space Launches Through 1970* (45th Space Wing History Office, First Edition, November 1991), Chapter III, Section 8 ff., https://ccspacemuseum.org/wp-content/uploads/histories/6555.pdf.

83 "Building Revives Brevard County," *OS*, November 21, 1971, 9-R, https://www.newspapers.com/image/221215327/.

84 *FT* KSC employment history spreadsheet.

85 "Brevard County Population & Economic Trends," Brevard Area Transportation Planning Annual Report, a supplement to *FT*, October 27, 1980, 8, https://www.newspapers.com/image/124903257.

86 *FT* KSC employment history spreadsheet. The spreadsheet shows 16,067 KSC-related jobs in 1985, then the number dropped to 13,664 in 1986. The job losses were temporary; in 1987, the number went up to 15,307. By 1988, the year shuttle returned to flight, the number was 16,559, more than at the time of the *Challenger* loss.

87 Anatoly Zak, "The Progress Cargo Ship," RussianSpaceWeb.com, updated November 30, 2023, http://www.russianspaceweb.com/progress.html.

88 NASA promoted the space shuttle as "reusable," but that was misleading. Some parts could be refurbished over several months.

89 Christopher Kraft et al., *Report of the Space Shuttle Management Independent Review Team*, February 1995, Executive Summary, viii, https://spp.fas.org/kraft.htm.

90 Jim Banke, "2 Companies Seek Shuttle Contract," *FT*, August 3, 1995, 1B, https://www.newspapers.com/image/175580949.

91 John Mintz, "Boeing to Buy Rockwell's Defense, Space Divisions," *WAPO*, August 2, 1996, https://www.washingtonpost.com/archive/business/1996/08/02/boeing-to-buy-rockwells-defense-space-divisions/5350b59e-c89f-4276-a04b-ee82c1d5b3b9/.

92 "United Space Alliance: The Space Operations Company," USA, http://www.unitedspacealliance.com/.

93 *FT* KSC employment history spreadsheet.

94 Wayne T. Price, "Economic Effect May Be 'Scary,'" *FT*, February 2, 2003, 7A, https://www.newspapers.com/image/178380219.

95 Paige St. John and Alisa LaPolt, "State Braces for Impact," *FT*, February 7, 2003, 1S–2S, https://www.newspapers.com/image/178396827. The article cited numbers from the Florida Space Authority.

Chapter 2. Something Old, Something New

1 President John F. Kennedy, "On Urgent National Needs," John F. Kennedy Presidential Library and Museum, https://www.jfklibrary.org/archives/other-resources/john-f-kennedy-speeches/united-states-congress-special-message-19610525. The website has both the complete text of the speech and an audio recording.

2 John M. Logsdon, *John F. Kennedy and the Race to the Moon* (Palgrave Macmillan, 2010), 61.

3 Logsdon, *Kennedy*, 79.

4 Logsdon, *Kennedy*, 29–31. The National Aeronautics and Space Act of 1958 created the Council, with the president as its chair. Kennedy decided to assign that responsibility to the vice president; Johnson as Senate majority leader had been instrumental in writing the original act and was intimately familiar with American space policy issues.

5 President John F. Kennedy, "Memorandum for Vice President," April 20, 1961, https://history.nasa.gov/Apollomon/apollo1.pdf, archived April 9, 2009, at the *WM*.

6 Vice President Lyndon B. Johnson, "Memorandum for the President: Evaluation of Space Program," April 28, 1961, http://history.nasa.gov/Apollomon/apollo2.pdf, archived April 9, 2009, at the *WM*.

7 Robert Seamans, "National and International Implications of NASA's Programs" (speech to the American Association for the Advancement of Science, Denver,

Colorado, December 30, 1961), 207, https://www.jfklibrary.org/asset-viewer/archives/JFKNSF/307/JFKNSF-307-001.

8 Joseph L. Myler, United Press International, "Billions Spent on Space to Seek Many Answers," *PBP*, January 1, 1962, 24, https://www.newspapers.com/image/133671279/.

9 Alvin Shuster, "Congress Wary on Cost, but Likes Kennedy Goals," *NYT*, May 26, 1961, 1, 13, https://timesmachine.nytimes.com/timesmachine/1961/05/26/issue.html.

10 "Public Hesitant on Extra Moon Funds," *MH*, May 31, 1961, 8-D, https://www.newspapers.com/image/619392105.

11 "Meaning for Apollo," *Project Apollo: A Retrospective Analysis*, 19, NASA, https://www.nasa.gov/wp-content/uploads/2023/04/sp-4503-apollo.pdf. The approximate cost in January 2025 dollars came from the CPI Inflation Calculator at the US Bureau of Labor Statistics website using $25.4 billion in January 1968 dollars. https://www.bls.gov/data/inflation_calculator.htm. The Planetary Society estimates the cost in 2020 dollars was $257 billion. "How Much Did the Apollo Program Cost?" The Planetary Society, https://www.planetary.org/space-policy/cost-of-apollo

12 Roger D. Launius, "Public Opinion Polls and Perceptions of US Human Spaceflight," *Space Policy* 19, no. 3 (2003): 163, https://www.sciencedirect.com/science/article/abs/pii/S0265964603000390.

13 Launius, "Public Opinion Polls," 165.

14 "Technology Triumphs, Morality Falters," Pew Research Center, July 3, 1999, https://www.pewresearch.org/politics/1999/07/03/technology-triumphs-morality-falters/.

15 CNN/ORC Poll 11, July 21, 2011, 2, http://i2.cdn.turner.com/cnn/2011/images/07/21/poll.july21.pdf.

16 Subpart 16.3—Cost-Reimbursement Contracts, Acquisition.gov, FAC Number 2024-05, May 22, 2024, https://www.acquisition.gov/far/part-16#FAR_Subpart_16_3. *Guidance on Using Incentive and Other Contract Types*, Department of Defense, March 2016, 9–12, https://www.acq.osd.mil/dpap/policy/policyvault/USA001270-16-DPAP.pdf. Joachim Hofbauer and Greg Sanders, "Defense Industrial Initiatives Current Issues: Cost-Plus Contracts," Center for Strategic & International Studies, 2008, https://csis-website-prod.s3.amazonaws.com/s3fs-public/legacy_files/files/media/csis/pubs/081016_diig_cost_plus.pdf.

17 Subpart 16.2—Fixed-Price Contracts, Acquisition.gov, https://www.acquisition.gov/far/subpart-16.2. Also see *Guidance*, and Hofbauer and Sanders.

18 McNaugher, Thomas L. "Weapons Procurement: The Futility of Reform." *International Security* 12, no. 2 (1987): 65, 67, 76, https://doi.org/10.2307/2538813.

19 J. C. Hunsaker, "Forty Years of Aeronautical Research," *Forty-Fourth Annual Report of the National Advisory Committee for Aeronautics* (US Government Printing Office, 1959), 3–4, https://digital.library.unt.edu/ark:/67531/metadc64173/m2/1/high_res_d/20050019296.pdf.

20 Hunsaker, "Forty Years," 5–6.

21 Hunsaker, "Forty Years," 7–8.

22 Hunsaker, "Forty Years," 10–11.

23 Hunsaker, "Forty Years," 8–9.

24 Hunsaker, "Forty Years," 21–22.

25 James H. Doolittle, "The Following Years: 1955–1958," *44th Annual NACA Report*, 29–30, also 35–36. Andrew Chaikin, "How the Spaceship Got Its Shape," *Air & Space Magazine*, November 2009, https://www.smithsonianmag.com/air-space-magazine/how-the-spaceship-got-its-shape-137293282/. Russell D. Howard et al., "Dream Chaser Commercial Crewed Spacecraft Review," American Institute of Aeronautics and Astronautics conference paper, April 12, 2011, https://www.researchgate.net/publication/271366758.

26 "Part III—Financial Report," *44th Annual NACA Report*, 94.

27 Constance McLaughlin Green and Milton Lomask, *Project Vanguard: The NASA History* (Dover Publications, 2009), 37–39. An earlier version, *Vanguard: A History* (NASA Scientific and Technical Information Division, SP-4202, 1970) is online at https://www.nasa.gov/wp-content/uploads/2023/03/sp-4202.pdf.

28 Walter Sullivan, "Light May Flash in Soviet's 'Moon,'" *NYT*, October 1, 1957, 1, 14, https://timesmachine.nytimes.com/timesmachine/1957/10/01/issue.html. UPI, "Tracking of Russian Moon Barred to Other Nations," *MH*, October 1, 1957, 3A, https://www.newspapers.com/image/619095052/. Alton L. Blakeslee, "Russians Might Win Race and Launch First Moon," *TD*, October 1, 1957, 11, https://www.newspapers.com/image/245084586/.

29 Paul Dickson, *Sputnik: The Shock of the Century* (Walker & Company, 2001), 17.

30 Dickson, *Sputnik*, 112–113.

31 AP, "Symington Asks Satellite Probe as Leaders Voice Alarm, Anger," *Daily Press*, October 6, 1957, 1A, 12B, https://www.newspapers.com/image/231327215/.

32 Green and Lomask, *Project Vanguard*, 49 et seq.

33 Green and Lomask, *Project Vanguard*, 209–210.

34 Franklin O'Donnell, *Explorer 1*, JPL, 2007, 22, 25, 43–44, https://www.jpl.nasa.gov/about/downloads/Explorer1.pdf, archived January 31, 2017, at the *WM*.

35 John M. Hightower, "Nation Regains Some of Prestige," *The Miami News*, February 1, 1958, 3A, https://www.newspapers.com/image/298215783/. "The Pulitzer Prizes," Pulitzer, https://www.pulitzer.org/winners/john-m-hightower.

36 Karl Hunziker, "Brevard Quietly Pleased," *Orlando Evening Star*, State Edition, February 1, 1958, 1, https://www.newspapers.com/image/289013727.

37 Yanek Mieczkowski, *Eisenhower's Sputnik Moment: The Race for Space and World Prestige* (Cornell University Press, 2013), 139–141. Walter A. McDougall, . . . *The Heavens and the Earth: A Political History of the Space Age* (The Johns Hopkins University Press, Paperback Edition, 1997) 148–150.

38 Mieczkowski, *Eisenhower*, 167. For a detailed discussion of ARPA's history, see *The Advanced Research Projects Agency, 1958–1974* (Richard J. Barber Associates, Inc., December 1975), https://apps.dtic.mil/sti/pdfs/ADA154363.pdf.

39 Mieczkowski, *Eisenhower*, 166–171.

40 Mieczkowski, *Eisenhower*, 173.

41 McDougall, *The Heavens and the Earth*, 174–176.

42 Mieczkowski, *Eisenhower*, 174–175.

43 Green and Lomask, *Project Vanguard*, 223.

44 "X-15 Hypersonic Research Program," NASA, updated February 28, 2014, https://www.nasa.gov/reference/x-15/.

45 Helen T. Wells et al., *Origins of NASA Names* (NASA Scientific and Technical Information Office SP-4402, 1976), 88, https://www.nasa.gov/wp-content/uploads/2023/03/sp-4402.pdf.

46 "From JPL to NASA," JPL, https://www.jpl.nasa.gov/who-we-are/history.

47 Loyd S. Swenson, Jr. et al., *This New Ocean: A History of Project Mercury* (NASA History Office SP-4201, 1998), 101 et seq., https://www.nasa.gov/wp-content/uploads/2023/02/sp-4201.pdf.

48 National Aeronautics and Space Act of 1958 (Unamended), Sec. 102(b), July 29, 1958, NASA, https://www.nasa.gov/history/national-aeronautics-and-space-act-of-1958-unamended/. The act has been amended many times over the years.

49 NASA Act of 1958 (Unamended), Sec. 102(c).

50 Dan Leone, "Wolf Asks NASA for List of All Space Act Agreements," *SN*, January 16, 2013, https://spacenews.com/wolf-asks-nasa-for-list-of-all-space-act-agreements/. "NASA's Use of Space Act Agreements," NASA Office of the Inspector General, Audit Report IG-14-020, June 5, 2014, https://oig.nasa.gov/wp-content/uploads/2024/02/ig-14-020.pdf.

51 Logsdon, *Kennedy*, 226.

52 "James E. Webb," NASA, https://history.nasa.gov/Biographies/webb.html. Richard Pearson, "James E. Webb Dies at 85," *The Washington Post*, March 29, 1992, https://www.washingtonpost.com/archive/local/1992/03/29/james-e-webb-dies-at-85/410b1d13-898e-4ac1-bb23-aaba70574f6b/.

53 McDougall, *The Heavens and the Earth*, 361.

54 W. Henry Lambright, *Powering Apollo: James E. Webb of NASA* (The Johns Hopkins University Press, 1995), 100.

55 NASA Administrator James Webb Memorandum to Vice President Lyndon Johnson, May 23, 1961, *Exploring the Unknown, Volume II: External Relationships*, ed. John M. Logsdon (NASA History Office, 1996), 493, https://history.nasa.gov/SP-4407/vol2/v2intro.pdf.

56 McDougall, *The Heavens and the Earth*, 376.

57 Piers Bizony, *The Man Who Ran the Moon: James E. Webb, NASA, and the Secret History of Project Apollo* (Thunder's Mouth Press, 2006), 73.

58 Seamans speech to AAAS, December 30, 1961, 208.

59 The final study was commissioned in September 1963. Logsdon, *Kennedy*, generally, 214–222.

60 Alexander MacDonald, *The Long Space Age: The Economic Origins of Space Exploration from Colonial America to the Cold War* (Yale University Press, 2017), 190. President John F. Kennedy, "Address to the U.N. General Assembly," September 20, 1963, US Department of State, https://2009-2017.state.gov/p/io/potusunga/207201.htm.

61 "A Historic Meeting on Spaceflight," updated December 2, 2013, NASA, https://history.nasa.gov/JFK-Webbconv/pages/audio.html. "Transcript of Presidential Meeting in the Cabinet Room of the White House," November 21, 1962, NASA, https://history.nasa.gov/JFK-Webbconv/pages/transcript.pdf.

62 "JFK Library Releases Recording of President Kennedy Discussing Race to the Moon," JFK Library press release, May 25, 2011, https://www.jfklibrary.org/about-us/news-and-press/press-releases/jfk-library-releases-recording-of-president-kennedy

-discussing-race-to-the-moon. The recording is on their website at https://www.jfklibrary.org/asset-viewer/archives/JFKPOF/MTG/JFKPOF-MTG-111-004/JFKPOF-MTG-111-004.

63 "NASA Langley Research Center's Contributions to the Apollo Program," NASA, https://www.nasa.gov/centers/langley/news/factsheets/Apollo.html, archived October 15, 2015, at the *WM*.

64 An advertisement for the lecture series was published in the *LAT*, March 13, 1960, 39, https://www.newspapers.com/image/380993326/.

65 The Nixon campaign released a space policy statement on October 25, 1960, while the vice president campaigned in Cincinnati, Ohio. It made no mention of the issues raised by Cordiner. "Statement of the Vice President of the United States on Space Exploration, Cincinnati, OH," The American Presidency Project, University of California Santa Barbara, https://www.presidency.ucsb.edu/node/273838.

66 Ralph J. Cordiner, "Competitive Private Enterprise in Space," in *Peacetime Uses of Outer Space*, ed. Simon Ramo (McGraw-Hill, 1961, reprinted by Hassell Street Press), 222, https://rjacobson.files.wordpress.com/2011/02/cordiner-article-1961.pdf.

67 "Text of the Address by President Eisenhower," White House press release, January 17, 1961, Dwight D. Eisenhower Presidential Library, Museum & Boyhood Home, https://www.eisenhowerlibrary.gov/sites/default/files/research/online-documents/farewell-address/1961-01-17-press-release.pdf.

68 AP, "Role in Moon Shot is Awarded to G.E.," *NYT*, February 10, 1962, 2, https://timesmachine.nytimes.com/timesmachine/1962/02/10/93829012.pdf.

69 John W. Finney, "Interest Conflict in Space Weighed," *NYT*, May 20, 1962, 46, https://timesmachine.nytimes.com/timesmachine/1962/05/20/140704202.pdf.

70 "G.E. Picks New Chief Executive to Succeed Ralph J. Cordiner," *NYT*, October 8, 1963, 59, 68, https://timesmachine.nytimes.com/timesmachine/1963/10/08/84800018.pdf. AP, "Orlando Not Affected in $75 Million Settlement by GE," *OS*, December 21, 1963, 7A, https://www.newspapers.com/image/223844669/.

71 Tom Inglis, "Ex-GE Boss' Beef Production Like Factory at Dundee Ranch," *The Tampa Times*, February 26, 1966, 9, https://www.newspapers.com/image/328943906/.

72 John M. Logsdon, *After Apollo? Richard Nixon and the American Space Program* (Palgrave Macmillan, 2015), 50.

73 NASA, *America's Next Decades in Space: A Report for the Space Task Group*, September 1969, https://ntrs.nasa.gov/api/citations/19710002945/downloads/19710002945.pdf.

74 NASA, *America's Next Decades*, 5.

75 NASA, *America's Next Decades*, 72–73.

76 NASA, *America's Next Decades*, 7.

77 NASA, *America's Next Decades*, 74–75.

78 Logsdon, *After Apollo*, 278–279. Richard Nixon, "Statement About the Future of the United States Space Program," The American Presidency Project, University of California Santa Barbara, https://www.presidency.ucsb.edu/documents/statement-about-the-future-the-united-states-space-program.

79 1969 *Space Task Group Report*, 49.

80 Logsdon, *After Apollo*, 288–289.

81 John Uri, *50 Years Ago: President Nixon Directs NASA to Build the Space Shuttle*, NASA, https://www.nasa.gov/feature/50-years-ago-president-nixon-directs-nasa-to-build-the-space-shuttle.
82 Craig B. Waff, "Project Echo, Goldstone, and Holmdel: Satellite Communications as Viewed from the Ground Station," *Beyond the Ionosphere: The Development of Satellite Communications*, ed. Andrew J. Butrica (NASA History Office SP-4217, 1997), 41 et seq., https://ntrs.nasa.gov/api/citations/19970026049/downloads/19970026049.pdf.
83 Bob Granath, "Telstar Opened Era of Global Satellite Television," NASA, July 10, 2012, https://www.nasa.gov/content/telstar-opened-era-of-global-satellite-television.
84 Communications Satellite Act of 1962, H.R. 11040, 87th Cong. (1962), https://www.govinfo.gov/content/pkg/STATUTE-76/pdf/STATUTE-76-Pg419.pdf.
85 Roger D. Launius, *Historical Analogs for the Stimulation of Space Commerce* (NASA History Program Office SP-2014-4554, 2014), 24–26, https://www.nasa.gov/sites/default/files/files/historical-analogs-ebook_tagged.pdf.
86 David J. Whalen, "Communications Satellites: Making the Global Village Possible," NASA, November 30, 2010, https://history.nasa.gov/satcomhistory.html.
87 J. D. O'Connell, "Policy Concerning U.S. Assistance in the Development of Foreign Communications Satellite Capabilities," August 25, 1965, *Exploring the Unknown, Volume III: Using Space*, ed. John M. Logsdon (NASA History Office, 1998), 92–93, https://www.nasa.gov/wp-content/uploads/2023/04/sp-4407-etuv3.pdf.
88 "Thirty Years of Ariane," European Space Agency, December 22, 2009, https://www.esa.int/About_Us/ESA_history/Thirty_years_of_Ariane.
89 Logsdon, *After Apollo*, generally 161–172.
90 President Jimmy Carter, "National Space Policy," Presidential Directive/NSC-37, Section 4f, May 11, 1978, Jimmy Carter Presidential Library & Museum, https://www.jimmycarterlibrary.gov/assets/documents/directives/pd37.pdf.
91 John M. Logsdon, *Ronald Reagan and the Space Frontier* (Palgrave Macmillan, 2019), 177.
92 President Ronald Reagan, "Commercialization of Expendable Launch Vehicles," National Security Decision Directive Number 94, May 16, 1983, https://www.reaganlibrary.gov/public/archives/reference/scanned-nsdds/nsdd94.pdf.
93 National Aeronautics and Space Administration Authorization Act of 1985, H.R. 5154, 98th Cong. (1984), Sec. 110, https://www.congress.gov/98/statute/STATUTE-98/STATUTE-98-Pg422.pdf.
94 NASA Authorization Act of 1985, Sec. 202(5).
95 Logsdon, *Reagan*, 388.
96 E. C. Peter Aldridge, Jr., "Assured Access: The Bureaucratic Space War," Robert H. Goddard Historical Essay, http://ocw.mit.edu/courses/aeronautics-and-astronautics/16-885j-aircraft-systems-engineering-fall-2005/readings/aldrdg_space_war.pdf.
97 President Ronald Reagan, "National Security Launch Strategy," National Security Decision Directive Number 164, February 25, 1985, https://www.reaganlibrary.gov/public/archives/reference/scanned-nsdds/nsdd164.pdf.
98 Logsdon, *Reagan*, 301.
99 John Pike, "The SLC-6 Saga," Federation of American Scientists, http://fas.org/spp/military/program/launch/sts_slc-6.htm, archived April 13, 2016, at the *WM*.

100 Harry F. Rosenthal, "Reagan Orders NASA To Halt Launch of Commercial Payloads," AP, August 18, 1986, http://www.apnewsarchive.com/1986/Reagan-Orders-NASA-To-Halt-Launch-of-Commercial-Payloads/id-7e6b76c27ec65f93f14fd7913cf95c48, archived March 6, 2014, at the *WM*.

101 "United States Space Launch Strategy," National Security Decision Directive Number 254, December 27, 1986, https://www.reaganlibrary.gov/public/archives/reference/scanned-nsdds/nsdd254.pdf.

102 According to the World Bank, US military spending in 1990 was $325.1 billion. By 1995, it was $295.8 billion. As a percentage of Gross Domestic Product, spending declined from 5.61 percent to 3.86 percent. https://www.macrotrends.net/global-metrics/countries/USA/united-states/military-spending-defense-budget.

103 Mark Albrecht, *Falling Back to Earth: A First Hand Account of the Great Space Race and the End of the Cold War* (New Media Books, 2011), 148, 186.

104 Cristina Chaplain, Government Accountability Office, "Evolved Expendable Launch Vehicle," March 5, 2014, 4, https://www.gao.gov/assets/gao-14-259t.pdf.

105 Frank C. Weaver, Office of Commercial Space Transportation, "LEO Commercial Market Projections," May 17, 1995, generally, http://commercialspace.pbworks.com/w/file/fetch/84014746/1995%20LEO%20Forecast.pdf.

106 "Worldwide Commercial Space Launches," US Department of Transportation Bureau of Transportation Statistics, https://www.bts.gov/content/worldwide-commercial-space-launches.

107 Kei Koizumi et al., *Past and Future: An Analysis of the FAA Commercial Space Transportation Forecasts*, George Washington University, May 8, 2017, Executive Summary, https://cpb-us-e1.wpmucdn.com/blogs.gwu.edu/dist/7/314/files/2018/10/Boll-Sloan-Solem-Past-and-Future_-An-Analysis-of-the-FAA-Commercial-Space-Transportation-Forecasts-2lj70q2.pdf.

108 "Lockheed Martin Files Lawsuit Against the Boeing Company," Lockheed Martin press release, June 10, 2003, https://news.lockheedmartin.com/2003-06-10-Lockheed-Martin-Files-Lawsuit-Against-The-Boeing-Company.

109 "Two Former Boeing Managers Charged in Plot to Steal Trade Secrets from Lockheed Martin," US Department of Justice press release, June 25, 2003, https://www.justice.gov/archive/criminal/cybercrime/press-releases/2003/branchCharge.htm.

110 Renae Merle, "U.S. Strips Boeing of Launches," *The Washington Post*, July 25, 2003, https://www.washingtonpost.com/archive/politics/2003/07/25/us-strips-boeing-of-launches/bfe387d0-549a-4ec1-93b5-49d4205293b4/.

111 "Air Force Lifts Boeing Suspension," US Air Force press release, March 4, 2005, https://www.af.mil/News/Article-Display/Article/134906/air-force-lifts-boeing-suspension/.

112 Not to be confused with United Space Alliance.

113 "Boeing, Lockheed Martin to Form Launch Services Joint Venture," United Launch Alliance press release, May 2, 2005, https://www.ulalaunch.com/about/news/2005/05/02/boeing-lockheed-martin-to-form-launch-services-joint-venture.

114 Chaplain, "Evolved Expendable Launch Vehicle," 4–5.

115 "Worldwide Commercial Space Launches," US DOT.

116 "The National Space Institute," National Space Society, https://space.nss.org/the-national-space-institute/.

117 Gerald K. O'Neill, "The Colonization of Space," *Physics Today* 27, no. 9 (1974): 32–40, reprinted by the National Space Society, https://space.nss.org/the-colonization-of-space-gerard-k-o-neill-physics-today-1974/.

118 David Brandt-Erichsen, "Brief History of the L5 Society," National Space Society, https://space.nss.org/brief-history-of-the-l5-society/.

119 "A Proclamation," March 28, 1987, National Space Society, https://space.nss.org/wp-content/uploads/NSS-Merger-Proclamation.pdf.

120 "History," Space Studies Institute, https://ssi.org/about/history/.

121 Elson Trinidad, "September 1980—Carl Sagan's 'Cosmos: A Personal Journey' Airs," KCET, September 9, 2014, https://www.kcet.org/kcet-50th-anniversary/september-1980-carl-sagans-cosmos-a-personal-journey-airs.

122 Carl Sagan, "The Adventure of the Planets," *The Planetary Report* 1, no. 1 (1980): 3, https://s3.amazonaws.com/planetary/assets/tpr/pdf/tpr-1981-v01n1_200424_191313.pdf.

123 "Our History: Space Frontier Foundation," Space Frontier Foundation, https://spacefrontier.org/our-history/

124 "History of NewSpace," Space Frontier Foundation, https://spacefrontier.org/history-of-newspace/, archived November 7, 2012, at the *WM*.

125 "What Is NewSpace?" Space Frontier Foundation, https://spacefrontier.org/what-is-newspace/, archived October 6, 2012, at the *WM*.

126 One example is a 2010 Near Earth LLC report, funded by NASA, that used the terms *NewSpace* and *OldSpace*, https://www.nasa.gov/sites/default/files/files/SupportingCommercialSpaceDevelopmentPart1.pdf, 13–14.

127 Andrew J. Butrica, "'La force motrice' of Reusable Launcher Development: The Rise and Fall of the SDIO's SSTO Program, From the X-Rocket to the Delta Clipper," Historical Seminar on Contemporary Science and Technology (National Air and Space Museum, November 18, 1999), https://www.hq.nasa.gov/office/pao/History/x-33/nasm.htm, archived November 18, 2004, at the *WM*. Greg Klerkx, *Lost in Space: The Fall of NASA and the Dream of a New Space Age* (Pantheon Books, 2004), generally, 101–110.

128 "Lockheed Martin X-33," NASA, February 17, 2016, https://www.nasa.gov/image-article/lockheed-martin-x-33/.

129 Steven Siceloff, "NASA Kills X-33 Program," *FT*, March 2, 2001, 1A, 3A, https://www.newspapers.com/image/178286729.

130 Elizabeth Douglass, "Futuristic Craft Taking Shape in Palmdale," *LAT*, July 20, 1998, D3, D9, https://www.newspapers.com/image/160164440.

131 Gary Letchworth, "X-33 Reusable Launch Vehicle Demonstrator, Spaceport and Range," AIAA Space 2011 Conference paper, September 27, 2011, 9, https://ntrs.nasa.gov/api/citations/20110016255/downloads/20110016255.pdf.

132 Chris Bergin, "X-33/VentureStar—What Really Happened," NASASpaceflight.com, January 4, 2006, https://www.nasaspaceflight.com/2006/01/x-33venturestar-what-really-happened/.

133 Ashlee Vance, *Elon Musk: Tesla, SpaceX, and the Quest for a Fantastic Future* (HarperCollins, 2015).
134 Vance, *Elon Musk*, 112.
135 Vance, *Elon Musk*, 116.
136 Vance, *Elon Musk*, 114.
137 Chris Anderson, "Elon Musk's Mission to Mars," *Wired*, October 21, 2012, https://www.wired.com/2012/10/ff-elon-musk-qa/.
138 Gwynne Shotwell, Makers.com video interview, August 6, 2018, https://www.facebook.com/watch/?v=1411901208918199.
139 "SpaceX Falcon 9 Rocket Flies Safely," ABC News, December 8, 2010, https://abcnews.go.com/Technology/spacex-falcon-launch-successful-privately-owned-spacecraft-return/story?id=12345299.
140 Anthony Ha, "Private Rocket Company SpaceX Gets $20M from The Founders Fund," *VentureBeat*, August 6, 2008, https://venturebeat.com/2008/08/06/private-rocket-company-spacex-gets-20m-from-the-founders-fund/.
141 Jeremy Hsu, "Strike Three for SpaceX's Falcon 1 Rocket," NBC News, August 3, 2008, https://www.nbcnews.com/id/wbna25990806.
142 "SpaceX Receives $20 Million Investment from Founder's Fund," SpaceX press release, August 4, 2008, http://www.spacex.com/press.php?page=47, archived August 6, 2008, at the *WM*.
143 Michael Braukus/J. D. Harrington, "NASA Establishes Commercial Crew/Cargo Project Office," NASA press release, November 7, 2005, https://www.nasa.gov/home/hqnews/2005/nov/HQ_05356_commercial_crew.html, archived December 22, 2005, at the *WM*.
144 "Commercial Orbital Transportation Services Demonstrations," Announcement Number COTS-01-05, NASA Johnson Space Center, January 18, 2006, generally, https://www.nasa.gov/pdf/225439main_COTS%20Final%20Announcement%20%28Amend%201%2C%20%202-17-06%29.pdf, archived March 18, 2010, at the *WM*.
145 Michael D. Griffin, "NASA and the Commercial Space Industry: Remarks on the Occasion of the X-Prize Cup Summit," October 19, 2006, 2–4, https://www.nasa.gov/pdf/161127main_mg_xprize.pdf, archived August 24, 2007, at the *WM*.
146 Alan Boyle, "SpaceX, Rocketplane Win Spaceship Contest," NBC News, August 18, 2006, https://www.nbcnews.com/id/wbna14411983.
147 Julie Bisbee, "Rocketplane Wins $207M Contract," *The Oklahoman*, August 19, 2006, 1A, 3A, https://www.newspapers.com/image/452597531/. Tariq Malik, "Competition Heats Up for NASA's Space Cargo Contract," Space.com, May 31, 2006, https://www.space.com/2444-competition-heats-nasa-space-cargo-contract.html.
148 "Rocketplane Kistler," NASA Commercial Crew and Cargo, https://www.nasa.gov/offices/c3po/partners/rpk/, archived September 20, 2016, at the *WM*.
149 U.S. Government Accountability Office, "Commercial Partners Are Making Progress, but Face Aggressive Schedules to Demonstrate Critical Space Station Cargo Transport Capabilities," June 16, 2009, 11, https://www.gao.gov/assets/gao-09-618.pdf.
150 "Inspiration4 Mission," SpaceX.com, September 30, 2021, https://www.spacex.com/updates/inspiration4/index.html, archived October 20, 2021, at the *WM*.
151 Stephen Clark, "Inspiration4 Booster Returns to SpaceX Hangar for Refurbishment,"

SpaceflightNow.com, September 23, 2021, https://spaceflightnow.com/2021/09/23/video-inspiration4-booster-returns-to-spacex-hangar-for-refurbishment/.

152 Vicky Stein and Scott Dutfield, "Inspiration4: The First All-Civilian Spaceflight on SpaceX Dragon," Space.com, January 5, 2022, https://www.space.com/inspiration4-spacex.html.

153 *Columbia Accident Investigation Board Report, Volume 1* (National Aeronautics and Space Administration, August 2003), 25, https://govinfo.library.unt.edu/caib/news/report/pdf/vol1/full/caib_report_volume1.pdf

154 *CAIB Report Volume 1*, 9.

155 *CAIB Report Volume 1*, 9.

156 "President Bush Offers New Vision for NASA," NASA press release, January 14, 2004, https://www.nasa.gov/history/vision-for-space-exploration/.

157 Edward C. "Pete" Aldridge et al., *A Journey to Inspire, Innovate, and Discover: Report of the President's Commission on Implementation of United States Space Exploration Policy* (US Government Printing Office, June 2004), 31, https://www.nasa.gov/pdf/60736main_M2M_report_small.pdf.

158 "Bringing the Vision to Reality," President George W. Bush White House archives, https://georgewbush-whitehouse.archives.gov/space/vision.html. Section C called for NASA to acquire crew and cargo transportation services to ISS. Section D proposed pursuing "commercial opportunities for providing transportation" and other services to ISS.

Chapter 3. Spaceport Florida

1 Lenore Beecken, "'Space Nut' Spreading the Word," *FT*, September 10, 1985, 2A, https://www.newspapers.com/image/176174067/.

2 Stephen Morgan profile, LinkedIn, https://www.linkedin.com/in/morganslusnr/.

3 "History," Space Studies Institute, https://ssi.org/about/history/.

4 Frank Yacenda, "Palm Bay Base of Space Studies Team," *Today*, July 4, 1984, 10A, https://www.newspapers.com/image/124852293.

5 Frank Yacenda, "Space Institute Seeking Backers," *Today*, August 4, 1984, 2B, https://www.newspapers.com/image/125036305/.

6 Ruth Rasche, "FIT 1st to Sign Up for Space Institute Effort," *OS*, August 7, 1987, C1, C7, https://www.newspapers.com/image/229039261/.

7 "In Orbit," *OS*, Central Florida Business Section, July 7, 1986, 2, https://www.newspapers.com/image/229570810/.

8 "History," Space Foundation, https://www.spacefoundation.org/who-we-are/.

9 Stephen L. Morgan and Freddie Garcia, "Advocating the Development of Space Enterprise: The Space Business Roundtable Movement Nationwide," presented at the 25th Space Congress, Cocoa Beach, Florida, April 1, 1988, 2-1–2-2, https://commons.erau.edu/cgi/viewcontent.cgi?article=2137&context=space-congress-proceedings.

10 Brian Bixler, "'Space Chamber of Commerce' Formed," *FT*, December 19, 1986, 1C, 17C, https://www.newspapers.com/image/173903776.

11 Brian Bixler, "Business Roundtable Sets Sights on New Enterprise for Brevard," *FT*, January 4, 1988, Business section, 6, https://www.newspapers.com/image/175024796/.

12 The National Commission on Space, *Pioneering the Space Frontier* (Bantam Books, May 1986), https://archive.org/details/pioneeringspacef00unit, archived September 26, 2011, at the *WM*.

13 Stephen L. Morgan, email message to author, February 3, 2022. Also Morgan, "The Florida Governor's Commission on Space: Its Impact on Space Enterprise," presented at the 25th Space Congress, Cocoa Beach, Florida, April 1, 1988, 2–39, https://commons.erau.edu/cgi/viewcontent.cgi?article=2141&context=space-congress-proceedings.

14 Todd Halvorson, "Governor Invests in Vision: Space City USA," *FT*, May 19, 1987, 1A, https://www.newspapers.com/image/174050271/.

15 Catherine Hinman, "Panel Will Boost Commercial Space Ventures," *OS*, D1, D5, https://www.newspapers.com/image/229145058/.

16 Michele Ostovar, "Presidential Directive on National Space Policy, February 11, 1988," NASA, https://www.nasa.gov/history/presidential-directive-on-national-space-policy-february-11-1988/.

17 Nick White, "Space Business Roundtable Takes on Booster Role," *FT*, February 21, 1988, 13A, https://www.newspapers.com/image/175028336/.

18 Morgan, *The Florida Governor's Commission on Space*, 2-43–2-44.

19 Todd Halvorson, "Martinez Gives Launch Biz a Boost," *FT*, March 26, 1988, 1A, https://www.newspapers.com/image/175057929/.

20 Ann Mittman, "Martinez Finds Spaceport Support from Nelson," *FT*, March 26, 1988, 1B, https://www.newspapers.com/image/175059135/.

21 Jeff Newell, "Early Governor's Race Focus Amazes Nelson," *PNJ*, January 10, 1988, 3B, https://www.newspapers.com/image/267572464/.

22 John C. Van Gieson, "Criticism Launched at Spaceport Plan," *OS*, April 14, 1988, D3, https://www.newspapers.com/image/230071237/.

23 Kirk Brown, "Gardner Questions Commercial Spaceport Plan," *FT*, April 20, 1988, 1B, https://www.newspapers.com/image/175033880.

24 The Space Coast Development Commission was funded by the City of Titusville as a partnership of the city and private business. It was separate from the county's Brevard Economic Development Council, a nonprofit private local development group that at one time was run by Brevard County.

25 Purvette Bryant, "Gardner Backs $500,000 Study for Spaceport," *FT*, April 24, 1988, 1B, https://www.newspapers.com/image/175049780.

26 Florida Governor's Commission on Space, *Steps to the Stars: Final Report to Governor Martinez*, July 7, 1988, UCF Libraries, https://ucf-flvc.primo.exlibrisgroup.com/permalink/01FALSC_UCF/faevv6/alma990222572870306596.

27 Space Commission, *Steps to the Stars*, 31.

28 Space Commission, *Steps to the Stars*, 32.

29 Space Commission, *Steps to the Stars*, 36.

30 Space Commission, *Steps to the Stars*, 38.

31 Space Commission, *Steps to the Stars*, 33.

32 Christopher Shove, "Regional Planning of Commercial Spaceports," *Journal of the American Planning Association* 68, no. 1 (2002): 87, https://www.tandfonline.com/

doi/abs/10.1080/01944360208977193. A PDF copy was provided by Chris Shove to this author.

33 Chris Shove, "Regional Planning," 88.

34 Chris Shove, "Regional Planning," 87.

35 Chris Shove, email message to author, March 5, 2022.

36 Catherine Hinman, "New Office to Plot State's Course in Space Commerce," *OS*, June 11, 1988, C1, https://www.newspapers.com/image/229519915.

37 Irene Klotz, "Spaceport Team Tackles Final Frontier," *FT*, July 10, 1988, 1A, https://www.newspapers.com/image/175057145/.

38 Irene Klotz, "Launch Complex Plan Draws 8 Competitors," *FT*, June 3, 1988, 4A, https://www.newspapers.com/image/175025985.

39 "Denver Firm to Plan Spaceport," *FT*, June 21, 1988, 4A, https://www.newspapers.com/image/175043043/.

40 Ann Mittman, "State to Study Spaceport Roads," *FT*, December 17, 1988, 7A, https://www.newspapers.com/image/177560510/.

41 Mittman, "State to Study . . ."

42 AP, "Poll: Martinez in Deep Trouble," *TD*, September 6, 1987, 1D, https://www.newspapers.com/image/247453176.

43 AP, "Fewer than 50% Approve of Martinez, New Poll Says," *FT*, September 6, 1988, 8B, https://www.newspapers.com/image/177589783/.

44 Dave Bruns, "Popularity Swings Way of Martinez," *TD*, April 6, 1989, 1C–2C, https://www.newspapers.com/image/247244839.

45 Michael Lafferty, "Volusia-Brevard Site Urged for Spaceport," *OS*, February 10, 1989, B1, https://www.newspapers.com/image/230618342. The leaked report was published by Irene Klotz, "Planners: Brevard Wins Spaceport," *FT*, February 10, 1989, 1A, https://www.newspapers.com/image/176131579/.

46 The page count comes from Irene Klotz, "Brevard Resists Backing County Spaceport Site," *FT*, February 22, 1989, 4A, https://www.newspapers.com/image/177611469.

47 United Engineers & Constructors, *Spaceport Florida Feasibility Study Executive Summary*, February 1989, included by Chris Shove in "Spaceport Florida and Other State-Sponsored Space Development Initiatives in Florida," presented at the 26th Space Congress, Cocoa Beach, Florida, April 27, 1989, https://commons.erau.edu/cgi/viewcontent.cgi?article=2091&context=space-congress-proceedings.

48 Charles D. Benson and William Barnaby Faherty, *Gateway to the Moon: Building the Kennedy Space Center Launch Complex* (University Press of Florida, First Paperback Printing, 2001), 96–98.

49 Benson and Flaherty, *Gateway*, 87–89. *NASA-Industry Program Plans Conference* (NASA SP-29, February 11–12, 1963), 57, https://ntrs.nasa.gov/citations/19630005481. The page has a map showing proposed sites for the "Nova Complex" near Shiloh.

50 UPI, "Misstart on Nova Causes Apprehension in Congress," *Orlando Evening Star*, February 26, 1963, 3A, https://www.newspapers.com/image/291083661/. Rex Newman, "U.S. Space Port Pumps Billions into Merritt Island," *OS*, November 28, 1963, 2E–3E, https://www.newspapers.com/image/223700412/.

51 "About Us," Merritt Island National Wildlife Refuge, https://www.fws.gov/refuge/merritt-island/about-us. "Interagency Agreement Between the National Aeronautics and Space Administration and U.S. Department of Interior Fish and Wildlife Service," updated July 25, 2012, https://www.fws.gov/uploadedFiles/NASAFWSAgreement2012.pdf, archived March 11, 2017, at the *WM*.

52 U.S. Fish and Wildlife Service, "History of the Shiloh Area," February 2014, https://www.fws.gov/uploadedFiles/ShilohHistory.pdf, archived February 6, 2017, at the *WM*.

53 *Spaceport Florida Feasibility Study*, 9–8.

54 "Progress on Range Reported," *The Pensacola Journal*, December 10, 1959, 10C, https://www.newspapers.com/image/263056489/.

55 *Spaceport Florida Feasibility Study*, 9–10.

56 *Spaceport Florida Feasibility Study*, 9–12–9–13.

57 *Spaceport Florida Feasibility Study*, 9–17.

58 *Spaceport Florida Feasibility Study*, 9–19.

59 *Spaceport Florida Feasibility Study*, 9–20–9–21.

60 *Spaceport Florida Feasibility Study*, 9–20.

61 Richard Burnett, "Martinez Wants $10 Million for Spaceport," *OS*, February 14, 1989, D1, D7, https://www.newspapers.com/image/230770167. Irene Klotz, "Martinez: Put $10 Million in Spaceport," *FT*, February 14, 1989, 1A–2A, https://www.newspapers.com/image/176205777/.

62 Richard Burnett, "Launch Industry Waits for Liftoff," *OS*, Central Florida Business section, February 20, 1989, 10, https://www.newspapers.com/image/230807555.

63 Burnett, "Launch Industry . . ."

64 Irene Klotz, "Scarborough Pushes Brevard for Spaceport Fla. Launch Site," *FT*, February 16, 1989, 4A, https://www.newspapers.com/image/177553285.

65 Catherine Liden, "Restructured BEDC Emerges with New Images, New Goals," *FT*, July 9, 1989, 2E, https://www.newspapers.com/image/177590074.

66 Cory Jo Lancaster, "Spaceport Discussion to Focus on Environmental Concerns," *OS*, February 17, 1989, D2, https://www.newspapers.com/image/230804630.

67 Irene Klotz, "Shiloh Development Faces Opposition," *FT*, February 18, 1989, 4A, https://www.newspapers.com/image/177572524.

68 Chris Shove, "Regional Planning of Commercial Spaceports," 89.

69 Irene Klotz, "Brevard Resists Backing County Spaceport Site," *FT*, February 22, 1989, 4A, https://www.newspapers.com/image/177611469.

70 Jeff Cole and Irene Klotz, "Commission Wants Spaceport Fla. All in Brevard, Despite Criticisms," *FT*, March 1, 1989, 1B–2B, https://www.newspapers.com/image/177221021.

71 Vincent Willmore, "Officials Lean Towards Shiloh for Spaceport Site," *FT*, March 15, 1989, 4A, https://www.newspapers.com/image/177225746.

72 Vincent Willmore, "Spaceport Becomes Political Issue," *FT*, April 3, 1989, 1A–2A, https://www.newspapers.com/image/177561344/.

73 Willmore, "Spaceport . . ."

74 Willmore, "Spaceport . . ."

75 John C. Van Gieson, "Spaceport Faces Political Test," *OS*, Central Florida Business section, April 3, 1989, 18, https://www.newspapers.com/image/230442211.

76 Irene Klotz and Vincent Willmore, "Cape Wins Boost for Spaceport," *FT*, April 8, 1989, 1A, https://www.newspapers.com/image/177596522.

77 John C. Van Gieson, "Spaceport Study Gets Slammed," *OS*, April 11, 1989, B3, https://www.newspapers.com/image/229966275.

78 *Spaceport Florida Feasibility Study*, 9–22.

79 Vincent Willmore, "Shiloh Out of Running; Spaceport Board Debated," *FT*, April 11, 1989, 6A, https://www.newspapers.com/image/177629286.

80 Vincent Willmore, "Spaceport Estimate Cut in Fourth but Should Be Enough," *FT*, April 28, 1989, 9B, https://www.newspapers.com/image/175899843.

81 Jane Musgrave, "House Panel Rejects New Taxes, Works to Pare Down Spending," *FT*, May 2, 1989, 7B, https://www.newspapers.com/image/177536427.

82 Vincent Willmore, "Budgets Put Squeeze on Spaceport," *FT*, May 4, 1989, 1A, https://www.newspapers.com/image/177543224.

83 John C. Van Gieson, "Spaceport Plans Fights for Survival in the Legislature," *OS*, May 7, 1989, D1, D3, https://www.newspapers.com/image/229939684/.

84 Bill Nelson telephone interview with author, December 29, 2023.

85 Van Gieson, "Spaceport Plans," D3.

86 Fred R. Shapiro, ed., *The Yale Book of Quotations* (Yale University Press, 2006), 86, https://archive.org/details/isbn_9780300107982/page/86/mode/2up, at IA. The passage notes that the quotation is sometimes attributed to German chancellor Otto von Bismarck, however earlier citations can be found attributing the quote to Saxe. Also *Professor Buzzkill History Podcast* (blog), May 1, 2018, https://professorbuzzkill.com/2018/05/01/bismarck-laws-and-sausages/.

87 Vincent Willmore, "New Spaceport Bill Jettisons Plans for Cape," *FT*, May 23, 1989, 1A, https://www.newspapers.com/image/175838781.

88 Vincent Willmore, "Panel Approves Senate Version of Spaceport Bill," *FT*, May 9, 1989, 1A, https://www.newspapers.com/image/177593248.

89 Chris Shove, email message to author, March 15, 2022.

90 Todd Halvorson, "Spaceport Would Change Area Economy," *FT*, June 4, 1989, 1B, https://www.newspapers.com/image/176076682.

91 Irene Klotz, "Martinez Signs Spaceport Florida Bill," *FT*, July 6, 1989, 1A, https://www.newspapers.com/image/177561618.

92 Florida Department of State, Division of Elections, Official Results, Governor & Cabinet, September 4, 1990, Democratic Primary Election, https://results.elections.myflorida.com/DetailRpt.Asp?ELECTIONDATE=9/4/1990&RACE=GOV&PARTY=DEM&DIST=&GRP=&DATAMODE.

93 Florida Department of State, Division of Elections, Official Results, Governor & Cabinet, November 6, 1990, https://results.elections.myflorida.com/Index.asp?ElectionDate=11/6/1990.

94 Irene Klotz, "Shove to Head Space Research Foundation," *FT*, June 27, 1989, 6A, https://www.newspapers.com/image/177620552/.

95 Chris Shove email, March 5, 2022.

96 "In Brief," *DP*, Inside Business Section, October 31, 1988, 8, https://www.newspapers.com/image/230922169/.

97 Stephen Morgan email message to author, March 18, 2022, also Stephen Morgan LinkedIn profile.

98 Ernest C. Gates, "Robb Campaigns for High-Technology Institute," *DP*, July 15, 1983, 5, https://www.newspapers.com/image/234592232/.

99 Ernest C. Gates, "High-Tech Center Gets Easy Approval in House," *DP*, March 10, 1984, 3, https://www.newspapers.com/image/238151540/.

100 Mark Di Vincenzo, "HU, W&M, Others Get Federal 'Space Grants,'" *DP*, September 20, 1989, B2, https://www.newspapers.com/image/236822599/.

101 "People Around Town," *The Observer*, Charlottesville, Virginia, March 15, 1990, 8, https://www.newspapers.com/image/613059121/.

102 "Wallops History," NASA, https://www.nasa.gov/wallops-history/.

103 AP, "Wallops Project Rejected," *DP*, October 25, 1994, C4, https://www.newspapers.com/image/236418526/.

104 AP, "Va. Could Be Satellite Launch Site," *DP*, March 13, 1995, C4, https://www.newspapers.com/image/233392899/.

105 Joe Taylor, "Commercial Rocket to Fly from Eastern Shore," *The Daily News Leader*, August 7, 1995, A3, https://www.newspapers.com/image/288763629/. "Rocket Explodes," *DP*, October 24, 1995, A1-A2, https://www.newspapers.com/image/237227134/.

106 "$pace$hot$," *DP*, April 25, 1996, C5, https://www.newspapers.com/image/236382610/.

107 AP, "State Space Authority Seeking Part of NASA's Wallops Program," *DP*, March 6, 1996, B3, https://www.newspapers.com/image/236381158/.

108 "Flight Center, NASA Agree on Launches," *DP*, March 8, 1997, C8, https://www.newspapers.com/image/237349617/.

109 "Liftoff for Spaceport," *DP*, December 25, 1997, A24, https://www.newspapers.com/image/237741616/?.

110 John Vandiver, "Wallops Woes," DT, Salisbury, Maryland, May 25, 2003, A1, A4, https://www.newspapers.com/image/282359986/.

111 Chris Guy, "Md., Va. Form Regional Spaceport," *Baltimore Sun*, December 5, 2003, 1D, 10D, https://www.newspapers.com/image/248542427/.

112 Stephen Furness, "Spaceport Receives $49 Million Contract," DT, November 10, 2004, 1, https://www.newspapers.com/image/282342915/.

113 Jim Hodges, "New Life Launched at Wallops Island," *DP*, December 17, 2006, A1, A5, https://www.newspapers.com/image/270425387.

114 Carol Vaughn, "Wallops Takes Lead in $45M Project," *DT*, June 10, 2008, A1, https://www.newspapers.com/image/283596998.

115 "NASA Orders Additional Cargo Flights to Space Station," NASA, March 25, 2022, https://www.nasa.gov/feature/nasa-orders-additional-cargo-flights-to-space-station. Jeff Foust, "Northrop Grumman Prepares for Final Flight of Antares with Russian and Ukrainian Components," *SN*, July 31, 2023, https://spacenews.com/northrop-grumman-prepares-for-final-flight-of-antares-with-russian-and-ukrainian-components/. As of this writing, Northrop Grumman was using SpaceX Falcon 9

rockets to launch Cygnus until their new booster, the Antares 330, is ready to launch from Wallops, targeting 2025.

116 Tamara Dietrich, "From Wallops to the Moon," *DP*, September 8, 2013, 6, https://www.newspapers.com/image/270303899/.

117 Jesse McKinley, "How Jerry Brown Became 'Governor Moonbeam,'" *NYT*, March 6, 2010, https://www.nytimes.com/2010/03/07/weekinreview/07mckinley.html. Mike Royko, "Gov. Moonbeam Has Landed," *LAT*, August 17, 1980, Opinion Section, 5, https://www.newspapers.com/image/387131024.

118 Leo Rennert, "Lobbying by Brown Boosts House Vote for Space Fund," *The Fresno Bee*, July 20. 1977, A4, https://www.newspapers.com/image/704031675/.

119 Miriam Pawel, *The Browns of California: The Family Dynasty That Transformed a State and Shaped a Nation* (Bloomsbury Publishing, 2018), 259.

120 Pawel, 259; W. E. Barnes, "Jerry Brown Hitches His Political Wagon to the Stars," *SFE*, August 14, 1977, A7, https://www.newspapers.com/image/461097331/.

121 "Drakes to Celebrate Anniversary at UCLA," *LAT*, August 11, 1977, View Section, 2, https://www.newspapers.com/image/383592281.

122 AP, "Jerry Brown's 'Era of Limits' Gives Way to Space Age," *The Daily Oklahoman*, September 6, 1977, 10N, https://www.newspapers.com/image/232238451/.

123 "Space Shuttle's First 'Flight' a Success," *SFE*, "This World" Section, August 21, 1977, 6, https://www.newspapers.com/image/461044393/.

124 "Above and Beyond," Historical Perspective, *Boeing Frontiers*, October 2011, https://www.boeing.com/news/frontiers/archive/2011/october/i_history.pdf, archived September 30, 2021, at the *WM*. Ralph Vartabedian, "A Shot Still Heard Round the World," *LAT*, July 26, 2013, A1 et seq., https://www.newspapers.com/image/203643815/.

125 George Alexander, "Proposed State Use of Satellite Detailed," *LAT*, December 23, 1977, Part I, 3, 28, https://www.newspapers.com/image/384053113/.

126 Leo Rennert, "No Pledges on Brown's Orbiter," *SB*, April 15, 1978, A12, https://www.newspapers.com/image/620926207/.

127 Leo Rennert, "Hughes Gets Blamed for State's Space Snags," *SB*, February 15, 1978, A1, A16, https://www.newspapers.com/image/620929432.

128 Jeff Raimundo, "Brown Claims Pleasure with Record," *SB*, September 13, 1978, A21, https://www.newspapers.com/image/620978476/.

129 "Orbiting Syncom 4 Conks Out," *FT*, September 17, 1985, 1A, https://www.newspapers.com/image/179253624/.

130 David Jackson, "Space Center Start-Up Funds to Be Obtained," *LR*, May 14, 1992, A1, https://www.newspapers.com/image/540764033.

131 David Jackson, "Spaceport Authority Concept Proposed," *LR*, January 13, 1993, A1, https://www.newspapers.com/image/540743043/. Michelle MacEachern, "Space Authority Bill on Wilson's Desk," *LR*, September 7, 1993, A1, https://www.newspapers.com/image/540632742/.

132 Phil Derkx, "Vandy Spaceport on Launch Pad?" *San Luis Obispo County Telegram-Tribune*, October 5, 1993, B3, https://www.newspapers.com/image/809574819.

133 Janene Scully, "California Space Authority Agrees to Dissolve," *SMT*, June 11, 2011, A3-A4, https://www.newspapers.com/image/483148841/.

134 Nick Wilson, "Central Coast Business Leaders Launch Initiative to Boost Jobs and Attract New Employers," *SLOT*, November 27, 2018, 1A, 6A, https://www.newspapers.com/image/716846523/. Kaytlyn Leslie, "Vandenberg Air Force Base Bids to Be Space Force Site," *SLOT*, November 24, 2019, 1A, 3A, https://www.newspapers.com/image/646048044/. "SLO County Flattened COVID Curve; Economic Recovery Next," REACH guest opinion column, *SLOT*, May 3, 2020, 5B, https://www.newspapers.com/image/661369137/.

135 *Commercial Space Master Plan, Vandenberg Space Force Base*, REACH et al., June 3, 3021, https://reachcentralcoast.org/wp-content/uploads/Commercial-Space-Master-Plan.pdf.

136 "For California's Central Coast, Space Is the Place," Fast Company, December 20, 2021, https://www.fastcompany.com/90707539/for-californias-central-coast-space-is-the-place.

137 "Space on the Agenda in Sacramento," REACH, May 31, 2024, https://reachcentralcoast.org/space-on-the-agenda-in-sacramento/

138 Sandra Irwin, "Space to Take Over West Coast Launch Pad Previously Used by ULA," *SN*, April 25, 2023, https://spacenews.com/spacex-to-take-over-west-coast-launch-pad-previously-used-by-ula/.

139 Isabella Leonard, "California Coastal Commission Hesitant to Support Increase on SpaceX Launches," *SBI*, April 16, 2024, https://www.independent.com/2024/04/16/california-coastal-commission-hesitant-to-support-increase-on-spacex-launches/.

140 Salvador Hernandez, "California Officials Reject More SpaceX Rocket Launches, with Some Citing Musk's X Posts," *LAT*, October 11, 2024, https://www.latimes.com/california/story/2024-10-11/la-me-spacex-coastal-commission. Nick Welsh, "Elon Musk Sues Coastal Commission Over Vandenberg Launches," *SBI*, October 22, 2024, https://www.independent.com/2024/10/22/elon-musk-sues-coastal-commission-over-vandenberg-launches/.

141 "Discover Our History," Mojave Air and Space Port, https://www.mojaveairport.com/history.html. Douglas Messier, "Stratolaunch's Roc Retracts & Extends Landing Gear During Fourth Test Flight," Parabolic Arc, February 25, 2022, http://www.parabolicarc.com/2022/02/25/stratolaunchs-roc-retracts-extends-landing-gear-during-fourth-flight-test/, archived January 11, 2024, at the *WM*. Douglas Messier, "Virgin Galactic's WhiteKnightTwo Returns to Mojave for Overhaul," Parabolic Arc, November 1, 2021, http://www.parabolicarc.com/2021/11/01/virgin-galactics-whiteknighttwo-returns-to-mojave-for-overhaul/, archived November 1, 2021, at the *WM*.

142 "Spaceport Debate Set in Alamogordo," *Deming Headlight*, July 27, 1992, 5, https://www.newspapers.com/image/558392200.

143 AP, "Anybody Want a Spaceport?" *EPT*, June 26, 1992, 1A, https://www.newspapers.com/image/431439102.

144 John Fleck, "New Mexico to Go After 'Spaceport,'" *AJ*, August 12, 1993, C3, https://www.newspapers.com/image/157732456/.

145 AP, "Space State?" *SFNM*, August 8, 1994, A7, https://www.newspapers.com/image/583621836/.

146 AP, "Firms Want Spaceport Go-Ahead," *AJ*, June 20, 1995, C3, https://www.newspapers.com/image/158146234/.
147 Eduardo Montes, "Launch of N.M. Spaceport Hits Snags," *AJ*, June 30, 1995, 3D, https://www.newspapers.com/image/158386398.
148 David Bennett, "NM Faces Tough Competition for Spaceport," *EPT*, August 15, 1997, 1B, 3B, https://www.newspapers.com/image/431900158/.
149 Hanson Scott, *The Albuquerque Tribune*, September 26, 1998, guest column, C1, https://www.newspapers.com/image/786355537/.
150 Miguel Navrot, "Air Force Rejects Experimental Craft," *AJ*, September 11, 2001, D3, https://www.newspapers.com/image/379318189/?.
151 "Governor Touts State's Aerospace Industry," *The Taos News*, February 19, 2003, B10, https://www.newspapers.com/image/418652188/.
152 Erica Molina, "Las Cruces Spaceport Gets Closer to Takeoff," *EPT*, September 25, 2003, 1B, 4B, https://www.newspapers.com/image/432243997.
153 Rene Romo, "New Mexico Wants to Host Spacecraft Competition," *AJ*, October 26, 2003, B1, B5, https://www.newspapers.com/image/443548224. For the history of the Kármán line, see S. Sanz Fernández de Córdoba, "100km Altitude Boundary for Astronautics," Fédération Aéronautique Internationale, June 21, 2004, https://www.fai.org/page/icare-boundary.
154 David Miles, "Gov. Betting Space Race a Boost to State's Economy," *AJ*, May 12, 2004, A1, https://www.newspapers.com/image/469640734.
155 Mark Prigg, "Holidays in Space," *Evening Standard*, London, England, September 27, 2004, 13, https://www.newspapers.com/image/726416360/.
156 Andrew Webb, "Gov. Pitches $100M for N.M. Spaceport," *AJ*, December 14, 2005, A1, A3, https://www.newspapers.com/image/206584798/. Webb, "Richardson, Virgin Chief Detail Plans for $225M Project," *AJ*, December 15, 2005, A1–A2, https://www.newspapers.com/image/206589098.
157 Andrew Webb, "On a World Stage," *AJ*, July 17, 2006, Business Outlook section, 1, https://www.newspapers.com/image/432332006/.
158 Walter Rubel, "Lawmakers Skeptical of Spaceport Proposal," *Carlsbad Current-Argus*, January 19, 2006, 1A, 10A, https://www.newspapers.com/image/507053065/.
159 Bill McCamley, Chairman, Doña Ana County Commission, *AJ* guest editorial, January 11, 2006, A11, https://www.newspapers.com/image/206275522/.
160 Barry Massey, "Governor Signs Bill to Fight Meth," *SFNM*, March 2, 2006, C-3, https://www.newspapers.com/image/207863248/.
161 "Voters to Decide on Tax Hike," *AJ*, February 6, 2007, C3, https://www.newspapers.com/image/206529093/.
162 "Spaceport Wins by 270 Votes," *AJ*, April 7, 2007, D3, https://www.newspapers.com/image/206252869/.
163 Rene Romo, "2nd County OKs Tax for Spaceport," *AJ*, April 23, 2008, C1, https://www.newspapers.com/image/206387842.
164 Rene Romo, "Voters Reject Spaceport Tax," *AJ*, November 6, 2008, F3, https://www.newspapers.com/image/207354937/.
165 Rene Romo, "Lease for Spaceport Signed," *AJ*, January 1, 2009, D1, https://www.newspapers.com/image/206203562.

166 Rene Romo, "Cheers to Space," *AJ*, October 18, 2011, A1–A2, https://www.newspapers.com/image/206308360.

167 National Transportation Safety Board, *Aerospace Accident Report: In-Flight Breakup During Test Flight, Scaled Composites SpaceShipTwo*, October 31, 2014, https://www.ntsb.gov/investigations/AccidentReports/Reports/AAR1502.pdf.

168 John Antczak, "VSS Unity Spaceship Lands in New Mexico," *AJ*, February 14, 2020, 12, https://www.newspapers.com/image/637959287.

169 Kevin Robinson-Avila, "A New Space Age," *AJ*, July 12, 2021, A1, https://www.newspapers.com/image/748647458.

170 Ryan Boetel, "Virgin Galactic Announces Layoffs," *AJ*, November 9, 2023, A1, A5, https://www.newspapers.com/image/1017857575/.

171 "Virgin Galactic Announces Fourth Quarter and Full Year 2024 Financial Results and Provides Business Update," Virgin Galactic press release, February 26, 2025, https://investors.virgingalactic.com/news/news-details/2025/Virgin-Galactic-Announces-Fourth-Quarter-and-Full-Year-2024-Financial-Results-and-Provides-Business-Update/default.aspx.

172 Megan Gleason, "Staying Aloft," *AJ*, June 17, 2024, Business Outlook section, Z10, Z12, https://www.newspapers.com/image/1081714573/

173 AP, "9 States Join in Spaceport Effort," *SFNM*, December 7, 2000, B-1, https://www.newspapers.com/image/583474748/.

174 Dennis Kintigh, "Spaceport America Looking More and More Like a Mirage," *AJ*, June 19, 2022, https://www.abqjournal.com/2509246/spaceport-america-looking-more-and-more-like-a-mirage.html.

Chapter 4. Scion of the Space Coast

1 Blanton McBride, "Per Person Charge Could Solve Playalinda Problem," *OS*, Brevard Sentinel edition, June 11, 1972, 4, https://www.newspapers.com/image/225041076/. According to Google Maps, Port St. Joe is about ten miles from Cape San Blas. The Nelson family tree can be traced on Ancestry.com, starting with Bill Nelson's great-great-grandfather, John "Jens" Edward Nelson, https://www.ancestry.com/genealogy/records/john-jens-edward-nelson-24-jpv1h.

2 Bill Nelson, "The Merritt Island Adventure," US Fish & Wildlife Service, November 17, 2014, YouTube video, 0:04:22, https://www.youtube.com/watch?v=22IlBi3cR2I. Bill Nelson, "US Space Programs Forum," Brookings Institution, October 1, 1986, C-SPAN video, 5:11:00, https://www.c-span.org/video/?150581-1/us-space-programs-forum.

3 "Nannie Merle Nelson," Ancestry.com, https://www.ancestry.com/genealogy/records/nannie-merle-nelson-24-7mfhl1. Her earlier generations were John Edward Nelson (1812–1891), John Nelson (1860–1923), John Wesley Nelson (1883–1960), and Nannie (1908–1967).

4 "Body of Hunt Is Located on Island Shores," *The Pensacola Journal*, May 6, 1937, 1, https://www.newspapers.com/image/352817683.

5 "Brief Illness Fatal to Attorney's Wife," *MH*, November 23, 1937, 2A, https://www.newspapers.com/image/617958510/.

6 "University Students Will Visit 28 States," *MH*, June 7, 1929, 5, https://www.newspapers.com/image/616572087/.

7 "Miss Falligant Wins University Contest," *MH*, April 14, 1929, 2, https://www.newspapers.com/image/616544195/. "University of Miami Loses Jury Debate," *The Miami Herald*, April 28, 1929, 10, https://www.newspapers.com/image/616658105/.

8 "21 Dade Boys Win Diplomas at Miami U," *Miami Daily News*, May 24, 1929, 15, https://www.newspapers.com/image/298593845/. Bill Nelson, phone interview with the author, November 18, 2023.

9 "C. W. Nelson Urges Real Tax Deductions," *MH*, April 8, 1932, 3, https://www.newspapers.com/image/617528663/. "Official Tabulation Ends in All Races," *The Miami Herald*, July 2, 1932, 5, https://www.newspapers.com/image/617420290/. Nelson lost 2,780 to 1,700.

10 "Nelson Candidate for Legislature," *Miami Daily News*, May 13, 1934, 8, https://www.newspapers.com/image/298712630/. "Miami's Own Whirligig," *Miami Daily News*, June 8, 1934, 1, https://www.newspapers.com/image/298613811/. "Both Nelson and Baumgardner were eliminated from the second primary . . ."

11 Jack Stark, "Wings Over Miami," *MH*, December 20, 1938, 16A, https://www.newspapers.com/image/617727932/.

12 "Obituaries," *OS*, Brevard Edition, September 26, 1957, 7, https://www.newspapers.com/image/222241782/.

13 "Florida Properties," *MH*, December 17, 1950, 14C, https://www.newspapers.com/image/618631982/.

14 Bill Pomeroy, Soil Conservationist, USDA, "Farm Plans Prepared for Owners of Ranch by Soil Conservation Service," *The Cocoa Tribune*, August 14, 1953, 2, https://www.newspapers.com/image/779792315/.

15 Bill Pomeroy, Soil Conservation News, "Ranch Ready for Heavy Rains," *Orlando Evening Star*, Brevard County Edition, April 7, 1955, 6, https://www.newspapers.com/image/292282422/.

16 Harold Cunningham, Assistant County Agent, "Bill Nelson Wins State 4-H Contest," *OS*, Brevard Edition, October 25, 1957, 1, https://www.newspapers.com/image/222145470.

17 "Obituaries," *OS*, Brevard County Edition, August 26, 1957, 7, https://www.newspapers.com/image/222241782/.

18 "Melbourne Boy Wins 4-H Speaking Contest," *The Cocoa Tribune*, May 5, 1958, 1, https://www.newspapers.com/image/779603912.

19 "State Winner," *OS*, Brevard Edition, June 17, 1959, 9, https://www.newspapers.com/image/223356696/.

20 Logan Owen, Jr., "Melbourne Youth Busy as Key Clubs' Proxy," *OS*, Brevard Edition, July 24, 1959, 11, https://www.newspapers.com/image/223381489/.

21 "Melbourne Boy Wins Key Honor," *OS*, July 7, 1959, Brevard County Edition, 1, https://www.newspapers.com/image/223372902/.

22 "Key Club President Gives Talk," *OS*, Lake-Marion-Sumter Edition, May 15, 1960, 4, https://www.newspapers.com/image/223840775/.

23 Bill Bunge, "Melbourne Grad Breaks Tradition," *MH*, Brevard Edition, May 17, 1960, 1C, https://www.newspapers.com/image/619656761/.

24 Homer Pyle, "Never a Dull Day in Normal Life of Busy Bill Nelson," *OS*, Brevard County Edition, May 27, 1960, 1A, https://www.newspapers.com/image/223484398/.

25 Orval Jackson, "Melbourne Youth to Tour Europe," *OS*, Brevard County Edition, September 30, 1959, 1, https://www.newspapers.com/image/223119415.

26 "Your Freedom 'Up to You,' Club Told," *MH*, Brevard Edition, January 6, 1960, 1C, https://www.newspapers.com/image/619292719.

27 Ethel P. Fish, "Personally Speaking," *MH*, June 4, 1960, 10A, https://www.newspapers.com/image/619359858.

28 "Bill Nelson Awarded Scholarship," *MH*, Brevard Edition, July 19, 1960, 1C, https://www.newspapers.com/image/619094712.

29 Logan Owen, "Melbourne Youth Nears Climax of Key Club International Presidency," *OS*, Brevard Edition, July 7, 1960, 12, https://www.newspapers.com/image/223795488/.

30 United Press International, "Bill Nelson Elected," *The Cocoa Tribune*, September 21, 1960, 5, https://www.newspapers.com/image/779632256/.

31 "UF Students from Area Win Posts," *OS*, March 12, 1961, 12D, https://www.newspapers.com/image/223693337/.

32 Bill Nelson phone interview, November 18, 2023. Andre Conn, "He's Taking Things as They Come," *Today*, January 15, 1971, 1D, 3D, https://www.newspapers.com/image/125367879/. "Senator Bill Nelson," Florida 4-H Hall of Fame, http://florida4h.org/foundation/FL4H/NelsonB.htm, archived March 1, 2012, by the *WM*.

33 Bill Nelson phone interview, November 18, 2023.

34 Amy Clark, "This Is Brevard," *Today*, August 24, 1971, 1D, https://www.newspapers.com/image/125403945/.

35 Andre Conn, "He's Taking Things . . ."

36 Amy Clark, "This Is Brevard," *Today*, November 9, 1971, 1D, https://www.newspapers.com/image/124753440/. Jean Yothers, "The Yothers Side," *OS*, November 22, 1971, 4D, https://www.newspapers.com/image/221227589/.

37 Peggy Marquette, "From Rose Petals to Rice: A Very Special Wedding," *OS*, Brevard Sentinel Edition, February 20, 1972, 8, https://www.newspapers.com/image/225317257/.

38 "Bill Nelson Plans Bid for Legislature," *Today*, April 19, 1972, 2B, https://www.newspapers.com/image/125163204/.

39 Amy Clark, "This Is Brevard," *Today*, April 20, 1972, 1D, https://www.newspapers.com/image/125164237.

40 "Bill Nelson Has Great Legislator Potential," *Today*, Editorials page, October 27, 1972, 8A, https://www.newspapers.com/image/125342961/. The editorial page published an endorsement for four state candidates—two Democrats and two Republicans. The paper also endorsed President Richard Nixon for reelection.

41 "Nixon Leader in Win Margin," *Today*, November 18, 1972, 2B, https://www.newspapers.com/image/125286935/. Nelson defeated David Vozzola by 49,456 to 21,460. Other newspapers, depending on when they published, showed slightly different results.

42 "Just Saying Thanks," *Today*, November 9, 1972, 1B, https://www.newspapers.com/image/124914334/.

43 Jerry Greene, "Happy He's a Politician, Nelson Feels There's Hope," *Today*, October 14, 1973, 3B, https://www.newspapers.com/image/124997804/.

44 Patsy Palmer, "Brevard Offered Help in Economic Planning," *Today*, February 23, 1974, 3B, https://www.newspapers.com/image/125151779/.

45 "Gillespie Proposes 'Solar Exemption,'" *Today*, April 20, 1974, 3B, https://www.newspapers.com/image/125153120/. "Tourism and Shuttle Seen as Unemployment Buffer," *Today*, February 21, 1975, 14A, https://www.newspapers.com/image/124806524/. "Solar Center Close," *Today*, May 10, 1975, 1B, https://www.newspapers.com/image/125181662/.

46 Charles Reid, "Tucker Move Slap in Face," *Today*, May 23, 1976, 6B, https://www.newspapers.com/image/125269218/.

47 Bill Kaczor, "Vogt, Wilson Named; Nelson Loses Post," *Today*, November 17, 1976, 1A, https://www.newspapers.com/image/125377587.

48 Robert Rothman, "House Attracts Nelson," *Today*, February 11, 1977, 2B, https://www.newspapers.com/image/125189199/.

49 Michael Moore, "Eastern Test Range Will Be Renamed," *Today*, January 18, 1977, 2B, https://www.newspapers.com/image/124851701/. "Eastern Test Range Put Under Vandenberg Rule," *SMT*, January 19, 1977, 2, https://www.newspapers.com/image/446510421.

50 Michael Moore, "Test Range Phaseout Under Study," *Today*, February 11, 1977, 1A, https://www.newspapers.com/image/125188847. Charles Overby, "Jobs at ETR Are Safe," *Today*, February 17, 1977, 1B, https://www.newspapers.com/image/125153037/. "Governor's Help Sought," *Orlando Sentinel Star*, Brevard Sentinel Edition, February 18, 1977, 1B, https://www.newspapers.com/image/225110246/.

51 Don Murray, "Gov. Askew Enters Fight to Save ETR," *Today*, May 4, 1977, 1B, https://www.newspapers.com/image/125364615/.

52 Barry Bradley, "No Test Range Closing Without Impact Study," *Today*, July 1, 1977, 1A, https://www.newspapers.com/image/125162772/.

53 Michael Moore, "Wake Up on ETR—Nelson," *Today*, July 13, 1977, 1B, https://www.newspapers.com/image/125174833/.

54 Hubert Griggs, "Nelson Fears ETR, Patrick Base Closing," *Orlando Sentinel Star*, August 4, 1977, 2C, https://www.newspapers.com/image/225940325/.

55 Dick Baumbach, "Reports Brighten ETR Fate," *Today*, April 14, 1978, 1A, 16A, https://www.newspapers.com/image/125171482/.

56 Baumbach, "Reports Brighten . . ."

57 B. J. Kuehn, "Test Range Reports Launch Hopes," *Today*, April 15, 1978, 1B, https://www.newspapers.com/image/125172648/.

58 David Wilkening, "Race Ends as Best Concedes," *Orlando Sentinel Star*, November 9, 1978, 1A, https://www.newspapers.com/image/225915802.

59 Chris Collins, "Gurney and Nelson Find Little to Debate," *Today*, October 28, 1978, 3B, https://www.newspapers.com/image/125229259/.

60 Chris Collins, "Getting Foot in the Door His First Task," *Today*, November 19, 1978, 1A, https://www.newspapers.com/image/125352893/.

61 Anne Groer, "Fuqua Gets Top Committee Spot; Nelson Wins 2nd Plum," *Orlando Sentinel Star*, January 19, 1979, 6A, https://www.newspapers.com/image/225926129/.

62 Dick Baumbach, "NASA Releases Tile Report; 'Incomplete,' Says Nelson," *Today*, April 14, 1979, 1A, 10A, https://www.newspapers.com/image/125436448/.

63 Dick Baumbach, "Shuttle to Launch Satellites," *Today*, April 28, 1979, 1B, https://www.newspapers.com/image/125079468/.

64 Dick Baumbach, "'81 May Be Year of Shuttle," *Today*, January 19, 1980, 8A, https://www.newspapers.com/image/125194717/.

65 Anne Groer, "Latest Bet on Early Shuttle Flight in 1981," *Orlando Sentinel Star*, January 30, 1980, 9A, https://www.newspapers.com/image/300925978.

66 Peter Larson and Anne Groer, "Cuts in NASA Budget Don't Clip Wings of Space Shuttle," *Orlando Sentinel Star*, April 1, 1980, 4A, https://www.newspapers.com/image/226514588/.

67 Bernard Wideman, "New Beach Access 'Long Road Ahead,'" *Today*, February 22, 1980, 1B, 5B, https://www.newspapers.com/image/124961250.

68 Peter Larson, "Beach Block," *Orlando Sentinel Star*, February 20, 1980, 1A, 5A, https://www.newspapers.com/image/227307129/.

69 Peter Larson, "Nature Group to Challenge Beach Closing," *Orlando Sentinel Star*, March 28, 1980, 1C, https://www.newspapers.com/image/226689396.

70 Jere Maupin, "Beach Shuttle Bus Proposed," *Today*, June 10, 1980, 1B, 3B, https://www.newspapers.com/image/124765322/.

71 Joan Reus, "Cutting Playalinda Access Riles Beach-Loving Group," *Today*, July 18, 1980, 3B, https://www.newspapers.com/image/125391051/.

72 Peter Larson, "Beach Bus Plan Launches Strong Titusville Protest," *Orlando Sentinel Star*, August 20, 1980, 1C, https://www.newspapers.com/image/226571921/.

73 Jere Maupin, "Playalinda Access 'People's Victory'?" *Today*, September 9, 1980, 1B, 3B, https://www.newspapers.com/image/124985230/.

74 Joan Reus, "Playalinda Beach Victory Turns Protest Into Party," *Today*, September 12, 1980, 1B, https://www.newspapers.com/image/124988757/.

75 Gregory Miller, "At 78, Nelson Foe Boasts of Experience," *Today*, September 14, 1980, 2B, https://www.newspapers.com/image/124990654/.

76 "How Area Voters Cast Their Ballots," *Today*, November 6, 1980, 2B, https://www.newspapers.com/image/125414603/.

77 *Today*, November 6, 1980, 1B, https://www.newspapers.com/image/125414587/.

78 Bart Slawson, "Where Have All the VIPs Gone?" *Today*, December 30, 1980, 10A, https://www.newspapers.com/image/125336687/.

79 David Bauman, "Nelson: Roll Out the Red Carpet for Launch Visit," *Today*, January 3, 1981, 1B, https://www.newspapers.com/image/125014373/.

80 Amy Clark, "VIPs: From John Denver to Wally Schirra, They Came and Waited, but in Vain," *Today*, April 11, 1981, 4A, https://www.newspapers.com/image/125035500/.

81 Chevon Thompson, "Nelson: Create Firm Space Policy," *Today*, April 16, 1981, 1B, https://www.newspapers.com/image/124887415/.

82 Ken Klein, "'Boll Weevils' Win Some Friends and Foes," *TD*, August 30, 1981, 1B, https://www.newspapers.com/image/246779016.

83 David Bauman, "Nelson Disappointed in Reaganomics," *Today*, November 12, 1981, 1B, 3B, https://www.newspapers.com/image/124819198/.

84 "We Recommend Nelson," *Today*, October 17, 1982, 18A, https://www.newspapers.com/image/124964745/.

85 "Election '82 Final Results," *Today*, November 4, 1982, 2B, https://www.newspapers.com/image/125119600/.

86 "NASA Facts: Solid Rocket Boosters and Post-Launch Processing," NASA, 2006, http://www.nasa-klass.com/Curriculum/Get_Oriented%202/Solid%20Rocket%20Boosters/RDG_SRB-Additional/SRB-processing.pdf.

87 Joan Deming and Patricia Slovinac, *Space Transportation System: Motor Vessels Liberty Star and Freedom Star*, Archaeological Consultants, Inc., 2013, 6–7, https://www.nasa.gov/sites/default/files/files/SRB_Historical_Narrative.pdf, archived February 17, 2017, at the *WM*.

88 J. Wyatt Emmerich, "Port Authority Agrees to Lease Land to United Space Boosters," *Today*, March 11, 1982, 18C, https://www.newspapers.com/image/125228646/.

89 Jane Shealy, "USBI Denies Plan to Store Harmful Waste," *Today*, November 19, 1982, 3B, https://www.newspapers.com/image/125367769/.

90 Jane Shealy, "Plant Protest Letter Mailed to KSC Boss," *Today*, November 23, 1982, 2B, https://www.newspapers.com/image/125375084/.

91 Frank Yacenda, "USBI Loses Bid for Canal Site," *Today*, January 21, 1983, 1B, https://www.newspapers.com/image/125132632/.

92 Michael Mecham, "Nelson Opposes USBI Move Plan," *Today*, February 3, 1983, 1B, https://www.newspapers.com/image/124917176/.

93 Anne Groer and James Fisher, "Plant for Boosters Staying in Brevard," *OS*, April 6, 1983, A1, A7, https://www.newspapers.com/image/227579406. Frank Yacenda, "Brevard Keeps Booster Work," *Today*, April 6, 1983, 2B–3B, https://www.newspapers.com/image/124808719.

94 Frank Yacenda and Dave Hodges, "A Big Booster for USBI," *Today*, August 18, 1984, 1A, 16A, https://www.newspapers.com/image/legacy/125119878/. Michael Lafferty, "Construction Takes Off on KSC Booster Facility," *Today*, January 31, 1985, 18C, https://www.newspapers.com/image/legacy/174072997/.

95 Benson and Faherty, *Gateway*, 82–85; 120–122.

96 Patricia Slovinac, *Mobile Launcher Platforms*, Historic American Engineering Record (HAER) FL-8-11-D, Archaeological Consultants, Inc., 2009, 11–12, https://forum.nasaspaceflight.com/index.php?action=dlattach;topic=28441.0;attach=386629.

97 Mike Toner, "Save Tower, NASA Urged," *MH*, February 17, 1983, 2C, https://www.newspapers.com/image/624240667/.

98 Frank Yacenda, "NASA Requests Estimate of Keeping Launch Tower," *Today*, February 25, 1983, 3B, https://www.newspapers.com/image/125216114/.

99 Michael Mecham, "Gantry May Be Saved," *Today*, April 13, 1983, 1B, 3B, https://www.newspapers.com/image/124833451/.

100 James Fisher, "Last Gantry of Apollo to be Saved," *OS*, September 28, 1984, C3, https://www.newspapers.com/image/228212614/.

101 "Abort Apollo Tower Plan," *OS*, July 30, 1989, G2, https://www.newspapers.com/image/230248390/.

102 Roger Launius, "Whatever Happened to the Apollo/Saturn Launch Towers?" *Roger Launius's Blog*, August 22, 2011, https://launiusr.wordpress.com/2011/08/22/whatever-happened-to-the-apollosaturn-launch-towers/. Launius is a former NASA chief historian.

103 John E. Naugle et al., "Informal Task Force for the Study of Issues in Selecting Private Citizens for Space Shuttle Flight," June 16, 1983, 1. A PDF copy of the report was provided by Alan Ladwig.

104 Naugle et al., "Private Citizens for Space Shuttle Flight," 2.

105 Naugle et al., "Private Citizens for Space Shuttle Flight," 6.

106 What did the shuttle mission designations mean?

STS is an acronym for *Space Transportation System*, the original name for the Space Shuttle program going back to the Nixon administration. Prior to the *Challenger* accident, STS was followed by three characters. Using STS-51L as an example, the *5* meant the year the mission was planned to fly, in this example *1985*. The second digit indicated the launch site—*1* for Kennedy Space Center, *2* for Vandenberg AFB. The letter was sequential for the year; *A* for the first mission, *B* for the second mission, etc. Several missions on paper were cancelled, which is why you might find gaps in the mission historical record.

After the *Challenger* accident, the US Air Force ended its interest in launching missions from Vandenberg, so the numbering system changed to a sequential number. STS-51L had been the twenty-fifth shuttle flight, so when the space shuttle resumed flight in 1988 the twenty-sixth mission was labelled *STS-26*.

107 Chris Dubbs and Emeline Paat-Dahlstrom, *Realizing Tomorrow: The Path to Private Spaceflight* (University of Nebraska Press, 2011), 76, 79.

108 Alan Ladwig, NASA Headquarters Oral History Project, interviewed by Sandra Johnson, Washington, DC, April 11, 2017, https://historycollection.jsc.nasa.gov/JSCHistoryPortal/history/oral_histories/NASA_HQ/Administrators/LadwigA/LadwigA_4-11-17.htm.

109 John Noble Wilford, "Garn, Head of Senate Space Panel, Is Chosen to Fly Aboard Shuttle," *NYT*, November 8, 1984, A1, https://www.nytimes.com/1984/11/08/us/garn-head-of-senate-space-panel-is-chosen-to-fly-aboard-shuttle.html.

110 Charlie Jean, "Bill Nelson Seeks Trip on Shuttle," *OS*, July 12, 1983, A1, A6, https://www.newspapers.com/image/229885858.

111 "Walker in Line for Ride on Future Space Shuttle," *Intelligencer Journal*, Lancaster, Pennsylvania, August 9, 1983, 5, https://www.newspapers.com/image/562720027/.

112 Alan Ladwig, *See You in Orbit? Our Dream of Spaceflight* (To Orbit Productions, 2019), 239.

113 Ladwig, *See You in Orbit*, 241.

114 Ladwig, *See You in Orbit*, 240.

115 "Space Shuttle Mission STS-51D Press Kit," NASA, April 1985, 5, https://historycollection.jsc.nasa.gov/JSCHistoryPortal/history/shuttle_pk/pk/Flight_016_STS-51D_Press_Kit.pdf.

116 Thomas H. Gorey, "Garn Will Get to Fly in the Space Shuttle," *The Salt Lake Tribune*, November 8, 1984, 1A, 2A, https://www.newspapers.com/image/613548943.

117 "More Bigwigs May Follow Garn Aloft," *WAPO*, December 18, 1984, https://www.washingtonpost.com/archive/politics/1984/12/18/nasamore-bigwigs-may-follow-garn-aloft/5ea7b706-5831-45ac-a8c8-48b3a5127b88/.

118 "Absentee Ballots Decide Two Races," *Today*, November 8, 1984, 20A, https://www.newspapers.com/image/125337251/.

119 Michael Mecham, "Nelson Is New Leader of Panel on Space," *Today*, February 6, 1985, 3A, https://www.newspapers.com/image/173869704/.

120 Chet Lunner, "Garn: Americans Support My Role on Shuttle Flight," *Today*, February 22, 1985, 1B, 3B, https://www.newspapers.com/image/173905833. A video copy of the episode was provided (courtesy of Senator Nelson) by the Special and Area Studies Collections Department, George A. Smathers Library, University of Florida. The Library intends to have the episode eventually available to the public on the University of Florida Digital Collections at https://ufdc.ufl.edu/.

121 Mary Thornton, "Jarvis: Bumped from Two Flights by Members of Congress," *WAPO*, January 29, 1986, https://www.washingtonpost.com/archive/politics/1986/01/29/jarvis-bumped-from-two-flights-by-members-of-congress/72da2efe-647c-4f91-92f4-d866cae51ef4/.

122 James Fisher, "Nelson Awaits *SLOT* to Go on Shuttle Mission," *OS*, August 8, 1985, A-1, A-14, https://www.newspapers.com/image/228796493/.

123 Chet Lunner, "Move Over Jake, It's Flying Bill," *FT*, August 31, 1985, 1A, https://www.newspapers.com/image/178772586/.

124 United Press International, "Nelson Ready for Space," *MH*, September 8, 1985, 12F, https://www.newspapers.com/image/631551463/.

125 Chet Lunner, "It's Official: Bill Aboard Future Flight," *FT*, September 7, 1985, 1A, https://www.newspapers.com/image/176120602/.

126 "Poll: Residents Split on Nelson Flight," *FT*, September 8, 1985, 2A, https://www.newspapers.com/image/176130479/.

127 Jim Runnels, "Hawkins to Women: Embrace Challenge," *OS*, September 8, 1985, B3, https://www.newspapers.com/image/228801603/.

128 Mary Voboril, "Florida Congressmen Disagree About Need for Shuttle Observers," *MH*, April 13, 1985, 12A, https://www.newspapers.com/image/legacy/631680803/.

129 "Bill Nelson: A Space Diary," *FT*, October 2, 1985, 11A, https://www.newspapers.com/image/179303343/.

130 Chet Lunner, "Holidays Look High for Nelson," *FT*, October 5, 1985, 1A, https://www.newspapers.com/image/179324742/.

131 "Space Shuttle Mission STS-51L Press Kit," NASA, January 1986, 29, https://historycollection.jsc.nasa.gov/JSCHistoryPortal/history/shuttle_pk/pk/Flight_025_STS-51L_Press_Kit.pdf.

132 David Axe, "NASA Veterans Baffled by Biden Pick of Bill Nelson to Lead Agency," *The Daily Beast*, March 19, 2021, https://www.thedailybeast.com/nasa-veterans-baffled-by-reported-biden-pick-to-lead-agency.

133 Andy Ott, "Greg Jarvis—Slipping the Surly Bonds of Earth," Hughes Aircraft Company, Our Space Heritage 1960–2000, May 16, 2015, https://www.hughesscgheritage.com/slipping-the-surly-bonds-of-earth-andy-ott/, archived January 26, 2021, by the *WM*.

134 Logsdon, *Reagan*, 272.

135 Bill Nelson with Jamie Buckingham, *Mission: An American Congressman's Voyage to Space* (Harcourt Brace Jovanovich, 1988), 52.

136 Chet Lunner, "High-Flying Nelson Shuttled Down to Earth," *FT*, January 19, 1986, 3A, https://www.newspapers.com/image/179393931.

137 George Haj, "Outer Space Assistance," *FT*, January 23, 1986, 6A, https://www.newspapers.com/image/179401300/. Chet Lunner, "Nelson Dubs His Mission in Orbit Unqualified Success," *FT*, January 24, 1986, 7A, https://www.newspapers.com/image/179402185/. James Fisher, "Shuttle Homework Over for McAuliffe," *OS*, January 24, 1986, A9, https://www.newspapers.com/image/229605731/.

138 Nelson, *Mission*, 178, 180.

139 Nelson, *Mission*, generally, 176–181.

140 "Space Shuttle Challenger: Reactions," C-SPAN, January 28, 1986, https://www.c-span.org/video/?125955-1/space-shuttle-challenger-reactions.

141 House Science and Technology Committee press conference, C-SPAN, January 28, 1986, https://www.c-span.org/video/?125956-1/shuttle-accident.

142 A "solid rocket motor" is part of a "solid rocket booster." The SRB is the motor plus the parachute and recovery systems atop the motor, and an aft skirt attached to the bottom of the motor. These additions were the responsibility of United Space Booster, Inc. (USBI). Thiokol was responsible for the motor, which was most of the SRB fuselage.

143 William P. Rogers et al., *Report of the Presidential Commission on the Space Shuttle Challenger Accident* (1986), generally Chapters IV and V. The "engineering hat" quote is on 93. The report is on the NASA at https://history.nasa.gov/rogersrep/genindex.htm.

144 Four days of the committee's hearings are archived on the C-SPAN at https://www.c-span.org/organization/?2146/Science-Space-Technology for the dates June 10, 11, 12, and 17, 1986. Nelson's questioning can be found throughout these videos.

145 Nelson, *Mission*, 291, 294.

146 Nelson, *Mission*, 295–296.

Chapter 5. Out of the Nest

1 Seth H. Lubove, "Who's #1?" *Florida Trend* Magazine, June 1989, 46 et seq. The State Library of Florida emailed a scan of the article.

2 James R. Hagy, "Bright Lights, Small Cities," *Florida Trend*, 55–56.

3 Steve Dick, NASA Chief Historian, "Summary of Space Exploration Initiative," NASA, https://history.nasa.gov/seisummary.htm. *Report of the 90-Day Study on Human Exploration of the Moon and Mars*, NASA, November 20, 1989, generally, https://history.nasa.gov/90_day_study.pdf.

4 "17 Promote Space Issues," *FT*, February 10, 1989, 4A, https://www.newspapers.com/image/176133154/.

5 Joyce Harris, "Space Coast on Brink of R&D Success," *FT*, October 15, 1989, 1E, 2E, https://www.newspapers.com/image/176622047.

6 Kirk Brown, "County Approves Courthouse Expansion," *FT*, January 11, 1989, 5B, https://www.newspapers.com/image/177639670/.

7 Catherine Liden, "County-Funded Council Goes Private in April," *FT*, March 30, 1989, 1C, https://www.newspapers.com/image/175901741/.

8 Jeff Cole, "Development Chief Steps Down," *FT*, April 5, 1989, 1A, https://www.newspapers.com/image/177571715/.

9 "Development Corporation Focuses on Healthy Growth," *FT*, September 24, 1989, 17A, https://www.newspapers.com/image/175844153/.
10 Charlie Jean, "Campaign Organized to Build Memorial to Fallen Astronauts," *OS*, February 15, 1986, A14, https://www.newspapers.com/image/229615291/.
11 Chris Adams, "Graham Signs Bill for Shuttle License Plates," *FT*, June 15, 1986, 7A, https://www.newspapers.com/image/174155363/.
12 Ann Mittman, "Officials Work Toward Creating Research Park," *FT*, February 22, 1987, 1B, https://www.newspapers.com/image/173921289/.
13 George Haj, "Space Coast Score Card," *FT*, June 6, 1987, 6B, https://www.newspapers.com/image/174530999/.
14 "Governor Appoints 5 to Research Authority," *OS*, November 25, 1987, B5, https://www.newspapers.com/image/229058278/.
15 Stephen L. Morgan, email message to author, October 20, 2022.
16 Catherine Hinman, "FIT Targets Space Studies," *OS* Central Business Section, January 18, 1988, 23, https://www.newspapers.com/image/228640563/.
17 Vincent Willmore, "Challenger License Revenue Fight Might Limit Benefits," *FT*, June 3, 1988, 4A, https://www.newspapers.com/image/175025985/. Vincent Willmore, "Disagreement Aborts Plans to Fund Research Institute with Tag Revenues," *FT*, June 4, 1988, 5A, https://www.newspapers.com/image/175030554/.
18 David White, "House Okays Using Challenger Tag Funds for Space Research in Brevard," *OS*, June 8, 1988, D3, https://www.newspapers.com/image/229519372/.
19 Ben Everidge phone interview with author, May 18, 2024.
20 Ann Mittman, "Moran Takes Job as Head of Space Research Institute," *FT*, December 21, 1988, 4A, https://www.newspapers.com/image/177598148/.
21 Ben Everidge phone interview. Irene Klotz, "Space Expert to Start Research Group," *FT*, January 7, 1989, 5A, https://www.newspapers.com/image/177600301/. Marcia Smith, "David Webb: In Memoriam," SpacePolicyOnline.com, October 5, 2016, https://spacepolicyonline.com/news/david-webb-in-memoriam/.
22 Irene Klotz, "Florida Can Be Research Model," *FT*, February 9, 1989, 4A, https://www.newspapers.com/image/176121353/.
23 Irene Klotz, "At Last, Spaceport to Become Reality," *FT*, June 8, 1989, 1B, https://www.newspapers.com/image/176121097/.
24 Irene Klotz, "Shove to Head Space Research Foundation," *FT*, June 27, 1989, 6A, https://www.newspapers.com/image/177620552/. Chris Shove email to author, October 21, 2022.
25 George H.W. Bush, "Remarks on the 20th Anniversary of the *Apollo 11* Moon Landing," The American Presidency Project, University of California Santa Barbara, https://www.presidency.ucsb.edu/documents/remarks-the-20th-anniversary-the-apollo-11-moon-landing.
26 Bernard Weinraub, "President Calls for Mars Mission and a Moon Base," *NYT*, July 21, 1989, A1, https://timesmachine.nytimes.com/timesmachine/1989/07/21/issue.html.
27 "Research Plans," *OS*, August 15, 1989, C1, https://www.newspapers.com/image/230241849/.

28 Richard Burnett, "Helicopter Program Waiting for Takeoff," *OS*, Central Florida Business section, August 21, 1989, 20, https://www.newspapers.com/image/230238803/.
29 Todd Halvorson, "Vice President Supports Lunar Research Facility," *FT*, August 27, 1989, 2A, https://www.newspapers.com/image/177602481/.
30 Richard Burnett, "Moonbase Lands Backer," *OS*, September 13, 1989, C1, C6, https://www.newspapers.com/image/230200161/.
31 Chris Shove, email message to author, March 15, 2022. Shove wrote that Moonbase was inspired by EPCOT's The Land Pavilion. "Living with the Land," EPCOT, https://disneyworld.disney.go.com/attractions/epcot/living-with-the-land/. Christy Lynch, "How Walt Disney World's Farm Grows the Most Magical Produce on Earth," Farm Flavor, May 2, 2018, https://farmflavor.com/florida/walt-disney-world-farm-grows-magical-produce-earth/.
32 Jack Snyder, "Moonbase 1 Project Picks a Site in Orange County," *OS*, October 4, 1989, D1, D4, https://www.newspapers.com/image/230438023/.
33 John Kennedy, email message to author, March 15, 2022.
34 John Kennedy, "Astronaut Foundation Questioned," *Fort Lauderdale Sun-Sentinel* (now the *SFSS*), November 19, 1989, 1A, 20A, https://www.newspapers.com/image/238043529.
35 Richard Burnett, "Space Foundation Director Quits Over Moonbase Rift," *OS*, December 21, 1989, C1, C6, https://www.newspapers.com/image/230758752/.
36 Mickey Higginbotham, "Shove Resigns from Memorial Foundation," *FT*, December 20, 1989, 5A, https://www.newspapers.com/image/176090287/.
37 Chris Shove, email messages to author, March 15, 2022, October 21, 2022. In a December 2023 phone interview, Bill Nelson said his position with the AMF was strictly honorary. He had no knowledge of the SRF, Moonbase 1, or Shove.
38 "Far-Out Use of Tag Money," *OS*, November 5, 1989, G2, https://www.newspapers.com/image/230752226/.
39 Mickey Higginbotham, "AMF Requests Audit to Explain Use of Money," *FT*, January 4, 1990, 1B, https://www.newspapers.com/image/177538736/.
40 Mickey Higginbotham, "Fight Over Tag Money Escalates," *FT*, January 11, 1990, 1B, https://www.newspapers.com/image/177588008/.
41 Cory Jo Lancaster, "2 Agencies Fight Over Challenger Tag Funds," *OS*, January 18, 1990, B1, B5, https://www.newspapers.com/image/230779741/.
42 "Reining in Challenger Tags," *OS*, February 4, 1990, G2, https://www.newspapers.com/image/232288882/.
43 Mickey Higginbotham, "Fla. Lawmakers Question Use of Money from Challenger Tags," *FT*, February 14, 1990, 1A–2A, https://www.newspapers.com/image/177773830.
44 Mickey Higginbotham, "Deratany Will Resign from FIT in Wake of Criticism," *FT*, February 21, 1990, 1A, https://www.newspapers.com/image/177564601/.
45 John Kennedy, "'Challenger' Tag Organizations Embroiled in Dispute," *Fort Lauderdale Sun-Sentinel*, March 11, 1990, 22A, https://www.newspapers.com/image/238325786/.
46 John A. Nagy and Vincent Willmore, "Brevard Wins Some Issues, Loses Some in

Divided Outcome," *FT*, June 3, 1990, 11B, https://www.newspapers.com/image/177544748/.

47 Mickey Higginbotham, "AMF Proposes Education Center," *FT*, April 3, 1990, 1B–2B, https://www.newspapers.com/image/177544180/.

48 "Center for Space Education," Astronauts Memorial Foundation, https://www.amfcse.org/about-cse.

49 I. K. Brown, "Spaceport Florida to Request AMF Land for Center," *FT*, June 7, 1990, 1B–2B, https://www.newspapers.com/image/177579659.

50 "As Decade of '90s Begins, Brevard Faces Challenges," *FT* editorial column, January 1, 1990, 12A, https://www.newspapers.com/image/177529180/.

51 Sarah Oates, "Spaceport Authority Taps 1st Director," *OS*, November 22, 1989, D1, https://www.newspapers.com/image/230746587/. "Spaceport Florida Names Director," *FT*, January 10, 1990, 10A, https://www.newspapers.com/image/177579542/.

52 I. K. Brown, "First Task: Turn Idea Into a Reality," *FT*, February 5, 1990, 1A–2A, https://www.newspapers.com/image/177680391.

53 Irene Klotz, "Commercial Launch Firms Face Uncertainty," *FT* Business Monday, October 23, 1989, 8–10, https://www.newspapers.com/image/175858678.

54 I. K. Brown, "First Task," 2A.

55 Vincent Willmore, "Authority Targets April '91 for Launches at Cape," *FT*, April 5, 1990, 1A, https://www.newspapers.com/image/177552400/.

56 Vincent Willmore, "Panel Approves Bill Lifting Limits on Spaceport Florida Launches," *FT*, April 19, 1990, 1B, https://www.newspapers.com/image/177540722/. Adam Yeomans, "Space Bill to Bolster Commercial Program," *OS*, April 19, 1990, C1, https://www.newspapers.com/image/230736327/.

57 Vincent Willmore, "$6 Million Project Gets a Boost," *FT*, April 27, 1990, 1B, https://www.newspapers.com/image/177609594/.

58 Lance Oliver, "Universities' Tracking Experiment Will Soar Aboard Shuttle," *OS*, May 4, 1990, D1, D7, https://www.newspapers.com/image/230558002/.

59 Spacehab Inc. provided space habitat microgravity experimentation equipment and services to NASA during the space shuttle era. The company is now known as Astrotech Corporation.

60 Sarah Oates, "Agreement Will Help University Research," *OS*, June 7, 1990, C1, C6, https://www.newspapers.com/image/230569897/.

61 I. K. Brown, "Old Pad to Launch New Rockets," *FT*, July 8, 1990, 1E, https://www.newspapers.com/image/177236803/. "Launch Complex 20," Cape Canaveral Space Force Museum, https://ccspacemuseum.org/facilities/launch-complex-20/.

62 I. K. Brown, "U.S. Company Seeks Approval to Operate Australian Spaceport," *FT*, February 9, 1990, 1A, 2A, https://www.newspapers.com/image/177710537/.

63 Sarah Oates, "Space Firms Await Words of Support," *OS*, July 15, 1990, D1–D2, https://www.newspapers.com/image/230579885/.

64 I. K. Brown, "Firm Gets OK to Manage Australian Spaceport," *FT*, August 22, 1990, 6A, https://www.newspapers.com/image/176603876/. "Whatever Happened to the Cape York Spaceport," State Library of Queensland, October 21, 2013, https://www.slq.qld.gov.au/blog/whatever-happened-cape-york-spaceport.

65 Kurt Loft, "Businesses Fear They're Missing Soviet Bandwagon," *The Tampa Tribune*, November 30, 1989, 1A, https://www.newspapers.com/image/338091297/. Sarah Oates and Richard Burnett, "Soviets Ask to Use Cape for Launches," *OS*, August 1, 1990, A1, A4, https://www.newspapers.com/image/231172671/. I. K. Brown, "Rocket Seller Wants Soviet Launches in U.S.," *FT*, August 24, 1990, 10A, https://www.newspapers.com/image/176625729/. Sarah Oates, "Profit at Heart of Soviet Request," *OS*, August 12, 1990, D1, D5, https://www.newspapers.com/image/230430988/.

66 "Commercial Space Launch Policy (NSPD-2)," September 5, 1990, Federation of American Scientists Space Policy Project, https://spp.fas.org/military/docops/national/nspd2.htm.

67 Sarah Oates, "Launch Policy May Benefit County," *OS* Brevard Extra Edition, September 25, 1990, F1, https://www.newspapers.com/image/230504655/.

68 Shirish Date, "Rocket Will Peek at the Sun," *OS*, July 10, 1991, A1, A4, https://www.newspapers.com/image/230836228/. Beth Dickey, Reuters, "Silent Rocket Tells No Secrets About the Sun," *OS*, July 12, 1991, A1, A14, https://www.newspapers.com/image/230604538/.

69 I. K. Brown, "FIT Gets Money from Challenger Tags," *FT*, March 21, 1991, 8A, https://www.newspapers.com/image/176805755/.

70 "Business Portfolio," *OS*, June 18, 1991, F2, https://www.newspapers.com/image/230796931/.

71 I. K. Brown, "TRDA Gives Grant for Space Law Center," *FT*, June 26, 1991, 1B, https://www.newspapers.com/image/175225115/.

72 Jim Banke, "Decay at the Cape," *FT*, January 3, 1991, 1A–2A, https://www.newspapers.com/image/176783856/.

73 Mark C. Cleary, *The Cape: Space Military Operations, 1971–1992* (Patrick AFB, Florida, 45th Space Wing History Office, 1994), Chapter IV, Section 3, https://spp.fas.org/military/program/cape/cape4-3.htm.

74 I. K. Brown, "Spaceport Florida Seeks $25 Million for Complex," *FT*, October 16, 1990, 1B, https://www.newspapers.com/image/176723055/. Sarah Oates, "Space Industry Wants Face Lift," *OS*, October 16, 1990, C1, C6, https://www.newspapers.com/image/230596322/.

75 John A. Nagy, "State Senate Approves Spaceport Legislation," *FT*, April 27, 1991, 2B, https://www.newspapers.com/image/177613234/.

76 Jennifer M. Gardner, Bureau of Labor Statistics, "The 1990–91 Recession: How Bad Was the Labor Market?" *Monthly Labor Review*, June 1994, 3 et seq., https://www.bls.gov/opub/mlr/1994/06/art1full.pdf.

77 Cox News Service, "Commercial Space Prospects Dim," *FT*, July 28, 1991, 1E, https://www.newspapers.com/image/175127686/.

78 I. K Brown, "Congress Mulls Spaceport Plan to Revive Cape Launch Complexes," *FT*, May 12, 1991, 1E, https://www.newspapers.com/image/177563230/.

79 Todd Halvorson, "Commercial Launch Sites Get U.S. Aid," *FT*, October 8, 1992, 1A, https://www.newspapers.com/image/176097416/.

80 Clyde Weiss, "Officials Seek Launch Funds," *LR*, May 11, 1993, A1, A5, https://www.newspapers.com/image/540767829/.

81 Sarah Oates, "6 States Link Up in Space Project," *OS*, August 1, 1990, D1, https://www.newspapers.com/image/231173494/.

82 States News Service, "States Line Up for Slice of Space Pie," *Helena Independent Record*, January 25, 1992, 1A, https://www.newspapers.com/image/392435429/.

83 Shirish Date, "Spaceport Florida May Fold," *OS*, January 5, 1992, A1, A6, https://www.newspapers.com/image/230622085/.

84 *Spaceport Florida Authority: Annual Report for 1997*, 11. Adam Yeomans, "Spaceport Florida Stays in Business with State Loan," *OS*, July 3, 1992, C1, C6, https://www.newspapers.com/image/231200657/. I. K. Brown, "$600,000 Grant Will Keep Spaceport Florida Airborne," *FT*, July 7, 1992, 1B, https://www.newspapers.com/image/177241376/.

85 AP, "1st State Rocket Soars," *FT*, August 23, 1992, 14B, 13B, https://www.newspapers.com/image/177246507/.

86 I. K. Brown, "Spaceport Florida Debuts 'Rockets for Schools,'" *FT*, September 27, 1992, 1E, https://www.newspapers.com/image/175192134/. I. K. Brown, "Students Get a Blast Out of Liftoff," *FT*, October 15, 1992, 6A, https://www.newspapers.com/image/176162517/. "Area Teachers Chosen for FSU Program," *FT*, June 24, 1992, 2B, https://www.newspapers.com/image/177236824/.

87 I. K. Brown, "Titusville Students Serve Up Shuttle Shrimp Test," *FT*, October 5, 1992, 1A, https://www.newspapers.com/image/176081686/. AP, "Shuttle Gives Lift to High Schools' Experiments," *Fort Pierce Tribune*, October 20, 1992, B4, https://www.newspapers.com/image/779589443/.

88 AP, "FSU Project Is Part of Rocket Launches," *TD*, December 12, 1993, 3E, https://www.newspapers.com/image/247615852/.

89 Catherine Liden, "Fla. Task Force Targets Peace Dividend Efforts," *FT*, May 23, 1992, 14C, https://www.newspapers.com/image/177234196/.

90 "Graham, Bacchus Push Retraining Plan in Defense Package," *FT*, June 4, 1992, 12C–11C, https://www.newspapers.com/image/177227099/.

91 AP, "Florida Develops Strategy to Snare Defense Dollars," *FT*, October 10, 1992, 7B, https://www.newspapers.com/image/176116305/.

92 Michael Schrage, "Clinton to Tap Technology's Power," *LAT*, November 5, 1992, D2, https://www.newspapers.com/image/177275638/. *Journal of Commerce*, "House Panel Developing Plan to Help Out U.S. Ship Builders," *MH*, December 4, 1992, 7B, https://www.newspapers.com/image/637602698/.

93 Deborah Mathis and Kirk Spitzer, "President Pledges Assistance," *FT*, March 12, 1993, 3A, https://www.newspapers.com/image/177547299/. *Congressional Record—Senate*, March 23, 1993, 5957, et seq., https://www.govinfo.gov/content/pkg/GPO-CRECB-1993-pt5/pdf/GPO-CRECB-1993-pt5-1-2.pdf.

94 Catherine Liden, "Defense Reinvestment Bill Refiled," *FT*, April 15, 1993, 12C, https://www.newspapers.com/image/176822827/.

95 SFA advertisement for word processors and technical writers in the *FT* classified ads section, April 10, 1993, 6F, https://www.newspapers.com/image/176779747/. SFA advertisement for industrial partners in *FT*, April 19, 1993, 3F, https://www.newspapers.com/image/176757120/.

96 Victoria Reid, "Getting Enterprise Florida Off the Ground," *FT*, May 3, 1992, 1E, https://www.newspapers.com/image/177558080/.
97 Richard Burnett, "Lurching to Convert Defense," *OS*, June 13, 1993, F1, F6, https://www.newspapers.com/image/231714765/.
98 Bill Bergstrom, "Tax Break Aims to Keep Defense Jobs in State," AP as published in *TD*, March 11, 1993, 5B, https://www.newspapers.com/image/247161335/. Catherine Liden, "McDonnell Optimistic Despite Bill," *FT*, April 3, 1993, 14C, https://www.newspapers.com/image/176709983/.
99 Jim Ash, "Sales Tax Proposal Would Boost Spaceport," *FT*, February 24, 1993, 2B, https://www.newspapers.com/image/176532508/. The bill's failure was mentioned in passing in *FT*'s "Tracking Your Lawmakers" under Rep. Charlie Roberts, April 4, 1993, 7B, https://www.newspapers.com/image/176718666/.
100 Fla. Stat. § 331.360 (1993), http://library.law.fsu.edu/Digital-Collections/FLStatutes/docs/1993/1993TXXVC331.pdf.
101 *SFA 1994 Annual Report*, 11.
102 Tedra Williams, "Spaceport May Modify Launch Pad," *FT*, July 23, 1993, 1B, https://www.newspapers.com/image/179097642/.
103 Jim Banke, "Commercial Launches Receive $2 Million Lift," *FT*, August 1, 1993, 1A, https://www.newspapers.com/image/179308266/. Also an editorial on page 8A.
104 Todd Halvorson, "Commercial Launch Pad Set for Cape," *FT*, February 24, 1994, 1A, https://www.newspapers.com/image/177590213/.
105 Jim Banke, "NASA, KSC Officials Say Future Is Looking Up," *FT*, January 30, 1994, 24A, https://www.newspapers.com/image/177032382/.
106 "Crime, Space, Economy Top Legislative Agenda," *FT* editorial, February 6, 1994, 10A, https://www.newspapers.com/image/177554621/.
107 Catherine Liden, "Group Promotes Canaveral Facilities," *FT*, March 24, 1994, 12C–11C, https://www.newspapers.com/image/178445171/.
108 Kathy Yampert, "Workshop Debates Future Launch Capabilities," *FT*, March 25, 1994, 20C, https://www.newspapers.com/image/178445825/.
109 Jim Ash, "Budget Benefits Brevard," *FT*, April 15, 1994, 1A–2A, https://www.newspapers.com/image/175837270/.
110 Jim Banke, "Cape Eyes Air Force Commercial Launch Grants," *FT*, April 24, 1994, 12E, https://www.newspapers.com/image/177602274/. Jim Banke, "Authority Wins Air Force Grant of $2.74 Million," *FT*, October 20, 1994, 7A, https://www.newspapers.com/image/177551804/.
111 Matthew Hoy, "Four Firms Get Launch Seed Funds," *LR*, October 23, 1994, A1, A15, https://www.newspapers.com/image/540770203/. AP, "Alaska Aerospace Firm Gets $750,000 Air Force Grant," *Daily Sitka Sentinel*, October 26, 1994, 8, https://www.newspapers.com/image/13044347/.
112 Jim Banke, "California Takes Shot at Commercial Space," *FT*, November 26, 1994, 1A, https://www.newspapers.com/image/177187313/.
113 John C. Van Gieson, "Governor Picks Spaceport Panel," *OS*, August 24, 1989, C1, C6, https://www.newspapers.com/image/230241767/. AP, "Spaceport Seeks Public Financing," *OS* February 9, 1995, C1, https://www.newspapers.com/image/234121978/.

Victoria Reid, "Spaceport Florida Looks for Money," *FT*, October 29, 1995, 1E–2E, https://www.newspapers.com/image/175194469/.

114 Seth Borenstein, "Contractors Come Down to Earth," *OS*, January 9, 1995, Florida Forecast '95 section, 60, https://www.newspapers.com/image/234166460/.

115 Jeffrey Gayner, "The Contract with America," "Implementing New Ideas in the U.S.," The Heritage Foundation, October 12, 1995, https://www.heritage.org/political-process/report/the-contract-america-implementing-new-ideas-the-us.

116 Phil Long, "The Sky Is Falling, or So It Seems Near Cape," *MH*, May 30, 1995, 1A, 4A, https://www.newspapers.com/image/639145068/.

117 Jim Banke, "167 Workers Retire Today from KSC," *FT*, March 31, 1995, 1A, https://www.newspapers.com/image/175809692/.

118 This and subsequent paragraphs draw primarily from two sources: "Has the Russian Space Quota Achieved Its Purpose?" Senate Committee on Governmental Affairs International Security, Proliferation, and Federal Services Subcommittee Hearing, July 21, 1999, https://www.govinfo.gov/content/pkg/CHRG-106shrg59455/pdf/CHRG-106shrg59455.pdf, and H. P. van Fenema, *The International Trade in Launch Services: The Effects of U.S. Laws, Policies and Practices on Its Development* (Universiteit Leiden, The Netherlands, 1999), https://scholarlypublications.universiteitleiden.nl/handle/1887/44957. Contemporary newspaper articles also provided insights into these events.

119 Van Fenema, *International Trade*, 183 et seq.

120 Van Fenema, *International Trade*, 12.

121 Van Fenema, *International Trade*, 12.

122 *Report of the Space Shuttle Management Independent Review Team*, February 1995, https://spp.fas.org/kraft.htm.

123 Jim Banke, "2 Companies Seek Shuttle Contract," *FT*, August 3, 1995, 1B, https://www.newspapers.com/image/175580949/.

124 Todd Halvorson, "NASA Hands Shuttle Contract to Team," *FT*, November 8, 1995, 1A, https://www.newspapers.com/image/175185192/.

125 Jeff Cole and Steven Lipin, "Boeing Agrees to Acquire Two Rockwell Businesses," *WSJ*, August 2, 1996, https://www.wsj.com/articles/SB838902218460054000.

126 Seth Borenstein, "Space Industry Has Up-And-Down Expectations for '96," *OS*, January 8, 1996, Florida Forecast '96 section, 36, https://www.newspapers.com/image/233517850/.

127 *SFA 1995 Annual Report*, 16–17.

128 AP, "Chiles Upset by Decision to Launch U.S. Satellites on Foreign Rockets," *OS*, February 9, 1996, D4, https://www.newspapers.com/image/233321351/. Michael Cabbage, "International Launch Plan May Lure Business," *FT*, March 4, 1996, 2A, https://www.newspapers.com/image/174949491/. "The Florida International Spaceport," undated SFA memo. Another undated SFA document by the same title included the letter from Governor Chiles to President Clinton, as well as the proposal to use SLCs 34, 37, and 41.

129 Todd Halvorson, "Treaties Allow Other Countries to Bid for Business," *FT*, March 4, 1996, 1A–2A, https://www.newspapers.com/image/174949488/.

130 MOU, October 18, 1996.
131 Jim Banke, "Weldon Ready to Bat for Space Agency," *FT*, November 20, 1994, 2A, https://www.newspapers.com/image/177182752/.
132 Frank Oliveri, "Ukrainian Rockets in Brevard?" *FT*, April 12, 1996, 1A, https://www.newspapers.com/image/175059399/.
133 Clyde Weiss, "WCSC Asks for Government Help," *LR*, June 13, 1996, A1, https://www.newspapers.com/image/540759792/.
134 Larry Wheeler, "Bill Boosts Development," *FT*, September 18, 1996, 2A, https://www.newspapers.com/image/175047693/. H.R.3936—Space Commercialization Promotion Act of 1996, 104th Congress (1995–1996), https://www.congress.gov/bill/104th-congress/house-bill/3936.
135 "Spending Items," *FT*, May 4, 1996, 6A, https://www.newspapers.com/image/174953288/.
136 Todd Halvorson, "Cape's New Complex May Mean Millions for County," *FT*, July 15, 1996, 1B–2B, https://www.newspapers.com/image/175082026/. Larry Wheeler, "Spaceport Gets Grant for $500,000," *FT*, August 6, 1996, 1B, https://www.newspapers.com/image/174714230/. "NASA Adds to Spaceport," *FT*, September 8, 1996, 1E, https://www.newspapers.com/image/175072125/.
137 "Hughes to Use Japanese Rockets for 10 Launches," *FT*, November 27, 1996, 2A, https://www.newspapers.com/image/175127675/.
138 Adam Bryant, "Boeing Offers $13 Billion to Buy McDonnell Douglas, Last U.S. Commercial Rival," *NYT*, December 16, 1996, A1, https://www.nytimes.com/1996/12/16/business/boeing-offering-13-billion-to-buy-mcdonnell-douglas-last-us-commercial-rival.html.
139 Seth Borenstein, "Florida to Build Air Force Rockets," *OS*, December 21, 1996, C1, C9, https://www.newspapers.com/image/234026614/.
140 Borenstein, "Florida to Build . . ." "Russian Engine May Go to Atlas," *FT*, April 17, 1994, 10E, https://www.newspapers.com/image/177546814/. "Business Memo," *SFSS*, November 7, 1995, 3D, https://www.newspapers.com/image/238827970/. "Pratt Engine Picked," *SFSS*, January 18, 1996, 3D, https://www.newspapers.com/image/238771792/. Richard Martin, "From Russia, with 1 Million Pounds of Thrust," *Wired* website, December 1, 2001, https://www.wired.com/2001/12/rd-180/.
141 Yvonne Gibbs, "NASA Armstrong Fact Sheet: X-34 Advanced Technology Demonstrator," NASA, https://www.nasa.gov/centers/armstrong/news/FactSheets/FS-060-DFRC.html.
142 "A Conversation with Roy Bridges," *FT*, January 10, 2023, 15A, https://www.newspapers.com/image/175053577/.
143 Jim Ash, "Local Spending Projects Fare Well as Session Ends," *FT*, May 3, 1997, 1A–2A, https://www.newspapers.com/image/175144837/. See also the sidebar on 2A, "List of Projects Sent to Chiles."
144 Robyn Suriano, "Commercial Pad Ready to Go," *FT*, May 30, 1997, 4B, https://www.newspapers.com/image/174188208/.
145 Robyn Suriano, "Probe Races to Moon," *FT*, January 7, 1998, 1A–2A, https://www.newspapers.com/image/174944204/. Michael Cabbage, "Taiwanese Satellite Soars Aloft," *OS*, January 27, 1999, A11, https://www.newspapers.com/image/234350163/.

James Dean, "Cape Revives Pad for Launch," *FT*, February 13, 2017, https://www.newspapers.com/image/272337388/. "Launch Complex 46," Cape Canaveral Space Force Museum, https://ccspacemuseum.org/facilities/launch-complex-46/.

146 Wayne Tompkins, "Spaceport Is Seeking 2nd Launch Complex," *FT*, October 19, 1997, 1A–2A, https://www.newspapers.com/image/174519728/. Todd Halvorson, "Israeli Government Considers Cape for Launches," *FT*, June 4, 1997, 7A, https://www.newspapers.com/image/174481695/. "Launch Complex 20," Cape Canaveral Space Force Museum, https://ccspacemuseum.org/facilities/launch-complex-20/. "Super Loki," Gunter's Space Page, http://www.astronautix.com/s/superloki.html.

147 Wayne Tompkins, "Job Gains Could Ease Delta Blues," *FT*, October 5, 1997, 1A–2A, https://www.newspapers.com/image/174067527/.

148 Todd Halvorson, "Florida's Dominance Is Gone," *FT*, Florida's Future in Space special section, December 14, 1997, 2, https://www.newspapers.com/image/174678623/.

149 Todd Halvorson, "Florida Faces Tough Race for Spaceships," *FT*, January 13, 1998, 1A, https://www.newspapers.com/image/174945818/.

150 Todd Halvorson, "Bid for Spaceship Faces Barriers," *FT*, March 15, 1998, 1A–2A, https://www.newspapers.com/image/175090595/.

151 Jim Ash, "Budget Plan Marks $1.6 Million for Florida to Woo VentureStar," *FT*, March 18, 1998, 1A, https://www.newspapers.com/image/174648314/.

152 Jim Ash, "Brevard Gets Nice Cut of Budget Pie," *FT*, April 22, 1998, 10B–9B, https://www.newspapers.com/image/178216439/.

153 Jim Ash, "VentureStar Officials Size Up KSC," *FT*, June 17, 1998, 1A, https://www.newspapers.com/image/178452815/. Jim Ash, "Lockheed Martin Officials Tour KSC," *FT*, June 18, 1998, 2A, https://www.newspapers.com/image/178453576/.

154 Michael Cabbage, "Rocket Ship of Future Can't Fly," *OS*, March 19, 2000, A1, A6, https://www.newspapers.com/image/235681218/.

155 Steven Siceloff, "NASA Kills X-33 Program," *FT*, March 2, 2001, 1A, 3A, https://www.newspapers.com/image/178286729/. Allen Li, "Critical Areas NASA Needs to Address in Managing Its Reusable Launch Vehicle Program," GAO, June 20, 2001, generally, https://www.gao.gov/assets/gao-01-826t.pdf.

156 QuoteInvestigator.com, October 20, 2013, https://quoteinvestigator.com/2013/10/20/no-predict/. The quote has been attributed to many other people over the years.

157 Wayne T. Price, "Spaceport Florida Chief Is Retiring," *FT*, September 29, 2000, 12C, https://www.newspapers.com/image/179391384/.

158 *SFA 2000 Annual Report*, Independent Auditors' Report, 2–4.

159 *SFA 2000 Annual Report*, Space Industry Expansion & Diversification Section, 5.

160 Wayne T. Price, "Hangar to Hold Equipment for Shuttle," *FT*, July 7, 2001, 8A, https://www.newspapers.com/image/178387015/.

161 Michael A. Cianilli, "Research and Preservation Office," KSC, https://columbia.nasa.gov/Research%20and%20Preservation%20Office.

162 "Destin Doesn't Want to Wait for Nature to Rebuild Dunes," *PNJ*, November 3, 1995, 5C, https://www.newspapers.com/image/267940886/. Edward Ellegood email to author, January 8, 2023, confirmed the site was never rebuilt. Ellegood was SFA's director of policy and program development.

163 "Bill Gives Boost to Commercial Space Units," *OS*, October 10, 1998, C1, C9, https://www.newspapers.com/image/234638918/. Commercial Space Act of 1998, Pub. L. 105-303, 112 Stat. 2843 (1998), https://www.congress.gov/bill/105th-congress/house-bill/1702/text.

164 Larry A. Strauss, "State Battles for Tech Funds," *FMNP*, September 30, 1997, 1D, https://www.newspapers.com/image/217550393/.

165 Larry S. Strauss, "Speakers Say Florida Must Prepare in Order to Compete," *FMNP*, September 30, 1997, 2D, https://www.newspapers.com/image/217550403/.

166 Mary Shedden, "Lesson in County's Growth," *FT*, October 27, 1996, 1A, 4A, https://www.newspapers.com/image/175125877/.

167 Mary Shedden, "Higher Education Teamwork Pays Off Economically," *FT*, October 27, 1996, 4A, https://www.newspapers.com/image/175125892/.

168 Joni James, "New Space Institute Gets 1st Major Contract," *OS*, January 26, 1997, A18, https://www.newspapers.com/image/234483343/.

169 "NASA Dish to UCF," *OS* Central Florida Business Section, 28, https://www.newspapers.com/image/235043350/.

170 Florida Space Institute, https://fsi.ucf.edu/.

171 Wayne T. Price, "Group Promotes Fla. Space Research," *FT*, August 28, 1999, 16, https://www.newspapers.com/image/178012649/.

172 "Teamwork, Innovation Key to Florida's Future in Space," *FT* editorial column, January 16, 2000, 10A, https://www.newspapers.com/image/176799328/.

173 Jim Ash, "Legislators Wrap Up Work," *FT*, May 6, 2000, 1A, 3A, https://www.newspapers.com/image/177219981/.

174 Steven Siceloff, "New Lab Keeps KSC on Science Front Line," *FT*, February 9, 2001, 3A, https://www.newspapers.com/image/178253726/.

Chapter 6. What's in a Name?

1 Chris Shove emails to author, January 20 and 21, 2023.

2 Reedy Creek Improvement District, https://www.rcid.org/, archived January 11, 2023, by the *WM*.

3 Chris Shove emails, January 20 and 21, 2023.

4 Adam Yeomans, "Seeking Better Way to Lure Business," *OS*, December 29, 1991, G1-G2, https://www.newspapers.com/image/231171032/. "Business Environment Gets a Boost," *OS*, March 27, 1992, B6, https://www.newspapers.com/image/231075767/.

5 Bill Cotterell, "State Agency Could Close," *TD*, January 8, 1995, 1C–2C, https://www.newspapers.com/image/247471022/.

6 Mark Silva and Tim Nickens, "They Won Some, Lost Some, Missed Some," *TD*, May 6, 1995, 2D, https://www.newspapers.com/image/247311471/.

7 AP, Adam Yeomans, "Chiles Agrees Public-Private Groups Need More Oversight," *TD*, October 25, 1995, 2C, https://www.newspapers.com/image/247247555/.

8 AP, "Oversight of Public-Private Groups Questioned," *FT*, November 28, 1995, 5B, https://www.newspapers.com/image/176852032/.

9 Bill Cotterell, "Senate Strikes Deal on Agency," *TD*, May 4, 1996, 2B, 6B, https://www.newspapers.com/image/247478122/. Bill Bergstrom, "Some Got All the Breaks," *TD*, May 7, 1996, 1B–2B, https://www.newspapers.com/image/247478728/.

10 Leslie Doolittle, "Privatization Is the Only Way to Save State Tourism, Industry Leaders Say," *OS*, March 5, 1995, H1, H4, https://www.newspapers.com/image/235401440/.

11 Marguerite M. Plunkett, "Agency to Push Tourism Growth," *PBP*, May 31, 1996, 12B, https://www.newspapers.com/image/135016488/. Kristin Vaughan, "'Visit Florida' Tourism Ads to Be Revealed," *PBP*, August 11, 1997, 2B, https://www.newspapers.com/image/133441061/.

12 Richard Burnett, "Lockheed Martin to Give Boost to Small High-Tech Ventures," *OS*, April 27, 1995, C13, https://www.newspapers.com/image/232695764/.

13 An example is a full-page advertising supplement placed in *The Boston Globe* by the Florida Division of Tourism on March 5, 1995, https://www.newspapers.com/image/440851297/. Many newspaper ads retrieved by this author from the 1990s via Newspapers.com referred readers to write the Florida Division of Tourism for more information about KSC and other Florida tourist destinations.

14 Cory Lancaster, "House OKs Bill to Replace Commerce Agency," *SFSS*, May 4, 1995, 1D, https://www.newspapers.com/image/238635421/.

15 Bill Cotterell, "Privatization of Commerce a Step Closer," *TD*, March 22, 1996, 1B, 6B, https://www.newspapers.com/image/247679861/.

16 Jean Gruss, "New 'Commerce' Plan Will Do It with Less," *Tampa Tribune*, May 8, 1996, Business & Finance section, 1, 4, https://www.newspapers.com/image/340337991/.

17 Jean Gruss, "Better Jobs, Higher Wages," *Tampa Tribune*, June 3, 1996, Business & Finance section, 8–9, https://www.newspapers.com/image/340191975/.

18 Fla. Stat. § 14.2015 (1997), www.leg.state.fl.us/Statutes/?StatuteYear=1997.

19 Rene Stutzman, "Film, TV Revenue Questioned," *OS*, February 26, 1997, B1, B4, https://www.newspapers.com/image/234323118/. Adam C. Smith, "Entertainment Promotions Council Called a 'Fiasco,'" *TBT*, March 5, 1998, 1E, 7E, https://www.newspapers.com/image/326820294/. Clay W. Cone, "State Film Panel Plans New Script for How Industry Is Promoted Here," *Naples Daily News*, August 30, 1999, 1E, 4E, https://www.newspapers.com/image/803276347/.

20 Barry Flynn, "Audit Criticizes Enterprise Florida," *OS*, May 18, 1999, B1, B6, https://www.newspapers.com/image/235504171/.

21 Barry Flynn, "Enterprise Florida Tries to Fix Woes," *OS*, August 18, 1999, B1, B7, https://www.newspapers.com/image/235516912/.

22 Kyle Parks, "Bush Plans to Overhaul Development Boards," *TBT*, January 30, 1999, 1E, 7E, https://www.newspapers.com/image/340072823/.

23 "Dr. J. Antonio Villamil," James Madison Institute, https://jamesmadison.org/bio/dr-j-antonio-villamil/.

24 Kyle Parks, "Incentive Fund Falls Short of Lofty Goals," *TBT*, May 5, 1999, 1E, 6E, https://www.newspapers.com/image/327108585/.

25 Beatrice Garcia, "The Governor as Recruiter: How Bush Helps State's Economy," *MH*, September 9, 1999, 1C, 8C, https://www.newspapers.com/image/618128460/.

26 The Florida Commercial Space Finance Corporation was renamed in 2003 to the Florida Aerospace Finance Corporation.

27 Michael Cabbage and Joni James, "$6.2 Million Gives Space Industry Big Boost," *OS*, May 1, 1999, C1, C9, https://www.newspapers.com/image/235497332/. Jim Ash, "Space, Dodger Bills Pass; Local Lawmakers Win Big in Legislature," *FT*, May 1, 1999, 1B–2B, https://www.newspapers.com/image/177637194/. Alisa LaPolt, "Legislation Would Help Keep Space Projects in State," *FT*, April 13, 1999, 5B, https://www.newspapers.com/image/177712811/.

28 Fla. SB 2540 (1999) (Filed), https://www.flsenate.gov/Session/Bill/1999/2540/.

29 Fla. SB 2540 (1999); Fla. Stat. § 331.460(3) (1999), http://www.leg.state.fl.us/Statutes/?StatuteYear=1999.

30 Wayne T. Price, "Fla. Space Industry Leaders Team Up for Funding," *FT*, November 18, 1999, 14C, https://www.newspapers.com/image/178202287/.

31 "Governor Bush Wants Transportation Projects on the Fast Track," Florida Department of Transportation press release, September 8, 1999, http://www.dot.state.fl.us/moreDOT/spenews/fasttrack.htm, archived January 15, 2000, by the *WM*. Letter from SFA Executive Director Edward O'Connor to Patrick AFB Brigadier General Donald Pettit, October 4, 1999. "Year 2000 Fast Track Transportation Initiative Project Application Form & Instructions." Wayne T. Price, "State Approves Grant for KSC Road," *FT*, November 23, 1999, 10C, https://www.newspapers.com/image/178203115/. Scott Blake, "Additional Attractions Planned for KSC," *FT*, October 25, 2000, 3A, https://www.newspapers.com/image/179367951/. Chris Kridler, "New Road Offers KSC Shortcut," *FT*, July 31, 2003, 1B, https://www.newspapers.com/image/178007597. "New Entrance and Parking Plaza," Kennedy Space Center Visitor Complex, December 13, 2018, https://www.kennedyspacecenter.com/blog/new-kscvc-entrance.

32 John A. Volpe National Transportation Systems Center, Research and Special Programs Administration, *Building on Florida's Strength in Space: A Plan for Action, Final Report* (US Department of Transportation, December 1999), Section II, 3–5, https://rosap.ntl.bts.gov/view/dot/9015/dot_9015_DS1.pdf.

33 Volpe Report, Section II, 7.

34 Volpe Report, Section III, 23–25.

35 Volpe Report, Section IV, 55–69.

36 Alisa LaPolt, "Space Business Plan Unfocused, Report Finds," *FT*, December 8, 1999, 8B–7B, https://www.newspapers.com/image/178198940/.

37 "Latest Report on Space Race Has Good Advice for Florida," *FT* editorial column, December 26, 1999, 10A, https://www.newspapers.com/image/178460123/.

38 John McCarthy, "Fla., Space Officials to Discuss Future of Commercial Launches," *FT*, January 11, 2000, 1A, https://www.newspapers.com/image/178207248/. Robyn Suriano, "Forum to Discuss Aerospace Industry's Future," *FT*, January 13, 2000, 6A, https://www.newspapers.com/image/178207656/. Robyn Suriano, "Politicians, Space Leaders Aim to Lure Launch Business," *FT*, January 15, 2000, 1A, 4A, https://www.newspapers.com/image/178208221/.

39 Dynamac Corporation and the Bionetics Corporation, *a Business Prospectus: Space Experiment Research and Processing Laboratory (SERPL)*, March 31, 2000, Executive Summary, E-3.

40 John McCarthy, "Research Center Could Boost Economy, KSC," *FT*, January 15, 2000, 4A, https://www.newspapers.com/image/178208227/.

41 Commercial Space Partnership Act of 1999, S.2316, 106th Cong. (1999), https://www.congress.gov/bill/106th-congress/senate-bill/2316.

42 Spaceport Investment Act of 1999, H.R.2289, 106th Cong. (1999), https://www.congress.gov/bill/106th-congress/house-bill/2289. Spaceport Investment Act of 1999, S.1239, 106th Cong. (1999), https://www.congress.gov/bill/106th-congress/senate-bill/1239/. Tom Breen, "Legislation Would Allow Leasing NASA-Owned Property," *FT*, March 16, 2000, 4A, https://www.newspapers.com/image/177189482/.

43 Jim Ash, "Legislators Wrap Up Work," *FT*, May 6, 2000, 1A, 3A, https://www.newspapers.com/image/177220016/.

44 Tom Breen, "Boeing Dedicates Delta 4 Plant," *FT*, September 12, 2000, 1A, 3A, https://www.newspapers.com/image/179375714/. "Horizontal Integration Facility," Space Florida, https://www.spaceflorida.gov/projects/horizontal-integration-facility.

45 Todd Halvorson, "State Sets Sights on New Rocket for Space Coast," *FT*, September 8, 1999, 1A, https://www.newspapers.com/image/178956284/. Sean Hao, "Beal Shutdown Disappoints," *FT*, November 22, 2000, 10C–9C, https://www.newspapers.com/image/179346059/. "Statement from Andrew Beal," Beal Aerospace, October 23, 2000, https://www.bealaerospace.com/statement.html.

46 Sean Hao, "Rocket Company, KSC Closer to a Deal," *FT*, July 7, 2000, 10C–9C, https://www.newspapers.com/image/179363805/.

47 David S. Langdon et al., "U.S. Labor Market in 2001: Economy Enters a Recession," *Monthly Labor Review*, US Bureau of Labor Statistics, February 2002, 29, https://www.bls.gov/opub/mlr/2002/02/art1full.pdf.

48 Frank Oliveri, "Companies Face Launch Declines," *FT*, January 19, 2001, 2A, https://www.newspapers.com/image/178247571/.

49 Miles O'Brien, "A Look at Shuttle Security After 9/11," CNN, January 17, 2003, https://www.cnn.com/2003/TECH/space/01/16/btsc.obrien.shuttle/. Denise Chow, "NASA Grapples with U.S. Space Security in Post-9/11 Era," *Scientific American*, September 8, 2011, https://www.scientificamerican.com/article/nasa-grapples-with-us-space-security/.

50 Marilyn Meyer and Steven Siceloff, "Spaceport Worker Faces Arrest Today," *FT*, March 30, 2001, 1A, https://www.newspapers.com/image/178262330/. R. Norman Moody, "Spaceport Finance Worker Turns Self In," *FT*, March 31, 2001, 3A, https://www.newspapers.com/image/178262812/. Brevard County Clerk of the Court website public records search, March 12, 2023; in-person public records search, March 15, 2023.

51 Frank Oliveri and Alisa LaPolt, "Spaceport Takes Credit Cards from Employees," *FT*, April 25, 2001, 1B–2B, https://www.newspapers.com/image/178270606/.

52 Alisa LaPolt, "Allegations of Theft Jolt Spaceport," *FT*, March 22, 2001, 1A–2A, https://www.newspapers.com/image/178259289/. LaPolt, "Legislators Question Spaceport Funding," *FT*, March 23, 2001, 1A–2A, https://www.newspapers.com/image/178259667/.

53 Alisa LaPolt, "Alpha Crew Lobbies Lawmakers," *FT*, April 3, 2001, 2A, https://www.newspapers.com/image/178268504/.

54 NASA eventually transferred the X-37 project to the Defense Advanced Research Projects Agency (DARPA) in 2004. At that point, it became a classified project. Mike Wall, "X-37B: The Air Force's Mysterious Space Plane," Space.com, August 30, 2021, https://www.space.com/25275-x37b-space-plane.html.

55 Steven Siceloff, "X-37 to Make Cape Canaveral Home," *FT*, April 4, 2001, 2A, https://www.newspapers.com/image/178268778/.

56 "X-37B," Boeing, https://www.boeing.com/space/x37b/index.page. Amanda Miller, "X-37B Space Plane Eclipses Its Record for Longest Flight," *Air & Space Forces Magazine*, July 10, 2022, https://www.airandspaceforces.com/x-37b-space-plane-eclipses-its-record-for-longest-flight/.

57 Alisa LaPolt, "Space Projects Set to Get $3.1 Million from State," *FT*, April 29, 2001, 1A, https://www.newspapers.com/image/178272156/.

58 *2001 Commercial Space Transportation Forecasts*, FAA and COMSTAC, May 2001, iii, https://www.faa.gov/about/office_org/headquarters_offices/ast/media/0501forcast.pdf.

59 *2002 Commercial Space Transportation Forecasts*, FAA & COMSTAC, May 2002, 41, https://www.faa.gov/about/office_org/headquarters_offices/ast/media/ForecastMay2k2GSO_NGSO.pdf.

60 *2002 Commercial Space Forecasts*, 1.

61 AP, "Gov. Bush Goes to California to Boost Florida in Space Race," *TD*, May 28, 1999, 9C, https://www.newspapers.com/image/249249654/. Victoria Reid, "Brevard Buzz," *FT*, August 29, 1999, 1E, https://www.newspapers.com/image/178013908/. David K. Rogers, "Fast Train Proposal Stops Next at Bush's Desk," *TBT*, November 20, 1999, 3B, https://www.newspapers.com/image/327287200/. Robyn Suriano, "Politicians, Space Leaders Aim to Lure Launch Business," *FT*, January 15, 2000, 1A, 4A, https://www.newspapers.com/image/178208221/. *SFA 1999 Annual Report*, 1.

62 2002 Florida Statutes, Title XXV, Chapter 331, Section 304, Paragraph (1), "Spaceport Territory," http://www.leg.state.fl.us/Statutes/?StatuteYear=2002.

63 "Florida Space Authority Christens New Name," *FT*, November 21, 2001, 1B, https://www.newspapers.com/image/178803037/.

64 Paige St. John and Steve Siceloff, "Space Coast Gets Shuttle Repair Work," *FT*, February 6, 2002, 1A–2A, https://www.newspapers.com/image/178373318/.

65 Paige St. John, "Lawmakers Consider $1 Lease for Hangar," *FT*, February 6, 2002, 2A, https://www.newspapers.com/image/178373334/. Paige St. John, "Space Contractors Upset with Authority," *FT*, February 8, 2002, 1A, 8A, https://www.newspapers.com/image/178376593/. Paige St. John, "Bush, Space Contractors Battle," *FT*, February 23, 2002, 1B–2B, https://www.newspapers.com/image/178250283/.

66 Paige St. John, "Moving NASA Work to Brevard Draws Ire," *FT*, February 28, 2002, 1C–3C, https://www.newspapers.com/image/178250985/. 2002 Florida Statutes, Title XXV, Chapter 331, Section 308, Paragraph (1), "Board of Supervisors," http://www.leg.state.fl.us/Statutes/?StatuteYear=2002.

67 Kelly Young, "Planners Focus on Spaceport," *FT*, April 28, 2002, 1C–2C, https://www.newspapers.com/image/178390393/.

68 *Cape Canaveral Spaceport Master Plan*, Team ZHA, July 2002.

69 Kelly Young, "Space Leaders Unveil Plan," *FT*, August 29, 2002, 1A, 5A, https://www.newspapers.com/image/178175121/.

70 2002 *Spaceport Master Plan*, 1–2.

71 2002 *Spaceport Master Plan*, III-5.

72 2002 *Spaceport Master Plan*, III-10.

73 2002 *Spaceport Master Plan*, III-7.

74 James Dean, "KSC Prospects Soar with Idea," *FT*, December 14, 2011, 1A–2A, https://www.newspapers.com/image/360465267/. Stratolaunch, www.stratolaunch.com.

75 "Swiss Company to Use Kennedy Space Center Shuttle Runway," WKMG-TV, March 14, 2014, https://www.clickorlando.com/news/2014/03/14/swiss-company-to-use-kennedy-space-center-shuttle-runway/. "Failed Space Flight Firm Was Backed by 'Phantom Bank,'" SWI, December 5, 2019, https://www.swissinfo.ch/eng/business/swiss-space-systems_failed-space-flight-firm-was-backed-by-phantom-bank-/45414356.

76 Jeff Foust, "Virgin Orbit Files for Bankruptcy," *SN*, April 4, 2023, https://spacenews.com/virgin-orbit-files-for-bankruptcy/. "Stratolaunch Expands Fleet with Virgin Orbit's Modified Boeing 747," Stratolaunch press release, May 25, 2023, https://www.prnewswire.com/news-releases/stratolaunch-expands-fleet-with-virgin-orbits-modified-boeing-747-301834904.html.

77 2002 *Spaceport Master Plan*, IV-6.

78 2002 *Spaceport Master Plan*, IV-7.

79 2002 *Spaceport Master Plan*, IV-7.

80 2002 *Spaceport Master Plan*, V-1.

81 *Final Report of the Commission on the Future of the United States Aerospace Industry*, November 2002, ix, https://www.nasa.gov/wp-content/uploads/2024/01/aerocommissionfinalreport.pdf.

82 Eric Berger, "Space Force Considers Merging Cape Canaveral with Kennedy Space Center," *AT*, October 8, 2020, https://arstechnica.com/science/2020/10/space-force-considers-merging-cape-canaveral-with-kennedy-space-center/.

83 John Kelly, "Aerospace Settles, Launches Quicken," *FT*, December 29, 2002, 6S, https://www.newspapers.com/image/178238995/.

84 "Boeing Launches Orbital Space Plane Design," Boeing press release, April 18, 2003, https://boeing.mediaroom.com/2003-04-18-Boeing-Launches-Orbital-Space-Plane-Design. Patrick M. McKenzie, OSP Business Development Manager, Lockheed Martin Space Systems, "Orbital Space Plane (OSP) Program," presented at the 54th International Astronautical Congress, October 1, 2003, https://ntrs.nasa.gov/api/citations/20030111797/downloads/20030111797.pdf. "X-37 Demonstrator to Test Future Launch Technologies in Orbit and Reentry Environments," NASA Facts FS-2003-5-65-MSFC, May 2003, https://www.scribd.com/document/49072037/NASA-Facts-X-37-Demonstrator-to-Test-Future-Launch-Technologies-in-Orbit-and-Reentry-Environments-2001.

85 "Columbia Accident Investigation Board Releases Final Report," August 26, 2003, at the CAIB archive, University of North Texas, https://govinfo.library.unt.edu/caib/news/press_releases/pr030826.html. Admiral (Retired) Harold W. Gehman, Jr. et al.,

Columbia Accident Investigation Board Report, Volume 1, August 2003, 25, https://govinfo.library.unt.edu/caib/news/report/pdf/vol1/full/caib_report_volume1.pdf.

86 CAIB Report, 13, 97.

87 CAIB Report, 102.

88 Alisa LaPolt, Victor Epstein, and Paige St. John, "Economic Impact Not as Bad as Challenger," *FT*, February 4, 2003, 1S, 3S, https://www.newspapers.com/image/178387752/. Wayne T. Price, "Survey Gauges Job-Market Impact," *FT*, February 4, 2003, 3S, https://www.newspapers.com/image/178387800/. Alisa LaPolt and V. J. Epstein, "Shuttle Grounding Could Slam Contractors," *FT*, February 5, 2003, 1S, 7S, https://www.newspapers.com/image/178390459/.

89 Alisa LaPolt et al., "Economic Impact . . ." February 4, 2003.

90 Wayne T. Price, "NASA Plans No Layoffs at KSC," *FT*, March 1, 2003, 1C, 3C, https://www.newspapers.com/image/177885495/.

91 Chris Kridler, "Authority's Retiring Director Optimistic About Spaceport," *FT*, March 11, 2003, 3A, https://www.newspapers.com/image/177907983/.

92 Rachael Lee Coleman, "Scott to Head Space Authority," *FT*, June 11, 2003, 1B, https://www.newspapers.com/image/177921867/. Winston E. Scott Biographical Data, NASA, August 2021, https://www.nasa.gov/sites/default/files/atoms/files/scott_winston_0.pdf.

93 Chris Kridler, "Air Force Hands Over Launch Station," *FT*, November 8, 2003, 3B, https://www.newspapers.com/image/178232910/.

94 Chris Kridler, "Lab NASA-Florida Team Effort," *FT*, November 19, 2003, 1B, 4B, https://www.newspapers.com/image/178397837/.

95 CAIB Report, 11, 27, 30.

96 President George W. Bush, "President Bush Announces New Vision for Space Exploration Program," January 14, 2004, George W. Bush White House Archives, https://georgewbush-whitehouse.archives.gov/news/releases/2004/01/20040114-3.html

97 President Ronald Reagan, "Address Before a Joint Session of the Congress on the State of the Union," January 25, 1984, Ronald Reagan Presidential Library & Museum, https://www.reaganlibrary.gov/archives/speech/address-joint-session-congress-state-union-january-1984.

98 "Strategy Based on Long-Term Affordability," NASA Fiscal Year 2005 Budget Chart, https://www.nasa.gov/pdf/54873main_budget_chart_14jan04.pdf, archived March 25, 2004, at the *WM*. Wayne Hale, "Killing Constellation," *Wayne Hale's Blog*, September 23, 2010, https://waynehale.wordpress.com/2010/09/23/6/.

99 Larry Wheeler, "NASA Lab Has a New Mission," *FT*, April 10, 2006, 1A, 5A, https://www.newspapers.com/image/179036116/.

100 National Research Council, *Review of NASA Plans for the International Space Station* (National Research Council, The National Academies Press, 2006), Executive Summary, 1, https://nap.nationalacademies.org/catalog/11512/review-of-nasa-plans-for-the-international-space-station.

101 Chris Kridler, "Lease Details Threaten KSC Park," *FT*, June 18, 2005, 1A, 5A, https://www.newspapers.com/image/178515559/. John Kelly, "Florida Suspends Research Park Plans," *FT*, September 16, 2005, 1A, 5A, https://www.newspapers.com/image/

179323405/. Chris Kridler, "Space Park Plans Stall After Loss of Partner," *FT*, November 13, 2005, 1A, 5A, https://www.newspapers.com/image/179113538/.

102 Fla. SB 1026 (2005), https://www.flsenate.gov/Session/Bill/2005/1026. Paul Flemming, "Budget, Growth Lead Day," *FT*, May 7, 2005, 1A, 3A, https://www.newspapers.com/image/178515315/. Chris Kridler, "Panel Tackles Uniting State's Space Interests," *FT*, June 11, 2005, 1A, 6A, https://www.newspapers.com/image/178511772/.

103 The agenda for the July 13, 2005, commission meeting was loaned by David Teek.

104 Wayne T. Price, "CEO: Florida Needs More Aerospace Talent," *FT*, July 14, 2005, 4A, https://www.newspapers.com/image/178518435/.

105 Ashlee Vance, *Elon Musk*, 102–108.

106 *Governor's Commission on the Future of Space and Aeronautics in Florida: Final Report*, January 2006. The report was archived in two parts on July 12, 2006, by the *WM*. Part 1 is at https://web.archive.org/web/20060712004855/http://www.myflorida.com/myflorida/government/governorinitiatives/space_commission/pdfs/final_report1.pdf. Part 2 is at https://web.archive.org/web/20060712005837/http://www.myflorida.com/myflorida/government/governorinitiatives/space_commission/pdfs/final_report2.pdf.

107 Governor's Commission Report, ES-2–ES-3.

108 Executive Order No. 05-120, State of Florida Office of the Governor, June 10, 2005. A Microsoft Word copy of the Executive Order was loaned by David Teek.

109 Governor's Commission Report, ES-1.

110 Governor's Commission Report, ES-3.

111 Governor's Commission Report, 3–17.

112 Futron Corporation, *Space Tourism Market Study: Orbital Space Travel & Destinations with Suborbital Space Travel*, October 2022, 2, https://www.americaspace.com/wp-content/uploads/space_docs/Space_Tourism_Market_Study_Futron_2002.pdf. Futron was 14,982 passengers short. The first American commercial suborbital flights occurred in 2021. Fourteen passengers flew that year on Blue Origin's New Shepard over three flights launched near Van Horn, Texas, including founder Jeff Bezos and actor William Shatner. Virgin Galactic flew one flight from New Mexico's Spaceport America with two pilots and four passengers, one of whom was founder Sir Richard Branson.

113 Governor's Commission Report, 2–12.

114 Paige St. John, "Jeb Seeks $55M for Space," *FT*, January 19, 2006, 1A, 7A, https://www.newspapers.com/image/179112972/.

115 Fla. HB1489 (2006), https://www.myfloridahouse.gov/Sections/Bills/billsdetail.aspx?BillId=33774&SessionId=42. The quote is from Fla. Stat.§ 331.3011(3) (2006).

116 Fla. Stat.§ 331.308(1)(f) (2006).

117 Fla. HB 1489, § 69 (2006).

118 Fla. HB 1489, § 68 (2006).

119 John Kelly and Scott Blake, "Lockheed Picks KSC as Assembly Site," *FT*, February 23, 2006, 1A, 3A, https://www.newspapers.com/image/179298539/.

120 Todd Halvorson, "KSC Assembly Historic," *FT*, September 1, 2006, 1A, 9A, https://www.newspapers.com/image/179155847/.

121 "Orion Takes Its Place at KSC," *FT*, January 31, 2007, 3B, https://www.newspapers.com/image/363036052/.

122 Chris Kridler, "Space Panel Resigned to Losing Chief," *FT*, July 28, 2006, 1A, https://www.newspapers.com/image/179093683/.

123 Wayne T. Price and Paul Flemming, "Space Firm Welcomes New Leader," *FT*, August 31, 2006, 1B, 4B, https://www.newspapers.com/image/179360720/.

124 Kohler's years of service on the Governor's Action Team were listed on his LinkedIn page, https://www.linkedin.com/in/steve-kohler-4828a922/. An example of his work is being directed by Governor Ridge to find a potential buyer for a closing Bethlehem Steel plant. Elliot Grossman and Mariella Savidge, "Bethlehem Steel Corp. Plans to Shut Down Coke Works," *The Morning Call*, Allentown, Pennsylvania, December 30, 1997, A1, A5, https://www.newspapers.com/image/276846632/.

125 Kris Hundley, "The New Race," *TBT*, November 12, 2006, 1D, 4D, https://www.newspapers.com/image/330689628/.

126 John Kelly, "Agency Shooting for More Than the Moon," *FT*, October 21, 2006, 1A, 6A, https://www.newspapers.com/image/179301070/.

127 "Pull Out the Stops," *FT* editorial column, December 4, 2006, 10A, https://www.newspapers.com/image/179177775/. "Fighting for Brevard," *FT* editorial column, December 15, 2006, 14A, https://www.newspapers.com/image/179182879/.

Chapter 7. The End of the Beginning

1 John Kelly, "Company to Launch Small Rockets for Less Cost," *FT*, December 12, 2002, 1A, https://www.newspapers.com/image/178239267/.

2 Ashlee Vance, *Elon Musk: Tesla, SpaceX, and the Quest for a Fantastic Future* (HarperCollins, 2015), 113–114.

3 Rachael Lee Coleman, "Tycoon Taps into Small Rockets," *FT*, June 15, 2003, 1B–2B, https://www.newspapers.com/image/177923564/.

4 *SFA 1997 Annual Report*, 4.

5 Josh Friedman, "Entrepreneur Tries His Midas Touch in Space," *LAT*, April 22, 2003, C1, C9, https://www.newspapers.com/image/189900016/. "TacSat-1 Heads for Orbit," FlightGlobal, October 20, 2003, https://www.flightglobal.com/tacsat-1-heads-for-orbit/51100.article.

6 John Vandiver, "Wallops Woes," *DT*, May 25, 2003, A1, A4, https://www.newspapers.com/image/282359986/.

7 Janene Scully, "New Rocket May Fly in December," *LR*, May 25, 2003, A1, A6, https://www.newspapers.com/image/540385152/.

8 Janene Scully, "Titan 4 Delay Bumps Falcon," *LR*, July 3, 2005, A1, A8, https://www.newspapers.com/image/540975783/.

9 Ashlee Vance, *When the Heavens Went on Sale: The Misfits and Geniuses Racing to Put Space Within Reach* (HarperCollins, 2023), 36–38.

10 "USAF/DARPA FALCON Program," Air-Attack.com, http://www.air-attack.com/page/32/USAF-DARPA-FALCON-Program.html, archived August 30, 2008, at the *WM*.

11 "Falcon HTV-2," DARPA, https://www.darpa.mil/about-us/timeline/falcon-htv-2.

12 Tim Buzza Oral Project History Interview, interviewed by Rebecca Wright, NASA Johnson Space Center Oral History Project Commercial Crew & Cargo Program Of-

fice, Hawthorne, California, January 15, 2013, https://historycollection.jsc.nasa.gov/JSCHistoryPortal/history/oral_histories/C3PO/BuzzaT/BuzzaT_1-15-13.htm.

13 Kelly Young, "New Rocket May Save Money," *FT*, July 9, 2003, 3B, https://www.newspapers.com/image/177924306/.

14 Keith Cowing, "SpaceX Falcon Launch Vehicle Unveiled in Washington D.C.," SpaceRef, December 4, 2003, https://spaceref.com/uncategorized/spacex-falcon-launch-vehicle-unveiled-in-washington-dc/. Jeff Foust, "The Falcon and the Showman," The Space Review, December 8, 2003, https://www.thespacereview.com/article/70/1

15 "March/April Update," SpaceX, http://www.spacex.com/, archived June 5, 2004, at the *WM*. Tony Reichhardt, "Son of Transhab," *Air & Space Magazine*, January 17, 2013, https://www.smithsonianmag.com/air-space-magazine/son-of-transhab-2210590/. "Space Inflatable Project Eyed by Bigelow Aerospace," Space.com, May 3, 2004, https://www.space.com/490-astronotes-2-15-2004.html.

16 Leonard David, "Bigelow Plans to Launch Genesis Pathfinder in Early 2006," *SN*, March 14, 2005, https://spacenews.com/bigelow-plans-launch-genesis-pathfinder-early-2006/. "Genesis Program," Bigelow Aerospace, https://www.bigelowaerospace.com/pages/genesis/.

17 Todd Halvorson, "Satellite Safely in Orbit," *FT*, February 4, 2005, 1B, 5B, https://www.newspapers.com/image/179018090/.

18 "SpaceX Announces the Falcon 9 Fully Reusable Heavy Lift Launch Vehicle," SpaceX, September 8, 2005, http://www.spacex.com, archived November 25, 2005, at the *WM*.

19 Michael Braukus/J. D. Harrington, "NASA Establishes Commercial Crew/Cargo Project Office," NASA press release, November 7, 2005, https://www.nasa.gov/home/hqnews/2005/nov/HQ_05356_commercial_crew.html, archived December 22, 2005, at the *WM*.

20 Todd Halvorson, "Cape Contends for SpaceX Launch," *FT*, December 28, 2005, 1A, 6A, https://www.newspapers.com/image/179332193/.

21 "Future at Hand," *FT* editorial column, August 23, 2006, 10A, https://www.newspapers.com/image/179313854/.

22 Jim Wolf, "SpaceX Gets Cape Canaveral Launch Pad," Reuters, April 26, 2007, https://www.reuters.com/article/space-usa-launch/update-3-spacex-gets-cape-canaveral-launch-pad-idUSN2622145920070426.

23 "SpaceX," archived February 27, 2007, at the *WM*.

24 Patrick Peterson, "Cape Conference Aims to Attract Space Industry," *FT*, October 31, 2007, 1C, https://www.newspapers.com/image/363034173/. Patrick Peterson, "Conference Attracts Space Entrepreneurs," *FT*, November 2, 2007, 3A, https://www.newspapers.com/image/363010911/.

25 Donna Balancia, "KSC Set to Host Space Expo," *FT*, September 13, 2007, 1B, 4B, https://www.newspapers.com/image/362984403/. Patrick Peterson, "Expo Reaches for Stars," *FT*, October 31, 2007, 1C, https://www.newspapers.com/image/363034173/.

26 S. V. Date, "Crist Rescinds 283 Bush Appointments," *PBP*, January 11, 2007, 2A, https://www.newspapers.com/image/227273378/. Jim Ash, "Space Florida Back in Action," *FT*, June 24, 2007, 1A, 6A, https://www.newspapers.com/image/362934097/.

"Crist Appoints Eight to Space Board," *FT*, August 9, 2007, 8B, https://www.newspapers.com/image/362987663/.

27 Paul Flemming, "Unions Pledge $250M for Jobs," *FT*, March 22, 2006, 1A, 5A, https://www.newspapers.com/image/179143687/. Paige St. John, "Space Bill Helps KSC," *FT*, April 22, 2006, p. 1A, https://www.newspapers.com/image/179009645/.

28 Jim Ash, "Space Initiative Retried," *TD*, March 8, 2007, 4B, https://www.newspapers.com/image/250021173/.

29 Fla. HB 7123 (2006) (Enrolled), https://www.myfloridahouse.gov/Sections/Bills/billsdetail.aspx?BillId=36885.

30 "Still More Setbacks," *FT* editorial column, June 27, 2007, 6A, https://www.newspapers.com/image/362941934/.

31 *FY 2008 Budget Estimates*, NASA, ESMD-16, https://www.nasa.gov/pdf/168652main_NASA_FY08_Budget_Request.pdf, archived February 14, 2007, at the *WM*.

32 Todd Halvorson, "KSC Chief Says More Tech Work Needed," *FT*, May 12, 2007, 1C–2C, https://www.newspapers.com/image/363020030/.

33 Wayne T. Price, "Florida Must Pursue Spaceport for Viability, Agency Head Says," *FT*, March 14, 2007, 1C–2C, https://www.newspapers.com/image/362961204/.

34 Todd Halvorson, "Cape Key Spot for Virgin Galactic," *FT*, February 7, 2006, 1A, 3A, https://www.newspapers.com/image/178497138/.

35 Governor's Commission Report, ES-3.

36 Scott Blake, "A Chance to Escape Earthly Constraints," *FT*, March 2, 2007, 1A, 3A, https://www.newspapers.com/image/362945887/. Todd Halvorson, "Space, Here I Come," *FT*, April 27, 2007, 1A, 3A, https://www.newspapers.com/image/362987773/.

37 "Space Florida and Zero Gravity Launch Research and Education Center," March 19, 2007, https://spacenews.com/space-florida-and-zero-gravity-corp-partner-to-launch-microgravity-education-and-research-center/.

38 Robert Block and Aaron Deslatte, "Critics Blast Space Florida as $50M Waste," *OS*, February 15, 2009, A1, A7, https://www.newspapers.com/image/268701693/.

39 "It's Called AWOL," *FT* opinion column, September 7, 2007, 12A, https://www.newspapers.com/image/362983146/.

40 "What Really Matters," *FT* opinion column, October 2, 2007, 10A, https://www.newspapers.com/image/362790444/.

41 "Poll Results," *FT* Community Conversations opinion page, September 17, 2007, 8A, https://www.newspapers.com/image/362987411/.

42 Bill Cotterell, "Crist's Influence Sought on Space," *FT*, October 10, 2007, 8B, https://www.newspapers.com/image/362808974/.

43 John Kennedy and Jim Stratton, "GOP Candidates Talk Tough," *OS*, October 22, 2007, A1, A12, https://www.newspapers.com/image/269863635/. "The Republican Debate on Fox News Channel," *NYT* transcript, October 21, 2007, https://www.nytimes.com/2007/10/21/us/politics/21debate-transcript.html.

44 John Heilemann and Mark Halperin, *Game Change* (HarperCollins, 2010), 98–99.

45 "Washington Post—ABC News Poll," *WAPO*, October 1, 2007, https://www.washingtonpost.com/wp-srv/politics/polls/postpoll_100307.html.

46 Joseph Carroll, "Clinton Maintains Large Lead Over Obama Nationally," Gallup, December 18, 2007, https://news.gallup.com/poll/103351/clinton-maintains-large-lead-over-obama-nationally.aspx.

47 Lori Garver, *Escaping Gravity: My Quest to Transform NASA and Launch a New Space Age* (Diversion Books, 2022), 79.

48 Hillary Clinton Speech and Q&A on Innovation, May 31, 2007, The American Presidency Project, University of California Santa Barbara, https://www.presidency.ucsb.edu/documents/hillary-clinton-speech-and-qa-innovation.

49 "Hillary Clinton's Agenda to Reclaim Scientific Innovation," Hillary for President, October 4, 2007, http://www.hillaryclinton.com/news/release/view/?id=3566, archived October 17, 2007, at the *WM*.

50 *SN* briefs, November 16, 2007, https://spacenews.com/briefs-120/.

51 Steve Robinson LinkedIn page, https://www.linkedin.com/in/steve-robinson-phd/. Steve Robinson phone interview, April 5, 2023.

52 "Barack Obama's Plan for Lifetime Success Through Education," Obama '08, http://www.barackobama.com/issues/pdf/PreK-12EducationFactSheet.pdf, archived January 2, 2008, at the *WM*.

53 Loretta Hidalgo Whitesides, "Obama Pits Human Space Exploration Against Education," Wired.com, November 21, 2007, https://www.wired.com/2007/11/obama-pits-huma/.

54 "Obama: Cut Constellation to Pay for Education," *Space Politics* (blog), November 20, 2007, http://www.spacepolitics.com/2007/11/20/obama-cut-constellation-to-pay-for-education/.

55 Marc Kaufman, "Clinton Favors Future Human Spaceflight," *WAPO*, November 23, 2007, https://www.washingtonpost.com/wp-dyn/content/article/2007/11/22/AR2007112201359.html.

56 John Zarrella and Patrick Oppmann, "Florida, Michigan Seek Exit from Democratic Penalty Box," CNN, March 6, 2008, https://www.cnn.com/2008/POLITICS/03/06/florida.michigan/.

57 Bill Cotterell, "Uninvited to Parties, Fla. Gets Serious Flirts," *FT*, January 27, 2008, 1A, 4A, https://www.newspapers.com/image/362899094/.

58 Rudy Giuliani, "Keeping America First in Space," *FT* guest opinion column, January 26, 2008, 11A, https://www.newspapers.com/image/362898286/.

59 Patrick Peterson, "Giuliani: Boost NASA's Budget," *FT*, January 25, 2008, 5A, https://www.newspapers.com/image/362897284/.

60 Eun Kyung Kim, Patrick Peterson, and John Kelly, "Candidates Vague on Space Plan," *FT*, January 23, 2008, 1A, 3A, https://www.newspapers.com/image/362896401/.

61 "America's Space Program," John McCain's campaign website, http://www.johnmccain.com/Informing/Issues/7366faf9-d504-4abc-a889-9c08d601d8ee.htm, archived January 30, 2008, at the *WM*.

62 Florida Department of State Division of Elections, January 29, 2008, Presidential Preference, https://results.elections.myflorida.com/Index.asp?ElectionDate=1/29/2008.

63 Brevard County Supervisor of Elections, Presidential Preference Primary, January 29, 2008, https://www.votebrevard.gov/Previous-Elections/2008-PPP-Election-Results.

64 Florida Department of State Division of Elections, January 29, 2008, Presidential Preference.

65 Brevard County Supervisor of Elections, Presidential Preference Primary, January 29, 2008.

66 Dan Nowicki, "McCain Clinches GOP Nomination," *Arizona Republic*, March 5, 2008, A1, A12, https://www.newspapers.com/image/126239567/.

67 Michael Finnegan and Mark Z. Barabak, "Clinton Dismisses Appeal to Step Aside," *LAT*, March 29, 2008, A1, A14, https://www.newspapers.com/image/193555261/.

68 Statement of Christina T. Chaplain, Director Acquisition and Sourcing Management, "Ares I and Orion Project Risks and Key Indicators to Measure Progress," GAO, April 3, 2008, 1, https://www.gao.gov/assets/gao-08-186t.pdf.

69 ISDC 2008, http://www.isdc2008.org/, archived May 13, 2008, at the *WM*. "Campaign 2008 and Space Policy," C-SPAN video, https://www.c-span.org/video/?205759-4/campaign-2008-space-policy.

70 Kate Snow and Eloise Harper, "Clinton Concedes Democratic Nomination; Obama Leads Party in Fall," ABC News, June 7, 2008, https://abcnews.go.com/Politics/Vote2008/story?id=5020581.

71 "The Buzz on Florida Politics," *St. Petersburg Times*, January 6, 2008, 3B, https://www.newspapers.com/image/330925092/.

72 William E. Gibson, "Clinton's Rally to End Democratic Boycott," *SFSS*, January 29, 2008, 8A, https://www.newspapers.com/image/286529731/. Hillary Clinton and Bill Nelson campaign event photo, *MH*, January 30, 2008, 1A, https://www.newspapers.com/image/656427637/.

73 AP, "Nelson Sees Mail-In Vote as Solution," *FT*, March 9, 2008, 10B, https://www.newspapers.com/image/362988437/.

74 John Kennedy, "Nelson Floats Plan for Counting Half of Dem Delegates," *OS*, March 15, 2008, A1, A3, https://www.newspapers.com/image/269795130/.

75 Tamara Lytle, "The Florida Compromise," *OS*, June 1, 2008, A10, https://www.newspapers.com/image/268370527/.

76 John McCarthy, "Nelson Is an Optimist in Brevard Visit," *FT*, February 21, 2008, 1B, 5B, https://www.newspapers.com/image/362982302/.

77 Robert Block, "Nelson: '08 Race Is Key for Space," *FT*, April 29, 2008, B1, B5, https://www.newspapers.com/image/268565557/.

78 Susanne Cervenka, "Obama Promises to Strengthen NASA," *FT*, May 22, 2008, 1A, 7A, https://www.newspapers.com/image/362841674/.

79 Patrick Peterson, "Brevard Out of Stumping Loop," *FT*, May 26, 2008, 1A, 5A, https://www.newspapers.com/image/362848476/.

80 Link to Launch advertisement, *FT*, June 18, 2008, 5A, https://www.newspapers.com/image/362943554/.

81 Dale Ketcham emails to author, June 16–19, 2023.

82 Jessica Raynor, "KSC Worker Rally Draws Hopeful Few," *FT*, June 24, 2008, 3A, https://www.newspapers.com/image/363039920/.

83 Link to Launch, http://www.linktolaunch.org/, archived May 14, 2010, at the *WM*.

84 Matt Reed, "Spaceflight Means More Than Profits," *FT*, June 24, 2008, 1B, https://www.newspapers.com/image/363039982/.

85 John Weinberg, Federal Reserve Bank of Richmond, "The Great Recession and Its Aftermath," Federal Reserve History, November 22, 2013, https://www.federalreservehistory.org/essays/great-recession-and-its-aftermath.

86 "Preparing for the Workforce Transition at Kennedy Space Center" hearing, US Senate Committee on Commerce, Science, & Transportation, June 23, 2008. Lynda Weatherman's submitted testimony is at https://www.commerce.senate.gov/services/files/FE882D07-381A-4DCD-8524-073274A53DDD. Steve Kohler's submitted testimony is at https://www.commerce.senate.gov/services/files/63004F22-FCF5-4DDA-A791-A32EDD68EA7E.

87 "Monthly Unemployment Rate in Brevard County, Florida, 2008–2010," Federal Reserve Bank of St. Louis, https://alfred.stlouisfed.org/series?seid=FLBREV3URN.

88 "Fire in the Belly," *FT* editorial column, November 18, 2007, 12A, https://www.newspapers.com/image/362987728/.

89 "People to Watch," *FT* editorial column, December 23, 2007, 8A, https://www.newspapers.com/image/363020868/.

90 Todd Halvorson, "Cargo Ship May Launch from Va.," *FT*, February 20, 2008, 6A, https://www.newspapers.com/image/362968459/?.

91 Todd Halvorson, "Space Florida to Endure Cuts," *FT*, May 10, 2008, 1A, 8A, https://www.newspapers.com/image/363016414/.

92 "Local Projects," *FT*, June 12, 2008, 6B, https://www.newspapers.com/image/362930814/.

93 Todd Halvorson, "Fla. Plans Duty-Free Launches," *FT*, August 8, 2008, 1A, 7A, https://www.newspapers.com/image/360690339/.

94 "Orbital Boost," *FT* editorial column, August 9, 2008, 8A, https://www.newspapers.com/image/360691469/.

95 Travis Griggs, "Beyond Medicine," *PNJ*, December 5, 2008, 1A, 2A, https://www.newspapers.com/image/271553109/.

96 Aaron Deslatte and Robert Block, "Space-Tourism Deal Spurs State Investigation," *OS*, January 24, 2009, A1, A10, https://www.newspapers.com/image/268761339/.

97 Paul Flemming, "Crist Calls for Inquiry into Director of Space-Tourism Medical Program," *PNJ*, January 27, 2009, 1A, 5A, https://www.newspapers.com/image/249952722/.

98 Robert Block and Aaron Deslatte, "Official's $150,000 Job Likely Broke Law," *OS*, April 11, 2009, A1, A15, https://www.newspapers.com/image/269068421/. Bill Cotterell, "Andrews Official Resigns," *PNJ*, April 15, 2009, 1A, 6A, https://www.newspapers.com/image/267993264/. Carlton Proctor, "No Probe of Ex-Director," *PNJ*, April 18, 2009, 1C–2C, https://www.newspapers.com/image/268008190/. AP, "Panel Finds Some Merit in Complaint," *PNJ*, March 4, 2010, 2B, https://www.newspapers.com/image/268769368/. In re Brice Harris, Fla. Amended Recommended Order (May 3, 2011), Case No. 10-2798EC (on file with Fla. Commission on Ethics, Complaint Number 09-061), 11, 23, https://ethics.state.fl.us/Documents/Orders/2009/09061%20Amended%20RO.pdf?cp=2024630. Carlton Proctor, "Ex-Andrews Employee Cleared," *PNJ*, May 4, 2011, 6D, https://www.newspapers.com/image/270767568/. Aaron Deslatte, "Ethics Panel Clears Tourism Ex-Official," *OS*, June 23, 2011, B9, https://www.newspapers.com/image/268468689/.

99 Aaron Deslatte and Robert Block, "Florida's Aerospace Agency Flounders," *OS*, January 30, 2009, B1, B6, https://www.newspapers.com/image/268785737/.

100 Steve Kohler, "Building a Better Spaceport," *FT* guest editorial column, February 8, 2009, 19A, https://www.newspapers.com/image/362884914/.

101 Aaron Deslatte, "Space Florida Chief Fights Back," *OS*, February 5, 2009, B1, B5, https://www.newspapers.com/image/268606725/.

102 "Review of Space Florida," Research Memorandum, the Florida Legislature Office of Program Policy Analysis and Government Accountability, January 30, 2009, 1. A PDF copy was provided via email to this author by OPPAGA.

103 "Review of Space Florida," 7.

104 Robert Block and Aaron Deslatte, "Critics Blast Space Florida as $50M Waste," *OS*, February 15, 2009, A1, A7, https://www.newspapers.com/image/268701693/.

105 Steve Kohler, "Firm's Achievements Will Ensure State's Status as Industry Leader," *OS* guest editorial column, February 19, 2009, A11, https://www.newspapers.com/image/268719124/.

106 Aaron Deslatte and Robert Block, "$43M? Space Florida Told to Fly a Kite," *OS*, February 20, 2009, A1, A8, https://www.newspapers.com/image/268726840/.

107 Robert Block and Mark K. Matthews, "Space Florida Builds Launchpad to Nowhere," *OS*, April 28, 2009, A1, A11, https://www.newspapers.com/image/269118806/.

108 Robert Block and Aaron Deslatte, "Space Florida Finds Way to Fly Lobbying Deal 'Under Radar,'" *OS*, April 29, 2009, A1, A10, https://www.newspapers.com/image/269121082/.

109 Carol Cratty, "Former NASA Official Accused of Steering Funds to Consulting Client," CNN, March 6, 2009, http://www.cnn.com/2009/CRIME/03/06/nasa.fund.indictment/.

110 Robert Block, "Ex-Board Member Got No-Bid Contract," *OS*, May 2, 2009, A1, A6, https://www.newspapers.com/image/269282366/.

111 Robert Block and Aaron Deslatte, "Space Florida President Resigns," *OS*, May 8, 2009, A1, A18, https://www.newspapers.com/image/269291887/.

112 According to OpenSecrets.org, the final amounts requested by Nelson differ from what was initially reported in the media. Nelson requested $1.1 million to Space Florida to renovate Launch Complexes 36 and 46, another $400,000 to Space Florida for the Thermal Vacuum Chamber, $1 million to the Brevard Workforce Development Board for "a job training initiative," $974,000 to the Center for Commercial Space Transportation, and $100,000 to TRDA for a Space Alliance Technology Outreach Program. https://www.opensecrets.org/members-of-congress/earmarks?cid=N00009926&cycle=2010&fy=FY10&name=bill-nelson.

113 Aaron Deslatte and David Damron, "Nelson Reverses, Asks $14M for Commercial Launchpad," *OS*, May 9, 2009, A4, https://www.newspapers.com/image/269294706/.

114 Patrick Peterson and Jim Ash, "Space Fla. Names Chief," *FT*, May 19, 2009, 8C, 7C, https://www.newspapers.com/image/363022684/. Robert Block, "Panel's Choice for New Leader Is Criticized," *OS*, September 16, 2009, B6, B8, https://www.newspapers.com/image/269529687/. Robert Block, "Space Florida Keeps Interim President," *OS*, September 18, 2009, B5–B6, https://www.newspapers.com/image/269537607/. "Space Florida President and CEO Frank DiBello Announces Upcoming Retirement After

14 Years Leading the Organization," Space Florida press release, March 15, 2023, https://www.spaceflorida.gov/news/space-florida-president-and-ceo-frank-dibello-announces-upcoming-retirement-after-14-years-leading-the-organization/.

115 "2008 Electoral College Results," National Archives, https://www.archives.gov/electoral-college/2008.

116 November 4, 2008, General Election Results, Florida Department of State, https://results.elections.myflorida.com/Index.asp?ElectionDate=11/4/2008.

117 General Election, November 4, 2008, Brevard County Supervisor of Elections, https://www.votebrevard.gov/Previous-Elections/2008-General-Election-Results.

118 General Election, November 7, 2000, Brevard County Supervisor of Elections, https://www.votebrevard.gov/Previous-Elections/2000-General-Election-Results.

119 General Election, November 2, 2004, Brevard County Supervisor of Elections, https://www.votebrevard.gov/Previous-Elections/2004-General-Election-Results.

120 Brevard County Supervisor of Elections email to author, July 6, 2023.

121 "Statement by John McCain on the 50th Anniversary of NASA," John McCain's campaign website, July 29, 2008, http://www.johnmccain.com/Informing/News/PressReleases/Read.aspx?guid=0064ea81-2ddd-4e64-ae9c-2fc032fc77ae, archived July 30, 2008, at the *WM*.

122 "NASA's 50th Anniversary: Statement from Sen. Obama," Obama '08, July 29, 2008, https://barackobama.com/2008/07/29/nasas_50th_anniversary_stateme.php, archived July 31, 2008, at the *WM*.

123 Patrick Peterson, "Obama Is Setting Stage for Stop Here," *FT*, July 31, 2008, 1A, https://www.newspapers.com/image/360556133/. "Charting the Cosmos," *FT* editorial column, August 1, 2008, 10A, https://www.newspapers.com/image/360684414/.

124 The I-4 Corridor refers to Florida's Interstate 4 highway, which runs from Tampa on the Gulf Coast northeast through the state to Daytona Beach on the Atlantic Coast. Voters along the Corridor tend to vote Democratic more than the rest of the state. Brevard County is not considered part of the I-4 Corridor. MCI Maps has an excellent discussion of the Corridor and its voting data during twenty-first-century presidential elections at https://mcimaps.com/floridas-infamous-i-4-corridor-and-its-politics/.

125 An audio recording of Barack Obama's August 2, 2008, Titusville town hall is on the author's Space SPAN YouTube channel at https://www.youtube.com/watch?v=Cyg4mL_rcR8.

126 Matt Reed, "Who Broke $40 Million Promise?" *FT*, May 1, 2011, 1B, 4B, https://www.newspapers.com/image/360536823/.

127 Robert Block, "Obama: Let's Go to Moon—Maybe Mars," *OS*, August 17, 2008, A4, https://www.newspapers.com/image/269207904/ "Advancing the Frontiers of Space Exploration," Obama '08, http://www.barackobama.com/pdf/policy/Space_Fact_Sheet_FINAL.pdf, archived October 28, 2008, at the *WM*. This is the first date for which the *WM* has the revised space policy.

128 "America's Space Program," John McCain's campaign website, http://www.johnmccain.com/Informing/Issues/7366faf9-d504-4abc-a889-9c08d601d8ee.htm, archived August 9, 2008, at the *WM*. This is the first date for which the *WM* has the revised space policy.

129 Jim Stratton and Robert Block, "McCain: No to Storm Fund, Yes to NASA Money," *OS*, August 19, 2008, A1, A10, https://www.newspapers.com/image/269210732/.

130 Jim Stratton, "McCain, Obama Spar on Just How Bad Economy Is," *OS*, September 16, 2008, A1, A10, https://www.newspapers.com/image/268466285/.

131 Robert Block, "McCain: Count Me in NASA's Corner," *OS*, October 18, 2008, A1, A4, https://www.newspapers.com/image/268734650/.

132 Robert Block, "Biden: NASA Will Soar Under Obama," *OS*, October 29, 2008, A3, https://www.newspapers.com/image/268791000/.

133 "Urgent Issues," GAO, November 6, 2008, https://www.gao.gov/press-release/gao-lists-top-urgent-issues-next-president-and-congress-unveils-new-transition-web-site. "Retirement of the Space Shuttle," GAO, http://www.gao.gov/transition_2009/urgent/space-shuttle.php, archived November 12, 2008, at the *WM*.

134 The $230 billion estimate was reported in 2006 by the GAO, based on NASA's January 2006 Exploration Systems Architecture Study. The cost estimate for ESAS through 2025 was $230 billion. Allen Li, "Exploration Cost and Schedule," GAO-06-817, July 17, 2006, https://www.gao.gov/assets/gao-06-817r.pdf.

135 "Shutting Down the Shuttle," *Wayne Hale's Blog*, NASA, August 28, 2008, https://blogs.nasa.gov/waynehalesblog/2008/08/28/post_1219932905350/.

136 Bill Gerstenmaier and Richard Gilbrech, "Workforce Transition Strategy Initial Report," NASA, March 31, 2008, 22, https://www.nasa.gov/wp-content/uploads/2015/01/220260main_workforce_transition_strategy_briefing.pdf. Patrick Peterson, "Job Loss Impact Will Reach Beyond KSC," *FT*, April 2, 2008, 1A, 6A, https://www.newspapers.com/image/363024948/.

137 Jeff Foust, "Space and the Financial Crisis," *The Space Review*, October 20, 2008, https://www.thespacereview.com/article/1236/1. "Supporting Commercial Space Development, Part 1: Support Alternatives Versus Investor Risk Perceptions & Tolerances," New Earth LLC, November 2010, 19, https://www.nasa.gov/sites/default/files/files/SupportingCommercialSpaceDevelopmentPart1.pdf, archived June 16, 2016, at the *WM*.

Chapter 8. A JFK Moment

1 Garver quoted Cordiner on October 20, 2011, in a speech at the International Symposium for Personal and Commercial Spaceflight; Alan Boyle, "NASA: Pay the Americans Now . . . or Pay the Russians Later," NBC News, October 20, 2011, https://www.nbcnews.com/science/cosmic-log/nasa-pay-americans-now-or-pay-russians-later-flna6c10402715. Garver quotes Cordiner in her book, *Escaping Gravity*, 57.

2 Vance, *When the Heavens Went on Sale*, 44. Vance writes that Pete Worden's father recalls the army air corps using the term during World War II.

3 Worden, S. P., "On Self-Licking Ice Cream Cones," *Seventh Cambridge Workshop on Cool Stars, Stellar Systems, and the Sun, ASP Conference Series*, Volume 26 (Astronomical Society of the Pacific, 1992), 600–601, https://adsabs.harvard.edu/full/1992ASPC . . . 26..599W.

4 "Obama Adds Three More to NASA Transition Team," Space.com, November 25, 2008, https://www.space.com/6161-obama-adds-nasa-transition-team.html.

5 "Former Administrator Michael Griffin," NASA biography, February 23, 2024, https://www.nasa.gov/about/highlights/griffin_bio.html.
6 Jason Davis, "'Apollo on Steroids': The Rise and Fall of NASA's Constellation Moon Program," The Planetary Society, August 1, 2016, https://www.planetary.org/articles/20160801-horizon-goal-part-2.
7 Rick Tumlinson, "Mike Griffin's Constellation Zombie," *SN*, October 27, 2010, https://spacenews.com/mike-griffins-constellation-zombie/.
8 Mark Matthews, "Nelson Tells Obama to Keep Griffin—for Now," *OS*, *The Write Stuff* (blog), November 7, 2008, http://blogs.orlandosentinel.com/news_space_thewritestuff/2008/11/nelson-tells-ob.html, archived November 10, 2008, at the *WM*.
9 Robert Block and Mark K. Matthews, "NASA Chief Griffin Wants to Keep His Job," *OS*, November 15, 2008, A10, https://www.newspapers.com/image/268462313/.
10 Garver, *Escaping Gravity*, 81.
11 Robert Block and Mark K. Matthews, "NASA Chief Griffin Bucks Obama's Transition Team," *OS*, December 11, 2008, A1, A8, https://www.newspapers.com/image/269507432/.
12 Robert Block, "A Message from the NASA Administrator," *OS*, *The Write Stuff* (blog), December 11, 2008, http://blogs.orlandosentinel.com/news_space_thewritestuff/2008/12/nasas-griffin-a.html, archived December 14, 2008, at the *WM*.
13 Scott Horowitz, "Keep Mike Griffin as the NASA Administrator," iPetitions, https://www.ipetitions.com/petition/KeepMike/.
14 Keith Cowing, "Vote to Keep Mike," NASA Watch, December 24, 2008, https://nasawatch.com/transition/vote-to-keep-mike/.
15 Jeffrey Kluger, "Does Obama Want to Ground NASA's Next Moon Mission?" *Time*, December 11, 2008, https://content.time.com/time/nation/article/0,8599,1866045,00.html.
16 Michael D. Griffin, NASA at 50 Oral History Project Edited Oral History Transcript, interview by Rebecca Wright, December 10, 2007, https://historycollection.jsc.nasa.gov/JSCHistoryPortal/history/oral_histories/NASA_HQ/NAF/GriffinMD/GriffinMD_9-10-07.htm.
17 James E. Webb, *Space Age Management: The Large-scale Approach* (McGraw-Hill, 1969), 131–132. The book is available on the IA at https://archive.org/details/spaceagemanageme0000webb.
18 Block and Matthews, "NASA Chief Griffin . . ."
19 H. George Frederickson, Henry J. Anna, and Barry Kelmachter, "Interview with Mr. James E. Webb," Syracuse/NASA Program Project Manager Research Group, May 15, 1969, 32. A PDF of the interview was provided by the NASA Program History Office, September 25, 2023.
20 Becky Iannotta, "Obama Asks Retired Air Force General to Run NASA," Space.com, January 14, 2009, https://www.space.com/6309-obama-asks-retired-air-force-general-run-nasa.html. George Cahlink, "CongressNow: Obama Enlists the Aid of 60 Generals and Admirals," *Roll Call* website, August 26, 2008, https://rollcall.com/2008/08/26/congressnow-obama-enlists-the-aid-of-60-generals-and-admirals/.

21 Garver, *Escaping Gravity*, 85.

22 Becky Iannotta, "Key U.S. Senator Cautions Obama on NASA Pick," Space.com, January 14, 2009, https://www.space.com/6313-key-senator-cautions-obama-nasa-pick.html. Todd Halvorson, "Nelson: Candidate Lacks NASA History," *FT*, January 15, 2009, 1A, 9A, https://www.newspapers.com/image/362903352/.

23 Robert Block, "NASA: Mike Griffin Out, Charlie Bolden In?" *OS, The Write Stuff* (blog), January 6, 2009, http://blogs.orlandosentinel.com/news_space_thewritestuff/2009/01/nasa-mike-griff.html, archived January 16, 2009, at the *WM*.

24 Mark K. Matthews, "Obama: NASA Is Adrift," *OS*, March 12, 2009, A1, A6, https://www.newspapers.com/image/268787092/. Todd Halvorson, "President Soon Will Hire Boss of NASA," *FT*, March 13, 2009, 1A, 7A, https://www.newspapers.com/image/362875249/.

25 National Aeronautics and Space Administration Authorization Act of 2008, Pub. L. No. 110-702, 122 Stat. 4779 (2008), https://www.congress.gov/bill/110th-congress/house-bill/6063. Section 101 (5) provided the funding. Section 611 (c) mandated the flight.

26 "Shuttle Program's Last External Tank Transported to VAB," NASA press release, https://www.nasa.gov/sites/default/files/atoms/files/shuttle_programs_last_external_tank_transported_to_vab.pdf.

27 Curtis Krueger, "Shuttle Shines in the Sunset," *TBT*, March 16, 2009, 1B, 7B, https://www.newspapers.com/image/331304077/.

28 Garver, *Escaping Gravity*, 85.

29 Lee Higgins, "Bolden on Course for Top NASA Job," *The State*, May 17, 2009, A1, A12, https://www.newspapers.com/image/754936103/.

30 "President Obama Announces More Key Administration Posts," Obama White House Archives, May 23, 2009, https://obamawhitehouse.archives.gov/the-press-office/president-announces-more-key-administration-posts.

31 Jay Barbree, "Obama Picks Ex-Astronaut to Lead NASA," NBC News, May 23, 2009, https://www.nbcnews.com/id/wbna30896443.

32 Mark K. Matthews and Robert Block, "Obama Picks Ex-Astronaut to Run NASA," *OS*, May 23, 2009, A1, A12, https://www.newspapers.com/image/269319464/.

33 Garver, *Escaping Gravity*, 91.

34 W. Henry Lambright, "Reflections on Leadership and Its Politics: Charles Bolden, NASA Administrator, 2009–17," *Public Administration Review* 77, no. 4 (July/August 2017): 619, https://www.jstor.org/stable/26648797.

35 Lambright, Bolden interview, 617, 619.

36 "U.S. Announces Review of Human Space Flight Plans," Obama White House Archives, May 7, 2009, https://obamawhitehouse.archives.gov/the-press-office/2015/11/16/us-announces-review-human-space-flight-plans.

37 Norman R. Augustine Biography, Bipartisan Policy Center, https://bipartisanpolicy.org/person/norman-r-augustine/.

38 "Meet the Committee," Review of U.S. Human Space Flight Plans Committee, https://www.nasa.gov/offices/hsf/members/index.html, archived June 14, 2009, at the *WM*.

39 "Upcoming Meeting Schedule," Augustine Committee, https://www.nasa.gov/offices/hsf/meetings/index.html, archived August 1, 2009, at the *WM*. Videos of the

Committee hearings are archived on this author's YouTube channel at https://www.youtube.com/@SpaceSPAN/videos.

40 Augustine Committee Cocoa Beach hearing video, July 30, 2009, https://www.youtube.com/watch?v=tE48GI-yJ3w.

41 Patrick Peterson, "Bolden to KSC: Stay the Course," *FT*, July 31, 2023, 1A, 3A, https://www.newspapers.com/image/362932393/.

42 Augustine Committee Cocoa Beach hearing video.

43 Review of US Human Spaceflight Plans Committee, *Seeking a Human Spaceflight Program Worthy of a Great Nation* (Augustine Committee Report), October 2009, 9, https://www.nasa.gov/wp-content/uploads/2015/01/617036main_396093main_hsf_cmte_finalreport.pdf.

44 Augustine Committee Report, 9.

45 Augustine Committee Report, 10–12.

46 NASA, *The Vision for Space Exploration*, February 2004, 19, https://www.nasa.gov/pdf/55583main_vision_space_exploration2.pdf.

47 NASA Authorization Act of 2008, Sec. 601(a).

48 Augustine Committee Report, 13.

49 Augustine Committee Report, 89.

50 Augustine Committee Report, 90.

51 Augustine Committee Report, 96.

52 *Options from the Review of US Human Space Flight Plans Committee, Before the Senate Science and Space Subcommittee*, 111th Cong. (2009), https://science.house.gov/2009/9//full-committee-hearing-options-and-issues-nasa-s-human-space-flight-program.

53 James E. Webb, "The Economic Impact of the Space Program," *Business Horizons*, Autumn 1963, 5 et seq., https://www.sciencedirect.com/science/article/pii/0007681363900454.

54 Webb, *Space Age Management*, 6–7.

55 Bizony, *The Man Who Ran the Moon*, 171.

56 Bizony, *The Man Who Ran the Moon*, 65.

57 James E. Webb, Lyndon B. Johnson Library Oral Histories, interview by T. H. Baker, April 29, 1969, 14, https://www.discoverlbj.org/item/oh-webbj-19690429-1-74-266.

58 "Labor Board Slates Machinists' Hearing at Cape Canaveral," *The Cocoa Tribune*, October 11, 1955, 2, https://www.newspapers.com/image/779679934/. "350 Pan American Workers Signify I.A.M. Preference," *The Cocoa Tribune*, February 5, 1957, 2, https://www.newspapers.com/image/779637022/. John Morton, "New Bid Is Slated by Union," *MH*, March 17, 1957, 2B, https://www.newspapers.com/image/618880229/. John Morton, "Machinists Withdraw from Fight," *MH*, April 8, 1957, 1B, https://www.newspapers.com/image/618627434/.

59 Henry Balch, "2,000 Honor Teamster Pickets at AFMTC; Construction Halts," *OS*, May 22, 1957, 1A, 4A, https://www.newspapers.com/image/222051280/. "1,100 Patrick Workers Return," *OS*, June 20, 1957, 3A, https://www.newspapers.com/image/222122359/.

60 Karl Hunziker, "Walkout Continues; No Pickets," *Orlando Evening Star*, May 20, 1958, 3, https://www.newspapers.com/image/290369081/. "Cape Strikers Vote to Work," *OS*, May 22, 1958, 1B, https://www.newspapers.com/image/222193339/.

61 “Injunction Blocks TWU Strike Against Pan Am,” *OS*, July 16, 1958, 1A, https://www.newspapers.com/image/222164679/. “TWU Strike at Cape Ends,” *OS*, July 19, 1958, 1A, https://www.newspapers.com/image/222166429/.

62 *IBEW Local 756 Commemorative History Book* (International Brotherhood of Electrical Workers Local No. 756, 2021). A copy was provided to this author by Matt Nelson, IBEW Local No. 756 business manager/financial secretary.

63 “TWU Local 525 Honored by AF for No-Strike,” *The Cocoa Tribune*, October 23, 1964, 1–2, https://www.newspapers.com/image/779640045/.

64 “Plumbers Withdraw Pickets at Cape,” *OS*, March 31, 1965, 1A, 18A, https://www.newspapers.com/image/223879761/.

65 UPI, “New Strike Hits Cape; Many Idle,” *MH*, September 17, 1965, 1A, https://www.newspapers.com/image/621894944/.

66 Sanders LaMont, “Machinists’ Strike Feared as Contract Deadline Nears,” *MH*, September 15, 1965, 1B, https://www.newspapers.com/image/621893978/.

67 George V. Hanna III, *Chronology of Work Stoppages and Related Events KSC/NASA and AFETR Through July 1965*, NASA John F. Kennedy Space Center, July 1967, https://ntrs.nasa.gov/api/citations/19670030427/downloads/19670030427.pdf.

68 William Z. Schenck, *Chronology of Work Stoppages and Related Events KSC/NASA and AFETR July 1965–July 1967*, NASA John F. Kennedy Space Center, October 1965, https://ntrs.nasa.gov/api/citations/19670031166/downloads/19670031166.pdf. Data charts are in Appendix A, although specific numbers are not provided.

69 John Quigley, “Machinist Union Strikes at Cape Kennedy,” *MH*, March 16, 1968, 2B, https://www.newspapers.com/image/621454354/. “Machinists Vote to Accept Offer,” *MH*, March 21, 1968, 1C, https://www.newspapers.com/image/621730451/.

70 “Picket Lines Go Up at Kennedy Space Center,” International Association of Machinists and Aerospace Workers press release, June 14, 2007, https://www.goiam.org/press-releases/picket-lines-go-up-at-kennedy-space-center/. Scott Blake, “Strikers Get a Hot Welcome from Weather on Picket Lines,” *FT*, June 15, 2007, 1C–2C, https://www.newspapers.com/image/362969120/. Scott Blake, “16% of Union Members Cross Picket Lines as USA Strike Continues,” *FT*, July 10, 2007, 1A, 5A, https://www.newspapers.com/image/362981427/. “Florida Machinists Ratify Space Center Contract,” IAMAW press release, November 6, 2007, https://www.goiam.org/news/imail-for-tuesday-november-06-2007/.

71 Steve Williams and Matt Nelson recorded interview with author at IBEW Local 756 union hall, Port Orange, Florida, September 21, 2023. Steve Williams email to author, September 24, 2023.

72 AP, “AFL-CIO Endorses Obama for President,” NBC News, June 26, 2008, https://www.nbcnews.com/id/wbna25392679.

73 Scott Carter, “The Flashback Files: Gators Nose Guard Robin Fisher,” Florida Gators, June 28, 2013, https://floridagators.com/news/2013/6/28/25920.

74 Robin Fisher State Farm Insurance Agent biography, https://www.statefarm.com/agent/us/fl/titusville/robin-fisher-j5j0l1ys000. Eleska Aubespin, “Fisher Lands 2nd Term at YMCA Helm,” *FT*, May 23, 2005, 1E, https://www.newspapers.com/image/178513345/. “State Farm Agent Sponsors Sportsmanship Program,” *FT*, October 1, 2006, 2C, https://www.newspapers.com/image/179157890/. Peter Kerasotis,

"Now It Feels Like We're Living on Space Coast," *FT*, July 26, 2005, 8A, https://www.newspapers.com/image/178522550/.

75 Robin Fisher telephone interview with author, July 18, 2023.

76 Primary Election, August 26, 2008, Democratic candidates for Board of County Commissioners, Brevard County Supervisor of Elections, https://www.votebrevard.gov/Previous-Elections/2008-Primary-Election-Results.

77 Brevard County Supervisor of Elections Office email to the author, July 19, 2023.

78 General Election, November 4, 2008, Brevard County Supervisor of Elections, https://www.votebrevard.gov/Previous-Elections/2008-General-Election-Results.

79 "Breaking Barriers," *FT* editorial, November 6, 2008, https://www.newspapers.com/image/360495961/. Fisher confirmed in his phone interview that he was the county's first African American commissioner.

80 Todd Halvorson, "Obama Reviews Shuttle's Finale," *FT*, December 14, 2008, 1A, 3A, https://www.newspapers.com/image/360491236/.

81 Rick Neale, "Effort Aims to 'Save Space,'" *FT*, September 28, 2009, 1A, 3A, https://www.newspapers.com/image/360413237/. SaveSpace.US, http://www.savespace.us/, archived October 1, 2009, at the *WM*.

82 John A. Torres, "Fisher Extends Letter Deadline," *FT*, October 31, 2009, 1B, 4B, https://www.newspapers.com/image/360415973/.

83 NASA FY 2011 Fact Sheet, OMB, February 1, 2010, https://web.archive.org/web/20100204060427/http://www.whitehouse.gov/omb/factsheet_department_nasa/, archived February 4, 2010, at the *WM*. "Fiscal Year 2011 Budget Estimates," NASA, February 1, 2010, https://www.nasa.gov/wp-content/uploads/2023/08/fiscal-year-2011-budget-overview.pdf.

84 *FT*, February 2, 2010, 1A, https://www.newspapers.com/image/360408262/.

85 "The Big Unknowns," *FT* editorial column, February 3, 2010, 10A, https://www.newspapers.com/image/360408555/.

86 Kris Hundley and Alex Leary, *St. Petersburg Times*, "Florida Feels Heat of NASA Cutback," *MT*, February 2, 2010, 1A–2A, https://www.newspapers.com/image/659363060/.

87 *The President's Fiscal Year 2011 Budget Proposal, Before the Senate Committee on the Budget*, 111th Cong. (2010), https://www.budget.senate.gov/hearings/the-presidents-fiscal-year-2011-budget-proposal.

88 *Key Issues and Challenges Facing NASA: Views of the Agency's Watchdogs, Before the House Committee on Science and Technology*, 111th Cong. (2010), http://science.house.gov/publications/hearings_markups_details.aspx?NewsID=2720, archived February 9, 2010, at the *WM*.

89 Consolidated Appropriations Act, 2008, Pub. L. No. 110-161, 121 Stat. 1919 (2007), https://www.congress.gov/110/plaws/publ161/PLAW-110publ161.pdf.

90 "NASA Space Shuttle Workforce Transition Strategy," July 2009 Update, NASA, 6, https://www.nasa.gov/pdf/433535main_Workforce%20Transition%20Strategy%203rd%20Edition.pdf, archived March 17, 2010, at the *WM*.

91 "Workforce Transition Office Opens," NASA, March 9, 2010, https://www.nasa.gov/centers/kennedy/news/workforce_transition.html, archived March 15, 2010, at the *WM*.

92 Consolidated Appropriations Act, 2010, Pub. L. No. 111-117, 123 Stat. 3143 (2009), https://www.congress.gov/111/plaws/publ117/PLAW-111publ117.pdf.

93 Robert B. Aderholt et al. letter to NASA Administrator Charles Bolden, February 12, 2010, http://www.posey.house.gov/UploadedFiles/LetterToBolden-CancellingConstellation-Feb15-2010.pdf, archived March 3, 2010, at the *WM*.

94 Robert Block, "At Crist's Space Summit, Crowd Faces New Reality," *OS*, February 19, 2010, B1, B9, https://www.newspapers.com/image/269261163/. Patrick Peterson, "'Tough Time for NASA,' Florida," *FT*, February 19, 2010, 1A, 3A, https://www.newspapers.com/image/360859238/. "Kosmas to Attend Florida Statewide Space Industry Summit," press release, February 17, 2010, archived at SpaceRef.com, https://spaceref.com/press-release/kosmas-to-attend-florida-statewide-space-industry-summit/.

95 bh Public Relations, "Governor Charlie Crist Statement On President Obama's NASA Policy," February 17, 2010, https://www.prlog.org/10537166-governor-charlie-crist-statement-on-president-obamas-nasa-policy.html.

96 "Straight-Talk Express," *FT* opinion column, February 21, 2010, 20A, https://www.newspapers.com/image/360409565/.

97 Fernando Rendon, "Rally Saturday for Space," *FT* guest column, February 25, 2010, 13A, https://www.newspapers.com/image/360410545/.

98 "Fact Sheet #66: The Davis-Bacon and related Acts (DBRA)," US Department of Labor Wage and Hour Division, https://www.dol.gov/agencies/whd/fact-sheets/66-dbra.

99 "McNamara-O'Hara Service Contract Act (SCA)," US Department of Labor Wage and Hour Division, https://www.dol.gov/agencies/whd/government-contracts/service-contracts. A list of Service Contract Act agreements at the Space Shuttle program's end can be found in a NASA document, "Test and Operations Support Contract Attachment J-08, Register of Service Contract Wage Determination," June 13, 2012, 12, https://www.nasa.gov/centers/kennedy/pdf/753746main_TOSC-attachment-J-08.pdf, archived February 17, 2017, at the *WM*.

100 "Space Act Agreement Between NASA and SpaceX for Commercial Orbital Demonstration Services Demonstration," signed August 18, 2006, https://www.nasa.gov/centers/johnson/pdf/189228main_setc_nnj06ta26a.pdf.

101 Joshua Finch, Office of Communications, NASA Headquarters, email to author, October 26, 2023. The email stated that Davis-Bacon is now known as the Wage Rate Requirements Statute.

102 Fernando Rendon phone interview with author, September 14, 2023.

103 Patrick Peterson, "'Worth Fighting For,'" *FT*, February 28, 2010, 1A, 11A, https://www.newspapers.com/image/360411292/.

104 The quotes from Bob Martinez and Richard Trumka are from a rally video posted March 4, 2010, on the IAMAW Machinists Union YouTube channel at https://www.youtube.com/watch?v=DSklyL2E3Q8.

105 Russian cosmonaut Sergei Krikalev flew on shuttle mission STS-60 in February 1994. American astronaut Norm Thagard launched on a Soyuz from Kazakhstan in March 1995, spent three months on the Russian space station Mir, and returned to KSC on the shuttle in July 1995.

106 "Our Dedication Mission," Vision4Space.com, https://www.goiam.org/publications/campaigns/vision4space/mission.htm.
107 White House Office of the Press Secretary, "President Obama to Host Space Conference in Florida in April," Obama White House Archives, March 7, 2010, https://obamawhitehouse.archives.gov/the-press-office/president-obama-host-space-conference-florida-april.
108 Video of the event is on the Space SPAN YouTube channel at https://www.youtube.com/watch?v=ofDqoF3Q8cA. A copy was provided by Eastern Florida State College.
109 "Barack Obama's Plan for Lifetime Success Through Education," Obama '08, http://www.barackobama.com/issues/pdf/PreK-12EducationFactSheet.pdf, archived on July 31, 2008, and August 7, 2008, at the *WM*.
110 Buzz Aldrin, "President Obama's JFK Moment," *Huffington Post*, February 3, 2010, http://www.huffingtonpost.com/buzz-aldrin/president-obamas-jfk-mome_b_448667.html, archived February 7, 2010, at the *WM*.
111 "John F. Kennedy Address at Rice University on the Space Effort," Rice University, September 12, 1962, https://www.rice.edu/kennedy. The May 25, 1961, Congress speech, "On Urgent National Needs," and the September 12, 1961, Rice University speech are often conflated.
112 Samuel Stebbins, "How Many People Were Born the Year You Were Born?" *USA Today*, June 12, 2020, https://www.usatoday.com/story/money/2020/06/12/how-many-people-were-born-the-year-you-were-born/111928356/. The number was determined by summing the number of births from 1970 through 2010. The total was 156,082,506.
113 Jeff Foust, "A Skeptic's Guide to Space Exploration," *The Space Review*, June 30, 2008, https://www.thespacereview.com/article/1160/1.
114 Syracuse University interview with Webb, 24–25.
115 Syracuse University interview with Webb, 21–24.
116 *Challenges and Opportunities in the NASA FY 2011 Budget Proposal, Before the Senate Science and Space Subcommittee*, 111th Cong. (2010), https://www.commerce.senate.gov/2010/2/challenges-and-opportunities-in-the-nasa-fy-2011-budget-proposal. A video of the hearing is available on the web page.
117 "An Orbital Grilling," *FT* editorial column, February 26, 2010, 12A, https://www.newspapers.com/image/360410741/. "Looking for Direction," *OS* editorial column, March 2, 2010, A10, https://www.newspapers.com/image/269216850/.
118 Andy Pasztor, "NASA Chief Bolden Seeks 'Plan B' for the Space Agency," *WSJ*, March 4, 2010, https://www.wsj.com/articles/SB10001424052748704541304575100000820591666., archived June 12, 2016, at the *WM*.
119 Kenneth Chang, "NASA Chief Denies Talk of Averting Obama's Plan," *NYT*, March 4, 2010, https://www.nytimes.com/2010/03/05/science/space/05nasa.html.
120 Michael L. Coats, Johnson Space Center Oral History Project, interview by Jennifer Ross-Nazzal, August 5, 2015, https://historycollection.jsc.nasa.gov/JSCHistoryPortal/history/oral_histories/CoatsML/CoatsML_8-5-15.pdf.
121 Coats interview, 28–29.

122 Mark K. Matthews, "Crist, Lawmakers 'Deeply Concerned' by NASA Job Losses," *OS*, March 17, 2010, A9, https://www.newspapers.com/image/269242347. Bart Jansen, "Nelson Sees Gains After Obama Talk," *FT*, March 17, 2010, 1A, 3A, https://www.newspapers.com/image/360412969/.

123 *Assessing Commercial Space Capabilities, Before the Senate Science and Space Subcommittee*, 111th Cong. (2010), https://www.commerce.senate.gov/2010/3/assessing-commercial-space-capabilities. A video of the hearing is available on the web page.

124 Todd Halvorson, "Nelson: Rocket to 'Cushion the Blow,'" *FT*, March 20, 2010, 1A, 3A, https://www.newspapers.com/image/360413516/.

125 Robin Fisher recalled in his July 2023 interview that Space Florida President Frank DiBello was one of the delegates. Larry Linkous recalled that Lynda Weatherman from the Economic Development Commission was also one of the delegates.

126 Dave Berman and James Dean, "Lobbyists Fighting for Space Funding," *FT*, March 21, 2010, 1B, 6B, https://www.newspapers.com/image/360413999/.

127 Dave Berman, "3 Astronauts, Lt. Governor to Address 'Save Space' Rally," *FT*, March 27, 2010, 1B, https://www.newspapers.com/image/360415686/.

128 SaveSpace.US, http://www.savespace.us/, archived March 21, 2010, at the *WM*.

129 Excerpts from the rally as of August 21, 2024, are on the SaveSpaceRally YouTube channel at https://www.youtube.com/@SaveSpaceRally/videos. The videos have been compiled into one video on the Space SPAN YouTube channel at https://www.youtube.com/watch?v=iaT-PYO8agE.

130 Robert Block, "Lt. Gov Wants Obama to Debate; Space Summit Venue Hunt Is On," *OS*, *The Write Stuff* (blog), March 24, 2010, http://blogs.orlandosentinel.com/news_space_thewritestuff/2010/03/lt-gov-wants-obama-to-debate-space-summit-venue-hunt-is-on.html, archived March 26, 2010, at the *WM*.

131 Todd Halvorson, "Orion Takes Its Place at KSC," *FT*, January 31, 2007, 3B, https://www.newspapers.com/image/363036052/.

132 Garver, *Escaping Gravity*, 108–110.

133 Lambright's Charlie Bolden interview, 618.

134 "A Bold Approach for Space Exploration and Discovery: Fact Sheet on the President's April 15th Address in Florida," Obama White House Archives, https://obamawhitehouse.archives.gov/sites/default/files/microsites/ostp/ostp-space-conf-factsheet.pdf.

135 "Direct Employment Resulting from Proposed Commercial Cargo and Crew Expenditures in the President's FY2011 Budget Request," The Tauri Group, April 9, 2010. The Tauri Group was purchased in 2019 by the Logistics Management Institute (LMI). Their analysis appears to no longer be online, but is in this author's archives. Alexandra Johnson, "11,800 Direct Jobs to Result from NASA's $6.1 Billion Commercial Spaceflight Investment, Independent Analysis Shows," Commercial Spaceflight Federation press release, April 13, 2010, http://www.commercialspaceflight.org/2010/04/11800-direct-jobs-to-result-from-nasas-6-1-billion-commercial-spaceflight-investment-independent-analysis-shows/, archived September 23, 2020, at the *WM*. Kenneth Chang, "President to Outline His Vision for NASA," *NYT*, April 13, 2010, https://www.nytimes.com/2010/04/14/science/space/14nasa.html.

136 Neil Armstrong, James Lovell, and Eugene Cernan letter obtained by Jay Barbree of NBC News, published online April 13, 2010, https://www.nbcnews.com/id/wbna36470363.

137 Mark Matthews, "Griffin, NASA Luminaries Urge Obama to Change Space Policy," *OS*, *The Write Stuff* (blog), April 13, 2010, http://blogs.orlandosentinel.com/news_space_thewritestuff/2010/04/griffin-nasa-luminaries-urge-obama-to-change-space-policy.html, archived April 17, 2010, at the *WM*.

138 "Protesters to Obama: Save Space Program Jobs," WESH 2 News YouTube channel, April 15, 2010, https://www.youtube.com/watch?v=E8n-lBRTaK0.

139 Dave Berman and Suzanne Cervenka, "Protestors' Flags, Signs Make Clear Messages," *FT*, April 16, 2010, 8A, https://www.newspapers.com/image/360420550/. Dave Berman, "But Downtown Appears Devoid of Hype Surrounding Obama's Visit," *FT*, April 16, 2010, 8A. See also the photo on this page. The Tea Party was a populist conservative movement of the early 2010s generally aligned with the Republican Party. Various newspaper reports and search engine results from this time list Space Coast Patriots as a Tea Party group.

140 Todd Halvorson, "President Makes Case; Promises Trip to Asteroid," *FT*, April 16, 2010, 1A, 3A, https://www.newspapers.com/image/360420506/.

141 "President Obama Tours SpaceX with Elon Musk [FOIA #22-16594-F]," Barack Obama Presidential Library YouTube Channel, posted June 7, 2023, https://www.youtube.com/watch?v=CKY4DwTDxkc. The FOIA request was by Forbes senior contributor Matt Novak, "Obama Library Releases Photos of Elon Musk with Former President After FOIA Request," Forbes, July 14, 2023, https://www.forbes.com/sites/mattnovak/2023/07/14/obama-library-releases-photos-of-elon-musk-with-former-president-after-foia-request/.

142 Halvorson, "President Makes Case . . ."

143 Malcolm Kirschenbaum phone interview with author, November 13, 2023.

144 A transcript of the speech is on the National Archives website at https://obamawhitehouse.archives.gov/the-press-office/remarks-president-space-exploration-21st-century. A video of the speech is on the NASA YouTube channel at https://www.youtube.com/watch?v=3rNn_cUrlmE.

Chapter 9. The Grand Compromise

1 Al Neuharth, "A 'Devastating' Plan: Obama Doesn't Get It; Space Is Last Frontier," *FT* opinion column, April 16, 2010, 17A, https://www.newspapers.com/image/360420595/.

2 John Seigenthaler, "Why They Mattered: Al Neuharth," *Politico* website, December 22, 2013, https://www.politico.com/magazine/story/2013/12/al-neuharth-obituary-101428/.

3 "Reaction to the Death of *USA Today* Founder Al Neuharth," *USA Today*, April 20, 2013, https://www.usatoday.com/story/news/nation/2013/04/20/al-neuharth-dies-reaction/2099043/.

4 Elon Musk, "Obama Made Right Choice," *FT* guest editorial column, April 16, 2010, 6A, https://www.newspapers.com/image/360420537/.

5 Eliot Kleinberg, "Obama Vows Commitment to NASA, but Space Workers Remain Skeptical," *PBP*, April 16, 2010, 1A, 17A, https://www.newspapers.com/image/206767433/.
6 "Independent Report Slams Space Privatization," International Association of Machinists and Aerospace Workers, July 29, 2010, https://www.goiam.org/news/independent-report-slams-space-privatization/.
7 "Nelson—Obama's KSC Trip, Speech Show President Listening and Backs Robust Space Program," Senator Bill Nelson, April 15, 2010, http://billnelson.senate.gov/news/details.cfm?id=323888&, archived April 21, 2010, at the *WM*.
8 Human Space Flight Capability Assurance and Enhancement Act of 2010, S.3068, 111th Cong. (2010), https://www.congress.gov/bill/111th-congress/senate-bill/3068.
9 Memo from Paul E. Damphousse to Senator Nelson, "Meeting with Sen. Hutchison," March 22, 2010, Special and Area Studies Collections, George A. Smathers Libraries, University of Florida, Gainesville, Florida.
10 Letter from Senator Bill Nelson et al. to President Barack Obama, March 4, 2010, Smathers Libraries.
11 *The Future of US Human Space Flight Before the US Senate Committee on Commerce, Science, and Transportation*, 111th Cong. (2010), http://commerce.senate.gov/public/index.cfm?p=Hearings&ContentRecord_id=54f5c39e-f62c-487f-b9ed-fd4be38d096f, archived June 9, 2010, at the *WM*. The hearing video is on the Space SPAN YouTube channel at https://www.youtube.com/watch?v=KCTj6rW-na4.
12 "Written Testimony of Neil A. Armstrong Before the Committee of Commerce, Science and Transportation, United States Senate, May 12, 2010," 2, http://commerce.senate.gov/public/?a=Files.Serve&File_id=9ea75dad-e89d-4ae6-b20a-805ad602dda2, archived June 10, 2010, at the *WM*.
13 "Written Testimony of Captain Eugene A. Cernan, USN (ret.), Commander, Apollo 17, Astronaut (ret.) Before the Committee of Commerce, Science and Transportation, United States Senate, May 12, 2010," 5, http://commerce.senate.gov/public/?a=Files.Serve&File_id=78b85667-581a-4082-814f-cd91c39f09dd, archived June 10, 2010, at the *WM*. Cernan's claim that the proposal "flows against the will of the majority of Americans" is not supported by polls at the time. A January 15, 2010, Rasmussen Reports poll found that 50 percent of respondents thought "the United States should cut back on space exploration given the current state of the economy," up from 44 percent in July 2009. Only 31 percent in the January 2010 poll disagreed with cutting space exploration. Rasmussen Reports, "50% Favor Cutting Back on Space Exploration," January 15, 2010, http://www.rasmussenreports.com/public_content/lifestyle/general_lifestyle/january_2010/50_favor_cutting_back_on_space_exploration, archived January 19, 2010, at the *WM*.
14 "SpaceX: Elon Musk's Race to Space," *60 Minutes* YouTube channel, March 18, 2012, https://www.youtube.com/watch?v=23GzpbNUyI4.
15 Steve Jurvetson Flickr post, July 11, 2012, https://www.flickr.com/photos/jurvetson/7547788856/.
16 Kenneth Chang, "Astronauts Attack Obama's NASA Plan," *NYT*, May 12, 2010, https://www.nytimes.com/2010/05/13/science/space/13nasa.html.

17 "Astronaut Legends Criticize Obama Space Plan," CBS News, May 12, 2010, https://www.cbsnews.com/news/astronaut-legends-criticize-obama-space-plan/.
18 "Astronauts Criticize NASA Moon Program Cuts Before Congress," PBS NewsHour YouTube channel, May 12, 2010, https://www.youtube.com/watch?v=6pGm4_25WTY.
19 Bart Jansen, "Space Pioneers Criticize Budget," *FT*, May 13, 2010, 1A, 3A, https://www.newspapers.com/image/360418000/.
20 Mark K. Matthews, "Ex-Astronauts Call Plan for NASA a 'Pledge to Mediocrity,'" *OS*, May 13, 2010, A4, https://www.newspapers.com/image/269301487/.
21 "History," Congressional Budget Office, https://www.cbo.gov/about/history.
22 "Senator Richard Shelby February 2023 Porker of the Month," Citizens Against Government Waste, https://www.cagw.org/porker-of-the-month/citizens-against-government-waste-names-senator-richard-shelby-february-2023.
23 Lori Garver and Michael Sheetz, "Op-ed: Unleashing the Dragon—the NASA Bargain Behind This Week's SpaceX Launch," CNBC, May 26, 2020, https://www.cnbc.com/2020/05/26/op-ed-the-nasa-bargain-behind-spacex-launch-demo-2.html. Orion was renamed the Multi-Purpose Crew Vehicle because Congress foresaw using it not only for deep space missions, but also as "an alternative means of delivery of crew and cargo to the ISS" if the commercial companies were "unable to perform that function." 2010 NASA authorization act, Sec. 303(b)(3). The MPCV appellation eventually was dropped in favor of Orion.
24 Mark K. Matthews and Robert Block, "Senate Shorts NASA on Time, Cash for New Rocket, Foes Say," *OS*, July 23, 2010, A1, A10, https://www.newspapers.com/image/269092245/.
25 Bill Nelson telephone interview with author, December 29, 2023.
26 Mark K. Matthews and Robert Block, "Plan Kills NASA Moon Mission," *OS*, July 16, 2010, A1, A19, https://www.newspapers.com/image/269074898/.
27 Robert Block, "Is Utah Emerging as Rival to KSC?" *OS*, July 12, 2010, A1, A9, https://www.newspapers.com/image/269068924/.
28 National Aeronautics and Space Administration Authorization Act of 2010, Pub. L. No. 111-267, 124 Stat. 2805 (2010), https://www.congress.gov/bill/111th-congress/senate-bill/3729.
29 National Aeronautics and Space Administration Authorization Act of 2010, H.R.5781, 111th Cong. (2010), https://www.congress.gov/bill/111th-congress/house-bill/5781.
30 "NASA Authorization Stalls in the House," *SN*, August 2, 2010, https://spacenews.com/nasa-authorization-stalls-house/.
31 NASA Authorization Act of 2010, Sec. 304.
32 *Space Launch System: Resources Need to Be Matched to Requirements to Decrease Risk and Support Long Term Affordability*, GAO, July 2014, 16, https://www.gao.gov/assets/gao-14-631.pdf.
33 SLS GAO report, 17.
34 SLS GAO report, 26.
35 SLS GAO report, 10.

36 "Florida's Space Workers and the New Approach to Human Spaceflight," NASA, April 15, 2010, https://www.nasa.gov/pdf/444595main_Fact%20Sheet%20revised.pdf, archived April 24, 2010, at the *WM*.
37 President Barack Obama, "Memorandum on the Task Force on Space Industry Workforce and Economic Development," May 3, 2010, https://www.govinfo.gov/content/pkg/DCPD-201000337/html/DCPD-201000337.htm.
38 Scott Powers, "Feds Gather Ideas for Jobs After Shuttle," *OS*, June 5, 2010, B4, https://www.newspapers.com/image/269078038/.
39 Task Force on Space Industry Workforce and Economic Development web page, http://www.nasa.gov/offices/spacecoasttaskforce/home/index.html, archived June 27, 2010, at the *WM*.
40 "U.S. Department of Labor Announces $15 Million Grant to Assist Workers in Florida Affected by End of Space Shuttle Program," US Department of Labor press release, June 2, 2010, https://www.dol.gov/newsroom/releases/eta/eta20100602.
41 James Dean, "Plan Strives to Ease Sting of Shutdown," *FT*, June 3, 2010, 1A, 3A, https://www.newspapers.com/image/360403545/.
42 Brevard Workforce LaunchNewCareers.com, https://web.archive.org/web/20100413172057/http://www.launchnewcareers.com/, archived April 13, 2010, at the *WM*.
43 Patrick Peterson, "Space Workers Face Daunting Job," *FT*, June 25, 2010, 8C–7C, https://www.newspapers.com/image/360423813/.
44 James Dean, "Locke Pledges Support," *FT*, August 4, 2010, 1B, 4B, https://www.newspapers.com/image/360423822/.
45 Presidential Task Force on Space Industry Workforce & Economic Development, *Report to the President*, August 15, 2010, http://www.nasa.gov/pdf/475699main_Space_Industry_Report_to_the_President.pdf, archived August 26, 2010, at the *WM*.
46 H.R.5781, Sec. 223(a)(1). The annual dollar amounts are specified in Sec. 101(5)(B) for FY11, Sec. 102(5)(C) for FY12, and Sec. 103(5)(B) for FY13.
47 Commerce, Justice, Science, and Related Agencies Appropriations Act, 2011, S.3636, 111th Cong. (2010), https://www.congress.gov/bill/111th-congress/senate-bill/3636.
48 Election 2010 Results, *NYT*, https://archive.nytimes.com/www.nytimes.com/elections/2010/results/house.html.
49 David M. Herszenhorn, "Senate Republicans Threaten Tax Dispute Blockade," *NYT*, December 1, 2010, https://www.nytimes.com/2010/12/02/us/politics/02cong.html.
50 Department of Defense and Full-Year Continuing Appropriations Act of 2011, Sec. 1321, Pub. L. No. 112-10, 125 Stat.120 (2011), https://www.congress.gov/bill/111th-congress/senate-bill/3729.
51 "Future of NASA Space Program," C-SPAN, September 14, 2011, https://www.c-span.org/video/?301537-1/future-nasa-space-program.
52 Betsy N. Riley, Benjamin S. Solish, Lauren Halatek, and Richard R. Rieber, *Early Career Hire Rapid Training and Development Program: Status Report* (JPL, CalTech, 2009), 1, presented September 2009 to the AIAA SPACE 2009 Conference & Exposition, https://www.researchgate.net/publication/268568928_Early_Career_Hire_Rapid_Training_and_Development_Program_Status_Report.

53 John McCarthy, "County Will Survive a Final Touchdown," *FT*, January 10, 2010, 1A, 7A, https://www.newspapers.com/image/360409401/.

54 Bridget Testa, "New Workforce Orbit—Relaunch at NASA," *Workforce Management*, August 17, 2009, https://workforce.com/news/new-workforce-orbitrelaunch-at-nasa.

55 "Last of Shuttle Layoffs Loom for USA Workers," WKMG-TV, January 9, 2012, https://www.clickorlando.com/news/2012/01/09/last-of-shuttle-layoffs-loom-for-usa-workers/.

56 Mary Williams Walsh, "Shuttle's End Leaves NASA a Pension Bill," *NYT*, June 14, 2011, https://www.nytimes.com/2011/06/15/business/15nasa.html.

57 Nell Greenfeldboyce, "NASA, SpaceX Aim to Launch Private Era in Orbit," NPR Morning Edition, May 18, 2012, https://www.npr.org/2012/05/18/152953776/nasa-spacex-aim-to-launch-private-era-in-orbit.

58 Noam Scheiber and Ryan Mac, "SpaceX Employees Say They Were Fired for Speaking Up About Elon Musk," *NYT*, November 17, 2022, https://www.nytimes.com/2022/11/17/business/spacex-workers-elon-musk.html. Michael Sainato, "UAW Wants to Unionize Tesla. It Faces a Tough and High-Profile Battle with Musk," *The Guardian*, December 11, 2023, https://www.theguardian.com/technology/2023/dec/11/uaw-tesla-elon-musk-unions.

59 SpaceX Intern Program web page, https://www.spacex.com/internships/.

60 Ainsley Harris, "This U.S. Air Force Commander Helps Elon Musk's Interns Launch SpaceX Rockets," *Fast Company*, April 9, 2018, https://www.fastcompany.com/40549782/this-us-air-force-commander-helps-elon-musks-interns-launch-spacex-rockets.

61 Neel V. Patel, "SpaceX Must Pay $4 Million for Thousands of Underpaid Employees," *Inverse*, May 11, 2017, https://www.inverse.com/article/31478-spacex-settles-underpaid-workers-lawsuit-for-4-million.

62 Marisa Taylor, "At SpaceX, Worker Injuries Soar in Elon Musk's Rush to Mars," Reuters, November 10, 2023, https://www.reuters.com/investigates/special-report/spacex-musk-safety/.

63 Loren Grush, "SpaceX Employees Draft Open Letter to Company Executives Denouncing Elon Musk's Behavior," *The Verge*, June 16, 2022, https://www.theverge.com/2022/6/16/23170228/spacex-elon-musk-internal-open-letter-behavior. National Labor Relations Board Case Number 31-CA-307555, Space Exploration Technologies Corp., https://www.nlrb.gov/case/31-CA-307555.

64 Daniel Wiessner, "US Court Blocks Transfer of SpaceX Lawsuit Against NLRB, for Now," Reuters, August 12, 2024, https://www.reuters.com/investigates/special-report/spacex-musk-safety/.

65 Marisa Taylor, "Exclusive: Injury Rates for Musk's SpaceX Exceed Industry Average for Second Year," Reuters, April 23, 2024, https://www.reuters.com/technology/injury-rates-musks-spacex-exceed-industry-average-second-year-2024-04-22/.

66 "Compare SpaceX vs United Launch Alliance." Glassdoor.com website, https://www.glassdoor.com/Compare/SpaceX-vs-United-Launch-Alliance-EI_IE40371-E146300.htm.

67 Christian Davenport and Brian Fung, "Elon Musk's SpaceX to Sue Government Over Space Launch Contract," *WAPO*, April 25, 2014, https://www.washingtonpost.com/business/economy/elon-musks-spacex-to-sue-government-over-space-launch-contract/2014/04/25/1001aa6e-cca6-11e3-95f7-7ecdde72d2ea_story.html.

68 Christian Davenport, "Elon Musk's SpaceX Settles Lawsuit Against Air Force," *WAPO*, January 23, 2015, https://www.washingtonpost.com/business/economy/elon-musks-spacex-to-drop-lawsuit-against-air-force/2015/01/23/c5e8ff80-a34c-11e4-9f89-561284a573f8_story.html.

69 Allard Beutel, "NASA Selects SpaceX to Begin Negotiations for Use of Historic Launch Pad," NASA press release, December 13, 2013, https://www.nasa.gov/news-release/nasa-selects-spacex-to-begin-negotiations-for-use-of-historic-launch-pad/. Dan Leone, "Blue Origin Files Formal Protest of Proposed Shuttle Pad Lease," *SN*, September 16, 2013, https://spacenews.com/37162blue-origin-files-formal-protest-of-proposed-shuttle-pad-lease/.

70 Stephanie Schierholz and Stephanie Martin, "NASA Chooses American Companies to Transport U.S. Astronauts to International Space Station," NASA press release, September 16, 2014, https://www.nasa.gov/news-release/nasa-chooses-american-companies-to-transport-u-s-astronauts-to-international-space-station/.

71 Commerce, Justice, Science, and Related Agencies Appropriations Bill of 2013, H.R.5326, H. Rept. 112-463, 71, 112th Cong. (2012), https://www.congress.gov/congressional-report/112th-congress/house-report/463/1?outputFormat=pdf. Frank Morring, Jr., "Apollo Commanders Back Call for Quick Commercial Crew Selection," *Aviation Week*, May 8, 2012, http://www.aviationweek.com/Article.aspx?id=/article-xml/asd_05_08_2012_p04-01-455625.xml, archived July 7, 2012, at the *WM*. Rep. Frank R. Wolf letter to Maj. Gen. (Ret.) Charles F. Bolden, Jr., May 31, 2012, http://wolf.house.gov/uploads/DOC053112-160324.pdf, archived July 30, 2012, at the *WM*. Ledyard King, "Nelson Opposes Plan to Speed Up Selection of Rocket Maker," *FT*, May 15, 2012, 2A, https://www.newspapers.com/image/360454311/. Eric Berger, "SpaceX Likely to Win NASA's Crew Competition by Months, for Billions Less," *AT*, April 8, 2019, https://arstechnica.com/science/2019/04/spacex-likely-to-win-nasas-crew-competition-by-months-for-billions-less/.

72 Paul K. Martin, *NASA's Management of the Commercial Crew Program*, Report No. IG-14-001 (NASA Office of the Inspector General, November 13, 2013), iv, https://oig.nasa.gov/office-of-inspector-general-oig/ig-14-001/.

73 *Commercial Crew OIG Report*, 6.

74 Stephen Clark, "NASA's Starliner Decision Was the Right One, but It's a Crushing Blow for Boeing," *AT*, August 24, 2024, https://arstechnica.com/space/2024/08/after-latest-starliner-setback-will-boeing-ever-deliver-on-its-crew-contract/. Marcia Smith, "Boeing's Starliner Losses Reach $2 Billion," SpacePolicyOnline.com, February 3, 2025, https://spacepolicyonline.com/news/boeings-starliner-losses-reach-2-billion/.

Chapter 10. In with the New

1 James Boswell, *Boswell's Life of Johnson* (Oxford University Press, 1791), 849, accessed at the IA, https://archive.org/details/in.ernet.dli.2015.459439/.

2 C. William Nelson, *The Impact of Cape Kennedy on Brevard County Politics* (Senior Essay, Department of Political Science, Yale University, May 4, 1965), 37–38. A PDF copy was provided by Yale University to this author and the University of Florida with the permission of Senator Nelson.

3 Sharada Dharmasankar and Bhash Mazumder, "Have Borrowers Recovered from Foreclosures during the Great Recession?" Chicago Fed Letter, No. 370, 2016, Federal Reserve Bank of Chicago, https://www.chicagofed.org/publications/chicago-fed-letter/2016/370.

4 Evan Cunningham, "Great Recession, Great Recovery? Trends from the Current Population Survey," *Monthly Labor Review*, US Bureau of Labor Statistics, April 2018, https://www.bls.gov/opub/mlr/2018/article/great-recession-great-recovery.htm.

5 Robert Block, "New Space Florida Leader Tackles Many Challenges," *OS*, January 15, 2010, A1, A13, https://www.newspapers.com/image/269140303/. Frank DiBello, "Fighting Fiercely for Space," *FT* guest opinion column, February 28, 2010, 17A, https://www.newspapers.com/image/360411362/.

6 V. O. Key, Jr., *Southern Politics* (Vintage Books, 1949), 82, 86, accessed at the IA, https://archive.org/details/southernpolitics0000keyv_d3i3/.

7 "Space Life Sciences Lab," Space Florida, http://www.spaceflorida.gov/slsl.php, archived December 9, 2007, at the *WM*; also archived January 5, 2009, at the *WM*.

8 Dynamac Corporation, "Environmental Assessment for Exploration Park—Phase 1," December 2008, http://netspublic.grc.nasa.gov/main/Final%20Space%20Exploration%20Phase%201%20EA.pdf.

9 Patrick Peterson, "Deal Made to Develop 1st KSC Park Building," *FT*, April 11, 2009, 10C, https://www.newspapers.com/image/362918040/.

10 Rick Neale, "Park Needs Infrastructure," *FT*, December 11, 2009, 8C, https://www.newspapers.com/image/360412393/.

11 Rick Neale, "Leftover Funds Hit Crossroad," *FT*, February 11, 2010, 1A, 3A, https://www.newspapers.com/image/360856060/. Rick Neale, "Park Passes Critical Test," *FT*, February 12, 2010, 1A, 3A, https://www.newspapers.com/image/360856446/.

12 "Spending Breakdown," *FT*, May 12, 2010, 2A, https://www.newspapers.com/image/360417715/.

13 Patrick Peterson, "Expectation High for Exploration," *FT*, June 26, 2010, 8C–7C, https://www.newspapers.com/image/360424797/. Patrick Peterson, "Site-Prep Work Starts at Exploration Park," *FT*, March 15, 2011, 6C–5C, https://www.newspapers.com/image/360541782/.

14 Fla. CS for HB 451 (2010) (Space Florida), https://www.myfloridahouse.gov/Sections/Bills/billsdetail.aspx?BillId=42806.

15 Fla. CS for HB 1389 (2010) (Space and Aerospace Infrastructure), https://www.myfloridahouse.gov/Sections/Bills/billsdetail.aspx?BillId=44191.

16 Fla. CS for HB 969 (2010) (Space and Aerospace Infrastructure), https://www.myfloridahouse.gov/Sections/Bills/billsdetail.aspx?BillId=43585.
17 Fla. CS for SB 1752 (2010) (Space and Aerospace Infrastructure), https://www.myfloridahouse.gov/Sections/Bills/billsdetail.aspx?BillId=43585.
18 Frank DiBello, "Legislature Gets a STAR," *TD* guest opinion column, May 28, 2010, 4B, https://www.newspapers.com/image/249891753/.
19 John McCarthy, "Space Florida Gets $400K," *FT*, October 1, 2010, 8C, https://www.newspapers.com/image/360527824/. *FT* corrected its original incorrect reporting of the funding source under "Corrections" on October 5, 2010, 2A, https://www.newspapers.com/image/360529723/.
20 Patrick Peterson, "Space Florida Balks at Scott's Funding Plan," *FT*, April 17, 2011, 1E, 5E, https://www.newspapers.com/image/360536039/. Jim Ash, "Deal Gives Scott Influence over Economic Development," *TD*, April 30, 2011, 1–2, https://www.newspapers.com/image/249887761/. Bill Kaczor, "Agency Aims to Bring New Jobs to State," *PNJ*, October 1, 2011, 3C, https://www.newspapers.com/image/270686396/.
21 Fla. *SB* 2000 (2011) (Appropriations), § 5, 1918B, 253, https://www.myfloridahouse.gov/Sections/Bills/billsdetail.aspx?BillId=46874&SessionId=66.
22 Fla. *SB* 634 (2012) (Spaceport Facilities), https://www.flsenate.gov/Session/Bill/2012/634/.
23 James Dean, "Scott Signs Space Upgrade Bill," *FT*, February 17, 2012, 8B, https://www.newspapers.com/image/360468333/.
24 *Florida Spaceport Improvement Program Project Handbook 2016 Edition* (FDOT Spaceport), 3. *Spaceport Improvement Program 2023–2024* (FDOT Spaceport), 3, 10, https://cdn.prod.website-files.com/66c8a3fe36eef11411f2b1e5/66c8a3fe36eef11411f2b577_SF0080.02DEL-FDOT-Spaceport-Handbook-Update-2023-230426%20(1).pdf.
25 "Commercial Crew and Cargo Processing Facility (C3PF)," Space Florida, https://www.spaceflorida.gov/facilities/c3pf/, archived April 22, 2024, at the *WM*. James Dean, "Boeing, KSC Talk Space Taxi," *FT*, July 15, 2011, 1A, 3A, https://www.newspapers.com/image/360539339/. James Dean, "Boeing Brings Jobs Back to Brevard," *FT*, November 1, 2011, 1A, 2A, https://www.newspapers.com/image/360457906/. *Space Florida 2014 Annual Report*, 17, https://cdn.prod.website-files.com/66c8a3fe36eef11411f2b1ef/66c8a3fe36eef11411f2b920_663eb32f2f0f49c7c5a2ced1_SFL2014-AnnualReport.pdf.
26 James Dean, "KSC Lands Secret Space Plane, Jobs," *Florida Today*, January 4, 2014, 1A–2A, https://www.newspapers.com/image/113498889/. "NASA Partners with X-37B Program for Use of Former Space Shuttle Hangars," NASA press release, October 8, 2014, https://www.nasa.gov/news-release/nasa-partners-with-x-37b-program-for-use-of-former-space-shuttle-hangars/. *Space Florida 2014 Annual Report*, 17.
27 Irene Klotz, "NASA Picks Florida Agency to Take Over Shuttle Landing Strip," NBC News, June 29, 2013, https://www.nbcnews.com/science/science-news/nasa-picks-florida-agency-take-over-shuttle-landing-strip-flna6c10489060. "Launch and Landing Facility (LLF)," Space Florida, https://www.spaceflorida.gov/facilities/llf/.

28 Space Florida Board of Directors Meeting Material, April 22, 2021, 14, https://cdn.prod.website-files.com/66c8a3fe36eef11411f2b1ef/66c8a3fe36eef11411f2b8c3_Space-Florida-Board-of-Directors-Meeting-April-22-2021-1.pdf.

29 "Governor Ron DeSantis Announces Terran Orbital Will Invest $300 Million in Florida to Construct the World's Largest, State-of-the-Art, Commercial Spacecraft Facility," Terran Orbital press release, September 27, 2021, https://terranorbital.com/governor-ron-desantis-announces-terran-orbital-will-invest-300-million-in-florida-to-construct-the-worlds-largest-state-of-the-art-commercial-spacecraft-facility/.

30 "Terran Orbital Receives $100 Million Investment from Lockheed Martin," Terran Orbital press release, October 31, 2022, https://terranorbital.com/terran-orbital-receives-100-million-investment-from-lockheed-martin/.

31 Space Florida Board of Directors Meeting Agenda, January 26, 2023, 21, https://cdn.prod.website-files.com/66c8a3fe36eef11411f2b1ef/66c8a3fe36eef11411f2b7ee_January-23-2023-BOD-Advance-Package-Revised-1.pdf.

32 "Amazon's Project Kuiper Completes Successful Tests of Optical Mesh Network in Low Earth Orbit," Amazon, December 14, 2023, https://www.aboutamazon.com/news/innovation-at-amazon/amazon-project-kuiper-oisl-space-laser-december-2023-update. "Amazon's Project Kuiper Will Open a New Satellite-Processing Facility at Kennedy Space Center," Amazon, July 21, 2023, https://www.aboutamazon.com/news/innovation-at-amazon/project-kuiper-kennedy-space-center-florida. Emre Kelly, "Amazon Network Expands to KSC," *FT*, July 23, 2023, 1A, 15A, https://www.newspapers.com/image/986020225/. Amazon and Blue Origin are owned by American billionaire Jeff Bezos. They operate as separate companies.

33 Mark K. Matthews, "NASA Seeks Facility Tenants Before Upkeep Cash Runs Out," *OS*, April 9, 2012, A1, A4, https://www.newspapers.com/image/269004269/. James Dean, "Right People, Right Place," *FT*, August 24, 2012, 1A, 3A, https://www.newspapers.com/image/360429755/. James Dean, "Swiss Organization Plans to Use Shuttle's Runway," *FT*, March 16, 2014, 14A, https://www.newspapers.com/image/113521482/. Rachael Nail, "Virgin Orbit Successfully 'Air-Launched' 7 Satellites," *FT*, July 2, 2021, 1A, 2A, https://www.newspapers.com/image/746185207/.

34 "NASA Selects Sierra Nevada Corporation's Dream Chaser® Spacecraft for Commercial Resupply Services 2 Contract," SNC press release, January 14, 2016, https://www.sncorp.com/news-archive/nasa-selects-sierra-nevada-corporation-s-dream-chaser-spacecraft-for-commercial-resupply-services-2-contract/. Scott Powers, "Commercial-Rocket Company Seeking State Incentive Package," *OS*, May 5, 2012, A6, https://www.newspapers.com/image/269529587/. "Space Florida, Sierra Nevada Corporation Announce Use Agreement for Launch and Landing Facility Dream Chaser® Spaceplane to Land at Cape Canaveral Spaceport," Space Florida press release, May 4, 2021, https://www.spaceflorida.gov/news/space-florida-sierra-nevada-corporation-announce-use-agreement-for-launch-and-landing-facility-dream-chaser-spaceplane-to-land-at-cape-canaveral-spaceport/.

35 Patrick Peterson, "State Gives $5M for Navy to Create 100 Jobs at Cape Canaveral Test Site," *OS*, August 3, 2012, A6, https://www.newspapers.com/image/268104899/. Tina Lange, "U.S. Navy, State of Florida and EDC of Florida's Space Coast Break

Ground on New Submarine Missile Test Facility," Space Florida press release, November 8, 2012, http://www.spaceflorida.gov/news/2012/11/08/u.s.-navy-state-of-florida-and-edc-of-florida-s-space-coast-break-ground-on-new-submarine-missile-test-facility, archived December 8, 2012, at the *WM*.

36 "Space Florida Welcomes Firefly Aerospace to Cape Canaveral Spaceport." Space Florida press release, February 22, 2019, https://www.spaceflorida.gov/news/space-florida-welcomes-firefly-aerospace-to-cape-canaveral-spaceport/.

37 "Missions," Firefly Aerospace, https://fireflyspace.com/missions/

38 Stephen Clark, "Relativity Space Obtains Air Force Approval for Cape Canaveral Launch Pad," *Spaceflight Now*, January 18, 2019, https://spaceflightnow.com/2019/01/18/relativity-space-obtains-air-force-approval-for-cape-canaveral-launch-pad/. Emre Kelly, "Relativity Launches First 3D-Printed Rocket," *FT*, March 24, 2023, 1A, 8A, https://www.newspapers.com/image/942888790/.

39 Dave Berman and Wayne T. Price, "Space Company Seeks $8M Incentive," *FT*, May 17, 2015, 3A, 20A, https://www.newspapers.com/image/105634640/. *Space Florida FY16 Annual Report*, 12, https://cdn.prod.website-files.com/66c8a3fe36eef11411f2b1ef/66c8a3fe36eef11411f2b924_663eb2ab48e419427366dff3_SFL2016-AnnualReport.pdf.

40 "Exploration Park Area Overview and Description," *Cape Canaveral Spaceport Development Manual, Version 1.4*, Chapter 3 "Exploration Park," Section 1.2, Space Florida, June 2024, https://cdn.prod.website-files.com/663258da01a14bbdf80faa99/66bd166b07b6e6066a8d9718_CCS%20Development%20Manual%20June%202024%20(version%201.4)%20240614.pdf.

41 James Dean, "State Pursues 'Major' Aerospace Deal," *FT*, January 29, 2016, 3A, 9A, https://www.newspapers.com/image/150687108/. James Dean, "OneWeb Coming to Space Coast," *FT*, April 18, 2016, 1A–2A, https://www.newspapers.com/image/177474172/. James Dean, "OneWeb Expects to Create 250 Positions with Satellite Factory at Exploration Park," *FT*, April 20, 2016, 1A, 5A, https://www.newspapers.com/image/177474460/.

42 National Aeronautics and Space Administration Authorization Act of 2005, Pub. L. No. 109-155, 119 Stat. 2895 (2005), https://www.congress.gov/109/plaws/publ155/PLAW-109publ155.pdf. Sec. 101 (5) provided the funding. Sec. 611 (c) mandated the flight.

43 "International Space Station: Significant Challenges May Limit Onboard Research," GAO, GAO-10-9, November 2009, 26–27, https://www.gao.gov/assets/gao-10-9.pdf.

44 2010 NASA authorization act, Sec. 504.

45 Joshua Buck, "NASA Seeks Nonprofit to Manage Space Station National Lab Research," NASA press release, December 2, 2010, https://www.nasa.gov/news-release/nasa-seeks-nonprofit-to-manage-space-station-national-lab-research/.

46 James Dean, "KSC Group Scores Station Deal," *FT*, July 14, 2011, 1A, 3A, https://www.newspapers.com/image/360539192/. James Dean, "Nonprofit Lands ISS Lab Deal," *FT*, September 10, 2011, 10C–9C, https://www.newspapers.com/image/360540364/. "NASA's Efforts to Maximize Research on the International Space Station," NASA OIG, IG-13-019, July 8, 2013, i, https://oig.nasa.gov/wp-content/uploads/2024/02/IG-13-019.pdf.

47 Jeanne L. Becker letter to Frank DiBello, February 29, 2012, posted March 16, 2012, on the SpaceRef, https://spaceref.com/space-commerce/resignation-letter-from-casis-executive-director-jeanne-l-becker-2/. James Dean, "Months In, Agency Chief Out," *FT*, March 6, 2012, 1A–2A, https://www.newspapers.com/image/360472441/.

48 Dan Leone, "Ohio Delegation Calls on NASA to Fire ISS Nonprofit," *SN*, April 5, 2012, https://spacenews.com/ohio-delegation-calls-nasa-fire-iss-nonprofit/.

49 Eli Rosenberg, "A Man Conned His NASA-Funded Lab into Paying for Trips to See Prostitutes and Escorts, Feds Say," *WAPO*, April 18, 2019, https://www.washingtonpost.com/science/2019/04/18/man-conned-his-nasa-funded-lab-into-paying-trips-see-prostitutes-escorts-feds-say/. Joseph Vockley, CASIS statement regarding Resnick incident, April 12, 2019, https://spaceref.com/press-release/statement-on-behalf-of-casis-iss-us-national-laboratory-regarding-resnick-indictment/. Emre Kelly, "Accused of 'Expensing' Prostitutes, Former ISS Executive Pleads Guilty," *FT*, February 20, 2020, 7A, https://www.newspapers.com/image/639940613/. "Former Chief Economist for the Center for the Advancement of Science in Space Pleads Guilty," US Attorney's Office, Middle District of Florida, press release, February 19, 2020, https://www.justice.gov/usao-mdfl/pr/former-chief-economist-center-advancement-science-space-pleads-guilty.

50 "ISS Utilization: Nanoracks Logistics Services for Small Satellites and ISS Deployment Systems," eoPortal, December 10, 2014, https://www.eoportal.org/other-space-activities/iss-Nanoracks-services#iss-utilization-Nanoracks-logistics-services-for-small-satellites-and-iss-deployment-systems. "CASIS and Nanoracks Close Deal to Use Commercial Research Platform in the Extremes of Space," Nanoracks press release, April 12, 2012, http://Nanoracks.com/wp-content/uploads/Release-01-Casis-Nanoracks-Commercial-Platform.pdf, archived March 27, 2015, at the *WM*.

51 "Space Florida and Nanoracks Announce International Space Station Research Competition," Florida Space Grant Consortium website, https://floridaspacegrant.org/nasa-opportunity/space-florida-and-Nanoracks-announce-international-space-station-research-competition/.

52 Gary Jordan and Brock Howe, "Commercial Airlock," *Houston We Have a Podcast* transcript, NASA Johnson Space Center, January 22, 2021, https://www.nasa.gov/podcasts/houston-we-have-a-podcast/commercial-airlock/. Eric Berger, "To Boost Commercial Activity, NASA May Add Private Airlock to ISS," *AT*, January 27, 2016, https://arstechnica.com/science/2016/01/to-boost-commercial-activity-nasa-may-add-private-airlock-to-iss/. CASIS, *International Space Station National Laboratory Annual Report for Fiscal Year 2024*, January 2, 2025, 11, 30, https://issnationallab.org/download/46129/?tmstv=1743709348.

53 "Voyager Space Holdings, Inc. Acquires Majority Stake of X.O. Markets, Parent of Nanoracks," Voyager Space Holdings press release, May 10, 2021, https://www.prnewswire.com/news-releases/voyager-space-holdings-inc-acquires-majority-stake-of-xo-markets-parent-of-Nanoracks-301287168.html.

54 "Made in Space and CASIS: Launch World's First 3D Printer to Space," ISS National Lab YouTube channel, September 18, 2014, https://www.youtube.com/watch?v=8ex481nHY7k. Hasti Afsarifard, "The First Uplink Tool Made in Space Is . . ." Made

in *Space* (blog), December 17, 2014, http://www.madeinspace.us/the-first-uplink-tool-made-in-space-is, archived December 25, 2014, at the *WM*.

55 "Redwire Acquires Made in Space, the Leader in On-orbit Space Manufacturing Technologies," Redwire press release, June 23, 2020, https://redwirespace.com/newsroom/redwire-acquires-made-in-space-the-leader-in-on-orbit-space-manufacturing-technologies/. "Redwire's 3D Bioprinter Successfully Installed on Space Station and Prepped for Upcoming Meniscus Print," *Redwire* (blog), February 27, 2023, https://redwirespace.com/newsroom/redwires-3d-bioprinter-successfully-installed-on-space-station-and-prepped-for-upcoming-meniscus-print/. "Redwire BioFabrication Facility Successfully Prints First Human Knee Meniscus on ISS, Paving the Way for Advanced In-Space Bioprinting Capabilities to Benefit Human Health," Redwire press release, September 7, 2023, https://redwirespace.com/newsroom/redwire-biofabrication-facility-successfully-prints-first-human-knee-meniscus-on-iss-paving-the-way-for-advanced-in-space-bioprinting-capabilities-to-benefit-human-health/. "Redwire Pioneering Biopharma Production in Space by Successfully Bioprinting Live Human Heart Tissue and Delivering Second Batch of PIL-BOX Pharmaceutical Crystal Experiments," Redwire press release, May 8, 2024, https://redwirespace.com/newsroom/redwire-pioneering-biopharma-production-in-space-by-successfully-bioprinting-live-human-heart-tissue-and-delivering-second-batch-of-pil-box-pharmaceutical-crystal-experiments/.

56 "Genes in Space," ISS National Laboratory, October 5, 2020, https://www.issnationallab.org/stem/partner-organizations/genesinspace-organization/. Amy Thompson, "From Science Fairs to Space: Student Experiments Help Launch New Era of Space-Based Research," ISS National Laboratory, June 22, 2022, https://www.issnationallab.org/upward-genes-in-space/.

57 "NASA Extends Cooperative Agreement, Affirms Successful CASIS Partnership Pushing the Limits of R&D through the ISS National Lab to Benefit Humanity," ISS National Laboratory website, December 8, 2025, https://issnationallab.org/press-releases/casis-to-manage-iss-national-lab-through-2030/.

58 Todd Halvorson, "NASA Proposes Commercial Pad," *FT*, February 17, 2008, 1A, 5A, https://www.newspapers.com/image/362964849/.

59 Jessica Raynor, "Residents Rail Against Launch Pad," *FT*, February 26, 2008, 1A, 5A, https://www.newspapers.com/image/362987844/. Patrick Peterson, "Fisherman, Birdwatchers Fear Pad Plan," *FT*, February 28, 2008, 1B, 6B, https://www.newspapers.com/image/362988753/. Patrick Peterson, "Air Station in Play for Launch Complex," *FT*, July 15, 2008, 1A, 7A, https://www.newspapers.com/image/360554841/.

60 Team ZHA, *Cape Canaveral Spaceport Master Plan July 2002*, ZHA Inc., prime consultant, IV-1, Exhibits IV-SC1–IV-SC3, https://www.spaceflorida.gov/wp-content/uploads/2022/07/Cape-Canaveral-Spaceport-Master-Plan-July-2002.pdf, archived March 24, 2023, at the *WM*.

61 *Agency Master Plan: Report on the 2011 Plan*, NASA, 36, https://www.nasa.gov/wp-content/uploads/2023/06/nasamp23mar12-lores.pdf.

62 *Kennedy Space Center Master Plan 2012–2032 Executive Summary*, 11, https://visualmedia.jacobs.com/KSC/Executive-Summary/NASA%20KSC%20MP%20Executive%20Summary.pdf.

63 Florida Lt. Gov. Jennifer Carroll letter to NASA Administrator Charles Bolden and USDOT Secretary Ray LaHood, September 20, 2012. A copy was also provided by Dale Ketcham of Space Florida.

64 James Dean, "Fla. Makes Play for Pad," *FT*, September 22, 2012, 1A, 5A, https://www.newspapers.com/image/360441654/. Mark K. Matthews, "Plan for New Space Coast Site Worries Some Environmentalists," *OS*, October 2, 2012, A3, https://www.newspapers.com/image/269713989/.

65 NASA Associate Administrator for Legislative and Governmental Affairs Seth Statler letter to Florida Lt. Gov. Jennifer Carroll, November 30, 2012, https://media2.spaceref.com/news/2013/11-30-12LtrStatler-NASA.pdf.

66 James Dean, "Interest Building in Shiloh," *FT*, February 27, 2013, 1A–2A, https://www.newspapers.com/image/226754133/.

67 Craig Pittman, "A Historic Complication," *TBT*, July 5, 2013, 1A, 3A, https://www.newspapers.com/image/364294354/.

68 Tina Lange, "Space Florida Initiates Environmental Study Process for Proposed Commercial Spaceport," Space Florida press release, July 15, 2023, http://www.spaceflorida.gov/news/2013/07/15/space-florida-initiates-environmental-study-process-for-proposed-commercial-spaceport, archived October 24, 2013, at the *WM*.

69 *Final Scoping Summary Report: Environmental Impact Statement for the Shiloh Launch Complex, Brevard and Volusia Counties, Florida* (FAA Office of Commercial Space Transportation, May 2014), generally, https://www.faa.gov/about/office_org/headquarters_offices/ast/environmental/nepa_docs/review/documents_progress/shiloh_launch_statement/media/FAA_ShilohEIS_Scoping_Summary_Report_Website.pdf, archived February 9, 2017, at the *WM*.

70 "CEO Musk: Project Is 'World's First,'" *The Brownsville Herald*, August 5, 2014, A1, A5, https://www.newspapers.com/image/282237709/. "Gov. Perry Announces State Incentives Bringing SpaceX Commercial Launch Facility, 300 Jobs to the Brownsville Area," Office of the Governor press release, August 4, 2014, archived August 6, 2014, at the *WM*. Catherine Donnelly, "Coalition Wins Boca Chica Closure Appeal," *Port Isabel-South Padre Press*, February 15, 2024, https://www.portisabelsouthpadre.com/2024/02/15/coalition-wins-boca-chica-closure-appeal/.

71 Jeff Foust, "Fish and Wildlife Service Documents Damage from Starship Launch," *SN*, April 26, 2023, https://spacenews.com/fish-and-wildlife-service-documents-damage-from-starship-launch/. Jeff Foust, "Federal Agencies Caught in Environmental Crossfire Over Starship Launches," *SN*, December 15, 2023, https://spacenews.com/federal-agencies-caught-in-environmental-crossfire-over-starship-launches/.

72 Eric Lipton, "Wildlife Protections Take a Back Seat to SpaceX's Ambitions," *NYT*, July 7, 2024, https://www.nytimes.com/2024/07/07/us/politics/spacex-wildlife-texas.html.

73 Space Exploration Technologies Corp., US EPA Consent Agreement, Docket No. CWA-06-2024-1768, https://www.epa.gov/system/files/documents/2024-09/spacex_cafo_cwa-06-2024-1768_txu09110_090624__0.pdf. Andrea Guzmán, "EPA Fines SpaceX over Environmental Violations in South Texas," *Beaumont Enterprise*, September 12, 2024, https://www.beaumontenterprise.com/culture/article/epa-fine-spacex-texas-water-19760415.php.

74 FAA, "Notice To Rescind a Notice of Intent to Prepare an Environmental Impact Statement," *Federal Register*, Document Citation 85 FR 19793, https://www.federalregister.gov/documents/2020/04/08/2020-07338/notice-to-rescind-a-notice-of-intent-to-prepare-an-environmental-impact-statement.

75 Ashlee Vance, *Elon Musk*, generally, 181–211. John Yembrick and Josh Byerly, "NASA Awards Space Station Commercial Resupply Services Contracts," NASA press release, December 23, 2008, https://www3.nasa.gov/home/hqnews/2008/dec/HQ_C08-069_ISS_Resupply.html.

76 Todd Halvorson, "Falcon 9 Nails Test, Rocket's Roar into Orbit Bodes Well for Obama's Plan to Privatize Space," *FT*, June 5, 2010, 1A, 6A, https://www.newspapers.com/image/360404145/.

77 John Kelly, "They Did It: Here's Why We Should Smile," *FT*, June 5, 2010, 7A, https://www.newspapers.com/image/360404176/.

78 Brian Berger, "SpaceX Sues Boeing and Lockheed Martin," Space.com, October 21, 2005, https://www.space.com/1701-spacex-sues-boeing-lockheed-martin.html. Space Exploration Technologies Corporation v. The Boeing Company and Lockheed Martin Corporation, US District Court, Central District of California, CV 05-07533 FMC (MANx), February 15, 2006, https://www.nasaspaceflight.com/docs/spacex_dismissal_20060216a.pdf, archived March 28, 2006, at the *WM*.

79 Jonathan McDowell, "Space Activities in 2024," Rev. 1.4, January 2, 2025, 4, 8, https://planet4589.org/space/papers/space24.pdf.

80 Richard Tribou, "SpaceX Launch Raises Total to 93 for the Year," *OS*, January 1, 2025, Section 1, 3, https://www.newspapers.com/image/1155838168/.

81 Robert Z. Pearlman, "SpaceX's Last Launch of 2024 Puts Starlink Satellites into Orbit," Space.com, December 31, 2024, https://www.space.com/spacex-starlink-launch-group-12-6.

82 Mike Wall, "400 Rocket Landings! SpaceX Notches Reuse Milestone," Space.com, January 23, 2025, https://www.space.com/space-exploration/launches-spacecraft/400-rocket-landings-spacex-notches-reuse-milestone.

83 Will Robinson-Smith, "SpaceX launches 3,000th Starlink Satellite in 2025 on Record-Setting 32nd Flight of Falcon 9 Booster," *Spaceflight Now,* December 7, 2025, https://spaceflightnow.com/2025/12/07/live-coverage-spacex-to-launch-3000th-starlink-satellite-in-2025-on-record-setting-32nd-flight-of-falcon-9-booster/.

84 "LOFTID: Demonstrating Technology for Large Inflatable Heat Shields," *ULA* (blog), October 27, 2022, https://blog.ulalaunch.com/blog/loftid-demonstrating-technology-for-large-inflatable-heat-shields.

85 "New Glenn Launches NASA's ESCAPADE, Lands Fully Reusable Booster," Blue Origin press release, November 13, 2025, https://www.blueorigin.com/news/new-glenn-launches-nasa-escapade-lands-fully-reusable-booster.

86 Stephen Clark, "SpaceX Wants to Take Over a Florida Launch Pad from Rival ULA," *AT*, February 17, 2024, https://arstechnica.com/space/2024/02/spacex-wants-to-take-over-a-florida-launch-pad-from-rival-ula/.

87 Jeff Foust, "SpaceX Launches a Debate on Monopolies," *The Space Review*, September 18, 2023, https://www.thespacereview.com/article/4653/1.

88 2010 NASA authorization act, Sec. 302(a).

89 John Brophy et al., *Asteroid Retrieval Feasibility Study* (Keck Institute for Space Studies, California Institute of Technology, Jet Propulsion Laboratory, April 2, 2012), https://kiss.caltech.edu/final_reports/Asteroid_final_report.pdf.

90 "NASA's Asteroid Initiative Benefits from Rich History," NASA press release, April 20, 2013, http://www.nasa.gov/mission_pages/asteroids/news/asteroid_initiative.html, archived April 11, 2013, at the *WM*.

91 Mark K. Matthews, "NASA: Asteroid Visit All We Can Afford," *OS*, April 25, 2013, A1, A6, https://www.newspapers.com/image/268249609/.

92 Jeff Foust, "NASA Closing Out Asteroid Redirect Mission," *SN*, June 14, 2017, https://spacenews.com/nasa-closing-out-asteroid-redirect-mission/.

93 Jeff Foust, "NASA Closing Out . . ."

94 "Next Space Technologies for Exploration Partnerships (NextSTEP) Broad Agency Announcement, Amendment 1," NASA BAA NNH15ZCQ001K, November 10, 2014, 3, https://www.nasa.gov/wp-content/uploads/2017/11/nextstep-baa_2014-11-17_am1.pdf.

95 "NASA Selects Six Companies to Develop Prototypes, Concepts for Deep Space Habitats," NASA press release, August 9, 2016, https://www.nasa.gov/news-release/nasa-selects-six-companies-to-develop-prototypes-concepts-for-deep-space-habitats/.

96 "NASA Awards Northrop Grumman Artemis Contract for Gateway Crew Cabin," NASA press release, June 5, 2020, https://www.nasa.gov/news-release/nasa-awards-northrop-grumman-artemis-contract-for-gateway-crew-cabin/.

97 *National Space Policy of the United States of America*, June 28, 2010, 11, https://www.nasa.gov/wp-content/uploads/2015/01/national_space_policy_6-28-10.pdf. President Donald J. Trump, "Reinvigorating America's Human Space Exploration Program," Space Policy Directive-1, December 11, 2017, https://www.govinfo.gov/content/pkg/FR-2017-12-14/pdf/2017-27160.pdf.

98 Jeff Foust, "White House Nominates Bridenstine as NASA Administrator," *SN*, September 1, 2017, https://spacenews.com/white-house-soon-to-nominate-bridenstine-as-nasa-administrator/.

99 William Harwood, "Trump Requests $1.6B in New NASA Funding for 2024 Moon Landing," CBS News, last updated May 14, 2019, https://www.cbsnews.com/news/trump-requests-new-nasa-funding-for-2024-moon-landing-today-2019-05-13/. The teleconference audio announcing Project Artemis is on the Space SPAN at https://www.youtube.com/watch?v=dSCImT4nHjA.

100 "NASA Awards Artemis Contract for Lunar Gateway Power, Propulsion," NASA press release, May 23, 2019, https://www.nasa.gov/news-release/nasa-awards-artemis-contract-for-lunar-gateway-power-propulsion/. Jeff Foust, "NASA Selects Maxar to Build First Gateway Element," *SN*, May 23, 2019, https://spacenews.com/nasa-selects-maxar-to-build-first-gateway-element/. The video of the Florida Tech event is on the Space SPAN YouTube channel at https://www.youtube.com/watch?v=8btkNgDi9GM.

101 "As Artemis Moves Forward, NASA Picks SpaceX to Land Next Americans on Moon," NASA press release, April 16, 2021, https://www.nasa.gov/news-release/as-artemis-moves-forward-nasa-picks-spacex-to-land-next-americans-on-moon/.

102 "NASA Selects Blue Origin as Second Artemis Lunar Lander Provider," NASA press release, May 19, 2023, https://www.nasa.gov/news-release/nasa-selects-blue-origin-as-second-artemis-lunar-lander-provider/.

103 Bart Jansen, "NASA Advocates Pushing Congress," *FT*, September 20, 2010, 1A–2A, https://www.newspapers.com/image/360534260/.

104 "NASA Human Space Exploration: Delay Likely for First Exploration Mission," GAO, GAO-17-414, April 2017, generally, https://www.gao.gov/assets/gao-17-414.pdf.

105 "Space Launch System: Cost Transparency Needed to Monitor Program Affordability," GAO, GAO-23-105609, September 2023, 1, https://www.gao.gov/assets/gao-23-105609.pdf.

106 "NASA: Assessment of Major Projects," GAO, GAO-23-106021, May 2023, 59, https://www.gao.gov/assets/gao-23-106021.pdf.

107 NASA Artemis II web page, accessed March 21, 2025, https://www.nasa.gov/mission/artemis-ii/. Kathryn Hambleton, "NASA's First Flight with Crew Important Step on Long-term Return to the Moon, Missions to Mars," August 27, 2018, https://www.nasa.gov/missions/artemis/nasas-first-flight-with-crew-important-step-on-long-term-return-to-the-moon-missions-to-mars/.

108 "NASA's Readiness for the Artemis II Crewed Mission to Lunar Orbit," NASA Office of the Inspector General, May 1, 2024, IG-24-011, iii, https://oig.nasa.gov/wp-content/uploads/2024/05/ig-24-011.pdf.

109 NASA Artemis III web page, accessed March 21, 2025, https://www.nasa.gov/mission/artemis-iii/.

110 Eric Berger, "NASA May Alter Artemis III to Have Starship and Orion Dock in Low-Earth Orbit," *AT*, April 19, 2024, https://arstechnica.com/space/2024/04/nasa-may-alter-artemis-iii-to-have-starship-and-orion-dock-in-low-earth-orbit/.

111 *The Diane Rehm Show*, National Public Radio, January 2, 2014. A recording of the show is on the Space SPAN YouTube channel at https://www.youtube.com/watch?v=n-WJKXYDR2Q.

112 Florida Department of State, Division of Elections, November 6, 2012, General Election, US Senator, https://results.elections.myflorida.com/Index.asp?ElectionDate=11/6/2012&DATAMODE=.

113 US Senator Brevard County 2012 election results, Brevard County Supervisor of Elections, https://www.votebrevard.gov/Previous-Elections/2012-General/2012-General-Election-Results.

114 Florida Department of State, Division of Elections, November 6, 2018, General Election, US Senator, https://results.elections.myflorida.com/Index.asp?ElectionDate=11/6/2018&DATAMODE=.

115 US Senator Brevard County 2018 election results, Brevard County Supervisor of Elections, https://enr.electionsfl.org/BRE/1955/Summary/.

116 The complete video of Bill Nelson's December 10, 2018, remarks on the Senate floor is on the Space SPAN YouTube channel at https://www.youtube.com/watch?v=pOLGwQbTW6Q.

117 "President Biden Announces His Intent to Nominate Bill Nelson for the National Aeronautics and Space Administration," Joe Biden White House Archives, March

19, 2021, https://bidenwhitehouse.archives.gov/briefing-room/statements-releases/2021/03/19/president-biden-announces-his-intent-to-nominate-bill-nelson-for-the-national-aeronautics-and-space-administration/.

118 Lori Garver, "New NASA Administrator Should Reject Its Patriarchal and Parochial Past," *Scientific American*, April 12, 2021, https://www.scientificamerican.com/article/bill-nelson-isnt-the-best-choice-for-nasa-administrator/.

119 *A Review of the President's Fiscal Year 2023 Funding Request for the National Aeronautics and Space Administration and the National Science Foundation, Before the Subcommittee on Commerce, Justice, Science, and Related Agencies, Committee on Appropriations*, 117th Cong. (2022), https://www.appropriations.senate.gov/hearings/a-review-of-the-presidents-fiscal-year-2023-funding-request-for-the-national-aeronautics-and-space-administration-and-the-national-science-foundation.

120 Miles O'Brien, "Years Late and Billions Over Budget, NASA's Most Powerful Rocket Finally Set for Takeoff," *PBS NewsHour* web site, September 2, 2022, https://www.pbs.org/newshour/show/years-late-and-billions-over-budget-nasas-most-powerful-rocket-finally-set-for-takeoff.

121 "Orlando International Airport Fact Sheet," 2020, Orlando International Airport, https://orlandoairports.net/site/uploads/Fact-Sheet-2020.pdf

122 "About Us," Brooklyn Navy Yard, https://www.brooklynnavyyard.org/mission/. "About the Yard," Navy Yard Philadelphia, https://navyyard.org/about/the-yard/. "Hunters Point Naval Shipyard," City of San Francisco, https://www.sf.gov/hunters-point-naval-shipyard.

123 James Dean, "Air Force Considers Privatizing Cape Canaveral Operations," *FT*, July 14, 2013, https://www.floridatoday.com/article/20130714/NEWS01/307130037/Air-Force-considers-privatizing-Cape-operations, archived July 18, 2013, at the *WM*.

124 "4.0 Spaceport Business Model," *Kennedy Space Center Future Development Concept: 2012–2031*, NASA, 17, https://www.nasa.gov/wp-content/uploads/2015/03/634026main_future-concept.pdf.

125 James Dean, "DiBello: Independent Authority Should Run Spaceport," *FT*, June 15, 2016, 11A, https://www.newspapers.com/image/192982040/.

126 *Cape Canaveral Spaceport Master Plan*, Space Florida, January 2017, 9, https://cdn.prod.website-files.com/66c8a3fe36eef11411f2b1e5/66c8a3fe36eef11411f2b5f1_sf-bod-approved-ccs-master-plan-02-01-17%20(2)-part-1.pdf.

127 "Commercial Space Transportation: Improvements to FAA's Workforce Planning Needed to Prepare for the Industry's Anticipated Growth," GAO, GAO-19-437, May 2019, 12–13, https://www.gao.gov/assets/gao-19-437.pdf.

128 *A National Spaceport Strategy: A White Paper Prepared for the Chief of Space Operations, United States Space Force*, Range of the Future Task Force, August 2020, 8–9, 22.

129 "Office of Spaceports," FAA web page, https://www.faa.gov/space/office_spaceports.

130 John W. Raymond, "US Space Force Range of the Future 2028 Strategic Intent," February 2020, 2–3, https://govtribe.com/file/government-file/fa882321r0009-ussf-range-of-the-future-2028-strategic-intent-dot-pdf.

131 *The Florida Spaceport System Maritime Intermodal Transportation Study, Feasibility Phase*, Space Florida, April 2024, 1–3, https://cdn.prod.website-files.com/663258da01a14bbdf80faa99/66a4290a20bcabc01ffb76f9_WharfStudy-Final-Compressed.pdf.

132 *Kennedy Space Center Economic Impact Study FY21*, NASA KSC and FIT, May 2022, 5, 7, 12, 13, https://www.nasa.gov/wp-content/uploads/2022/06/ksc_economic_impact_report_fy2021.pdf.

INDEX

Stephen C. Smith is a child of the Apollo era. He grew up in Southern California at a time when the local economy heavily depended on the aerospace industry, military and civilian. He witnessed the layoffs in the early 1970s as the end of Project Apollo coincided with the détente between the United States and Soviet Union.

After a career that included public policy and political consulting, Stephen and his wife, Carol, moved to Florida's Space Coast to be part of the next generation of US spaceflight. For ten years, Stephen worked at NASA's Kennedy Space Center, delivering educational lectures, leading public and private tours, and escorting retired astronauts. He had a front row seat for the arrival of the NewSpace era, as NASA and the Department of Defense transitioned from government to commercial launch services, lowering costs for taxpayers and introducing innovation to the US aerospace industry.

Stephen retired in 2021 to write about this experience, based not only on his own observations but also questions from the public during his ten years of service at KSC. This is his first academic publication, although he's been a freelance writer for many years.

In 1978, Stephen graduated from the University of California, Riverside, with a bachelor's degree in political science. In 1985, he earned a master's degree in public administration from California State University, Long Beach.

You can find Stephen on social media. He's on CounterSocial and Bluesky at @WordsmithFL. You can read his columns on Substack at TheSpacePundit.com.

www.ingramcontent.com/pod-product-compliance
Lightning Source LLC
Chambersburg PA
CBHW031914160426
42812CB00096B/674
* 9 7 8 1 6 8 3 4 0 6 5 6 3 *